James Knight

Orthopaedia

Or, a practical treatise on the aberrations of the human form

James Knight

Orthopaedia

Or, a practical treatise on the aberrations of the human form

ISBN/EAN: 9783337370282

Printed in Europe, USA, Canada, Australia, Japan

Cover: Foto ©berggeist007 / pixelio.de

More available books at **www.hansebooks.com**

ORTHOPÆDIA

OR

A PRACTICAL TREATISE

ON THE

ABERRATIONS OF THE HUMAN FORM.

BY

JAMES KNIGHT, M. D.,

MEMBER OF THE MEDICO-CHIRURGICAL FACULTY OF MARYLAND, THE DISTRICT
MEDICAL SOCIETY OF OHIO, AND OF THE COUNTY MEDICAL SOCIETY OF
NEW YORK; PHYSICIAN AND SURGEON IN CHARGE OF THE
HOSPITAL OF THE NEW YORK SOCIETY FOR THE
RELIEF OF THE RUPTURED AND CRIPPLED,
NEW YORK CITY, ETC., ETC.

NEW YORK:
G. P. PUTNAM'S SONS,
23D STREET AND 4TH AVENUE.
1874.

Entered, according to act of Congress, in the year eighteen hundred and seventy-four,
By G. P. PUTNAM'S SONS,
In the office of the Librarian of Congress, at Washington.

CONTENTS.

CHAP.		PAGE.
I.	REMARKS ON DEFECTIVE PHYSICAL FORMATION	7
II.	IMPAIRMENT OF TISSUES RESULTING IN CONTORTIONS	37
III.	GENERAL REMARKS ON THE TREATMENT OF TALIPES	60
IV.	INFANTILE PARALYSIS	88
V.	ELECTRICITY AS A THERAPEUTIC AGENT IN THE TREATMENT OF PARALYSIS	115
VI.	CONTRACTION OF THE HANDS, FINGERS AND TOES	163
VII.	LATERAL CURVATURE OF THE SPINE — TORTICOLLIS	173
VIII.	RACHITIS	189
IX.	HERNIA — PROCIDENTIA UTERI — ECTROPION VESICÆ — RELAXED ABDOMEN	203
X.	VARICOSE VEINS — BURSÆ — GANGLION	240
XI.	PATHOLOGICAL CONSIDERATION OF DISEASES OF THE JOINTS	253
XII.	DISEASES OF THE BONES — NECROSIS	289
XIII.	TONICS, AND THEIR EFFECT UPON THE SYSTEM	334

ORTHOPÆDIA.

INTRODUCTORY REMARKS.

The attainment of practical information on the subject of aberrations of the human form, and the tendency thereto, is, at the present day, considered by the medical profession, both at home and abroad, to be one of the essential qualifications of a general practitioner. If he does not assume the treatment of this class of ailments, he should, at least, possess such information as would enable him to diagnosticate in its incipiency the pathological condition that would tend to deformity of the person, and thus refer the patient to an experienced orthopædic surgeon. To the afflicted, this is of inestimable importance, impressing them with confidence in treatment, as well as in the high order of the attainments of their medical adviser.

This knowledge is most difficult to attain in an ordinary general practice, the patients being so sparsely distributed, and institutions for their treatment so limited. As a means of imparting information upon this important branch of medical education, a careful revision of a series of years of successful practice is presented, with a desire that it may be considered, in a limited degree, as a favorable succedaneum to that of a more practical acquirement, and with the conviction that it is a duty due to each other from all associates in a liberal profession, to impart to the best of their ability such supposed acquirement of knowledge as long-continued labor, under favorable circumstances, would yield to diligent inquiry.

Our public and private practice has been most largely devoted in professional service to the relief of crippled children. The first effort at devising surgico-mechanical appliances was made at the dispensary, and for the relief of reducible hernia, with a view to obviate much of the suffering inflicted by the ordinary trusses then in use. The result was a very decided success, and an incentive to the construction of appliances for the restoration of impaired powers of locomotion in children laboring under deformities both congenital and the sequelæ of infantile paralysis. Other deformities, resulting from constitutional impairment, led to a careful study of the pathological conditions tending to the various phases of deformity of body and limbs, such as caries of the spine, terminating in spinal or psoas abscess, morbus coxarius in its several stages, involving the destruction of the joint and often loss of life, and synovitic invasion of the joints. These latter ailments were carefully considered, and a careful regime with proper sanitary regulations and conditions were found to be the primary requisites of a proper treatment. Such conditions, we concluded, were only attainable in the highest degree, in a hospital of proper construction. With this impression we introduced the initiatory efforts in our own dwelling — tending to success far exceeding expectation — resulting in the organization of a society, which was favored with most liberal donations, enabling us to organize at our discretion a Hospital for the Relief of the Ruptured and Crippled. The indigent were received free of charge, and the payments from others contributed to the support of the house. Upwards of 5,000 *in-patients* have received treatment during the ten years.

Other ailments incidental to, and requiring mechanical support, as procidentia uteri, varicose veins and consequent ulceration, have also received careful study, and the treatment, with suitable appliances, represented in this work by wood engravings, has been attended with much success, these appliances being largely supplied to out-door patients.

The treatment of deformity has been one of the greatest difficulties encountered by the general practitioner in medicine, even to the abandoning of all attempts at efforts to redress curable deformities in children, especially that of clubfoot. The seeming impossibility of treating deformities by regular practitioners in medicine, even in this advanced state of medical science, has caused orthopædic surgery to be assumed by mere adventurers, destitute of medical knowledge, to their great pecuniary advantage and the discredit of the medical profession.

There is another unfavorable circumstance, much to be regretted, as it seriously discourages those competent to attain a knowledge of orthopædic surgery. It is in being deceived by the exaggerated laudations of inventions of no particular merit, devised by practitioners in medicine, and some of great notoriety. All appliances, however skilfully constructed, are only auxiliary means in the treatment of deformity, and may be used to greater advantage by the inventor than by even those quite as familiar with the treatment of these ailments; each having peculiarly devised or modified apparatus for the treatment of their patients. To give promise of success in the treatment of deformity, it is a most essential acquirement to the student, to have sufficient practical knowledge to make a correct diagnosis of the case, then to carefully consider the indications that present themselves, and rely upon his own judgment as to the means that will most readily apply to the redressing of the abnormal condition. Even then he may fail in his first attempts in treatment, but it will not be so discouraging, inasmuch as this has been his own device, tending necessarily to improvement. It would have been discouraging, however, had he failed in the treatment relied upon for some highly-extolled appliance — the device of some reputable practitioner, and supposed to be perfect. The success in treatment is the result of practical skill, and not wholly in the well-designed apparatus, which will only meet certain indications for a time, and fail in the progress of treatment. From this fact, we became convinced of the impossibility of attaining success in orthopædic practice, by an entire reliance upon the devices of surgical appliances, however skilfully constructed. All apparatus must be modified, and the judgment of the practitioner exercised, as to the modifications that may be required from time to time, as successful progress is made in the treatment. Hence, the roller and its application claims important consideration, and practical skill in making it available in the redressing of deformity. Without knowledge of the application of the roller, or what is in some cases of greater utility in treatment, perfectly fitting, enveloping devices of woven fabric made to lace over large abscess developments, we are without the most efficient means for redressing deformity, especially of the feet and limbs. Complicated apparatus only serves to confuse and embarrass the practitioner, especially when not devised by himself, as he will construct or modify to meet various indications that are apparent in nearly every recurring case, and which cannot be considered by the practitioner who has constructed an apparatus for general use, a fair sample of which can be found in the surgical instrument makers' shops. The various engravings of apparatus presented in this work are only intended to impress ideas in regard to the construction of modified forms to meet presenting indications. Many novel forms are introduced to meet the results of existing anomalies presenting in the treatment of deformity, such as have failed to invite sufficient attention, more especially in the incipiency of ailments tending to an increase of the abnormal condition, as lateral curvature of the spine, depressed arch of the foot (which tends to valgus), contracted toes, and many other tendencies to contortion of parts of the body and limbs, which a very simple appliance would have restored to normal form. These modified appliances are the result of the peculiar necessities of cases presented for treatment. In this we have been greatly favored — having had the supervision of 26,448 patients within the past ten years in the "Hospital for the Relief of the Ruptured and Crippled," and previously a private practice of thirty years, limited mainly to ailments treated of in this work. During this period we have carefully studied the pathological condition of each patient and the result of treatment, availing ourselves of the observations and inventions of eminent practitioners in this and foreign countries, though not always in the actual form of the inventor's design, nor strictly to the letter as to treatment — personal experience and judgment tending to modification. Hence, the subject-matter of this work may be considered as the varied and consolidated experience of many practitioners, modified to our judgment. Nearly every written authority upon the subject herein treated has been carefully considered and made applicable to our practice, however much we may claim as our share of practical attainment in the treatment of this class of ailments, which we have greatly restricted from adventurous treatment to that of the expectant. The results have proved most favorable in our practice — seventy-five per cent of the ordinary conditioned patients laboring under synovitic disease having been restored to self-sustaining ability, the diseases having been arrested and the limbs restored to usefulness.

For this success in treatment we are greatly indebted for the advice of that most eminent surgeon, the late Professor Valentine Mott, the first Consulting Surgeon to the Hospital for the Relief of the Ruptured and Crippled, whose invaluable experience in the treatment of these constitutional chronic diseases was freely imparted; and for many years previous to the opening of this institution, was made available to the relief of vast numbers of suffering humanity.

The subject-matter of this volume extends through a large range of topics, all of which have received careful consideration, as they presented themselves in practice. It has also been supplemented by recent contributions from eminent living authors, of the medical fraternity, the list of whom, only, would include nearly all of our compeers who, as specialists, had directed their attention to the treatment of aberrations of the human form. To these authors we tender our most grateful acknowledgments for the assistance they have afforded us.

This work, from absolute necessity (our time for literary labors being necessarily very limited), is greatly condensed; yet we trust it will be perfectly comprehended by the reader. For valuable assistance in the preparation of this work for the press we are greatly indebted to Dr. VIRGIL P. GIBNEY, our Senior Assistant in the Hospital.

CHAPTER I.

REMARKS ON DEFECTIVE PHYSICAL FORMATION.

Congenital aberration of the human form anciently attributed to mental and physical impressions on the *enciente*.— The laws of Lycurgus in relation to malformed offspring. — The undue severity of these laws naturally prompted efforts for cure.— Modern theories and investigations with deductions therefrom.— R. W. Tamplin's opinion as to the cause of congenital deformities of the feet.— Unfavorable position of the feet in children and adults results in persistent deformity.— Intra-uterine life, and its susceptibility to change in health and physical formation.— The dogmas of Hippocrates, Paré, Petit, etc., in reference to the causes of congenital deformities.— The opinions of MM. Serres, Geoffrey, Saint Hilaire and Roux, compared with the diverse views of Tiedman on the subject.— Guerin's theories on the causes of fœtal deformity, as presented and illustrated by him before the Academy of Sciences at Paris. — Our own professional experiences as to defective physical formation, illustrated by the history and diagnosis of patients.— Arrest of development and contortion the inevitable result of abnormal nervous energy, as demonstrated in the histories of several recent cases.— Diagnosis and treatment of the case of Maggie K.— Marked and permanent improvement after a year's treatment. — Recapitulation of facts deduced from this case.— Extraordinary instance of uniform arrest of development.— History of patient.— Cases of angular bending of lower third of tibia and fibula, without retraction of the muscles.— Case of *primipara*, band encircling right leg below knee.— Case of club-foot, with cinctures in limbs.— Blemishes; difficulty in assigning origin ; the varied forms they assume.— Curious practices in horse-breeding.— Abnormal marks and appearances on newly-born children, their origin.— CONGENITAL LUXATION.— Luxation of the hip.— Knowledge as to treatment of this affliction extremely limited.— Congenital luxation in the earlier periods of gestation; its effect on the articulating cavities.— Its protracted development.— Peculiar symptoms.— Disease generally hereditary — Dupuytren's case of an entire family afflicted with congenital luxation.— Doubtful whether the cause is arrest of development of the cotyloid cavity or, simply muscular retraction.— ANCIENT TREATMENT OF DEFORMITY.— Talipes treated by Hippocrates 500 B. C., but not again made the subject of research for 1600 years.— Ambrose

Paré's Treatise on Talipes.—Construction of artificial limbs and surgical apparatus first described.—Hildanus' extension apparatus and splint.—Description of an extension splint for contracted knee-joints.—Glisson's treatment of deformity by extension, A.D. 1651.—Extension by weights and cords not *novel*.—Method of extension and counter-extension by Scultetus.—Shoes for the cure of talipes described by Circœus.—Treatment of spinal curvature and torticollis by Isaac Mincius and Heister.—First application of the term "Orthopœdia" to deformity in children, by Mr. Andrys.—Establishment of an institution for the cure of talipes by Venel and Tiphaisne.—Publication of the results of Venel's treatment by Dr. Ehremann and coadjutors.—Scarpa's shoe.—His celebrated monograph "Sulle Piedi Ferti."—The application of elastic force absolutely essential in all apparatus intended to overcome muscular contraction.—ANCIENT TREATMENT OF SPINAL DEFORMITIES.—Treatment by suspension and lever.—Paré's treatment for cleft palate.—Paré's metallic corset the earliest piece of mechanism for the relief of the distorted spine.—Le Vasher's apparatus for the relief of caries of the cervical vertebræ.—Sheldrake's apparatus for spinal curvature.—Darwin an advocate of this treatment.—SURGICAL MEANS OF REDRESSING DEFORMITY.—Section of Tendo-Achilles as a method of treatment for talipes.—Molinelli's opposition thereto.—Thelsemius, of Frankfort, the first promulgator of the principle of division of tendons as a treatment for pedal deformity.—Operation demonstrating the results of his treatment.—Experiment of Saratorius in a case of talipes equinus.—Michaelus, of Marburg, his successful treatment by partial severance.—Opinion of Benjamin Bell thereon.—Indifference of the practitioners of that day to the question or remedy of deformity.—Subcutaneous division of muscles and tendons.—John Hunter's advocacy of the system.—Classification of injuries sustained by the body.—Delpech's application of Hunter's principles in medical practice.—Stromeyer's initiatory operations and discoveries in scientific tenotomy.—Duval, of Paris, his heroic treatment of club-foot and cognate affections.—The new system enthusiastically adopted by Dieffenbach.—The uniform success of the new treatment for club-foot on the European continent.—Dr. Detmold its most successful and skilful operator.—Simple reliance on the operation, without subsequent careful treatment, not only delusive and disappointing to the patient, but fatal as to its ultimate success.

FROM the earliest psychological history of mankind to the records of the present generation, congenital aberration of the human form has invited the serious consideration of not only the illiterate, but the wisest men of their day; this affliction being one of the most deplorable that can be conceived of, not only to the victim himself, but to the parents of the unfortunate malformed, or blemished, child. The subject was long involved in obscurity, and early investigations resulted only in speculation, and probable surmise, as to the unfavorable influences that should have so impressed the *enciente*. Indeed, it is but recently that physiology has been brought to bear on fœtal development, culminating in little more than mere conjectural and seemingly plausible solutions of the cause of these lamentable occurrences.

The idea most commonly entertained, even now, is that mental and physical impressions of a decidedly pleasurable, or alarming, character made upon the *enciente* (as that of great desire, bodily injury, and frightful or extraordinary scenes), are causes of these abnormal conditions of development in the fœtus. This impression is of the most ancient origin; as we are informed in the Bible that, 1745 years before the Christian era, Jacob was in possession of information on this subject, and made it available to his advantage. There it is stated that he, while in charge of Laban's flocks, experienced an irresistible attachment for his employer's daughter, and expressed the desire of making her his wife. To this Laban assented, and entered into a covenant with him, wherein it was agreed that, after a certain period of time, all of the parti-colored cattle of the future increase of his flocks should be given to Jacob, who practiced the device of placing peeled rods in the watering places where the cattle came to drink. "And they conceived before the rods, and brought forth cattle, ringed, streaked, speckled and spotted;" * and by the consequent increase he became wealthy.

This supposed impression upon the fœtus, an idea entertained and strenuously supported by the ancients, it is most reasonable to suppose, influenced the Greeks in their submission to the barbarous laws of Lycurgus, which condemned to destruction all their deformed children, that the supposed baleful influence exercised upon the *enciente* by the sight of them might be thus avoided. Now, as then, this idea obtains credence among parents; and the birth of a deformed child is considered as a more sad affliction than the death of their most favored offspring; and, though not openly acknowledged, yet the impression is secretly entertained, that deficiency, contortion, or blemish in an infant, has been caused by some unfavorable sight, or injury, or by some extraordinary desire not gratified during gestation, and in this way an injury inflicted on the fœtus. Notwithstanding these prevalent notions as to the cause, it is very reasonable to suppose that treatment for deformities in children

---

* Gen., chap. xxx, 37th to 43d verses.

dates from a very early period; for parental sympathy, rendered even more active by the desire to save the little unfortunates from the execution of this obnoxious law, would naturally prompt the concealment of their condition, and induce efforts for their cure.

The more seemingly rational theories of the present age, deduced from physiological investigations, and a consideration of the primary formative process are:

1st. That of an arrestation of development, emanating from the nervous centres; or, as some believe the blood-vessels to precede the nerves, by their impairment affecting the peripheral appendages of the primary nucleus, and determining the normal or abnormal development of the fœtus.

2d. That, if some of the nervous centres, essential to functional force in the nutrient vessels, be unfavorably impressed, or partially interrupted in their functions, the inevitable result would be abnormal development.

3d. That the encircling of the limbs of the fœtus by adventitious bands has resulted in their amputation.

4th. That unfavorable positions of the fœtus *in utero*, influenced by the spherical boundary of the latter, is a primary cause of distortion, more especially of the extremities.

R. W. Tamplin, F. R. C. S. E., Surgeon to the Royal Orthopœdic Hospital, London, in his fifth lecture of a course delivered in that institution, and republished at Philadelphia, thus speaks of the cause of congenital deformities of the feet:

"These consist of *talipes varus*, *talipes valgus* and *talipes calcaneus*. I have never met with *talipes equinus*. You will recollect, I stated that, in my opinion, it was position, and position alone, which caused these malpositions (or rather the permanent extreme natural position, for this is its real character) during uterine existence; and if you observe the character of these three deformities, I think you will have little difficulty in reconciling the possibility, if not the probability, of this being the case. Take the first mentioned, *talipes varus;* there are many positions in which the extremities of the child may be kept while in utero, which would better adapt it to the surrounding interior, *provided the feet were inverted;* and, it so happens, that the greater number by far of congenital cases consist of double varus. And where it does not consist of double varus, nothing is more easy to account for than one or the other extremity being so placed that the foot should grow in that position, while its fellow may be free and unconfined. Then, if you refer to congenital varus, this also may be easily imagined — that the extremities or extremity may be so situated, that the flat surface of the feet press, more or less, on the walls of the uterus — if not constantly, sufficiently to influence them during the growth of the bones and ligaments; and, be it remembered, that it is only the ligaments and muscles that we find affected, the bones retaining, in either of the three forms, their natural proportions and relative size, and it must be evident that it would not require any force, but merely a constant position, to produce this effect, at least it is so to my mind."

It is an undisputed fact, that the unfavorable position of the feet of children and adults has, by long continuance, resulted in persistent deformity, and that so diverse in character as to produce the several varieties alluded to by Surgeon Tamplin.

Thorough investigation of intra-uterine life has well determined that the fœtus is liable to many affections, which may cause various changes, not only in health, but in physical formation; the facts furnishing a basis for the deduction of inferences as to malformations. The great father of medicine, Hippocrates, asserted that children, while contained within the organ of gestation, were subject to unfavorable impressions from falls, blows or pressure exerted on the abdominal walls of the mother; Paré expressed similar views in regard to congenital deformity; J. L. Petit believed in the alteration of the germ, or an aberration of the formative power. Dupuytren, in an arrest of the development of the deficient or deformed parts of the fœtus; Breschel and M. Sause ascribed as a cause the existence of certain articular maladies; Chaussier,. Delpech and Guerin, a primitive alteration in the nervous centres, and, from extended observations, attributed certain congenital deformities, as that of talipes and other contortions, to convulsions occurring in the fœtus.

The opinions of M. Serres, Geoffrey, Saint Hilaire and Roux are, that the evolution of the organs in the fœtal state proceeds in strict ratio with their supply of blood; consequently, that atrophy, imperfection, or absence of the organs generally, is attributable to the imperfection or absence of their nutrient arteries; while Tiedman asserts* that the nervous system is developed

---

* "London Medical and Physical Journal," July, 1828.

before any other part of the body, that congenital malformations or anomalies depend upon the imperfections, or the want of certain portions of the nervous system, and that they control the development of the embryo to the determining of the form and disposition of the organs; hence, deformities have their first cause in the irregular development of this system.

The absence or impairment of nervous centres having been considered by Guerin as a cause of fœtal deformity, this gentleman presented supposed evidence to sustain this opinion; and read a paper before the Academy of Science, Paris, presenting specimens of monstrosities having the attendant contortion of the limbs. He asserted that there was always to be observed a perfect relation between the absence of the nervous centres (or a portion of them, the result of which is muscular contraction) and that of dislocating the joints in the fœtus. This noted Orthopædist premises his argument in favor of this theory by the following remarks:

"The observation of certain monstrosities presents to us, in a striking manner, the combination of these four orders of facts, namely: that in them may be observed at the same time, a material lesion of the nervous centres, the retraction of the totality, or of the greater portion of the muscular system and of concomitant luxations, a vigorous relation between the seat, the extent, and the degree of the nervous lesion, and the seat, extent, and degree of retraction; and, finally, a relation of direction, of extent, and of degree between the retraction and the dislocation which it produces." For illustration, he presented an anencephalus, in which there was not only contraction of all of the muscles, producing curvature of the spine, but actual dislocations of the joints. He further asserted that, under certain circumstances, an arrest of development is the result of this deficiency in the nervous centres.

As the perfection of the nervous centres is essential to the perfect formation of the fœtus, we must infer that some pathological change had taken place at an early period of gestation to the impairment and destruction of the nervous centres; hence, an arrest of development and contortion are the results of abnormal nervous energy.

As an aid to the adequate consideration of this subject, of so much interest to all, we have introduced several engravings representing deformed children; two of them laboring under various congenital malformations, and now living subjects, and one simi-

larly afflicted that died of scarlet fever at the age of four years —
the drawings having been made from photographs of the patients.

The first of these cases (fig. 1) is a striking illustration of arrest
and deficiency of development, together with retraction of the
muscles. The history of the case is as follows:

Maggie K——, aged twenty months, was brought by her mother
from her residence in the country for treatment, or rather to obtain

Fig. 1.

our opinion as to the possibility of enabling the child to walk. We
could, however, only give a favorable opinion as to the ultimate
accomplishment of the desired object.

After a very careful examination, we advised an effort to be
made to relieve the child from its utterly dependent condition — a
sitting posture maintained with difficulty; the upper limbs being so
markedly deficient as to not only be unavailable in any effort at
locomotion, but of no assistance to the child in keeping an upright
position. We, at the same time, informed the mother that the
severe surgical treatment required for the extension of the lower

limbs would possibly endanger the child's life, but that its deplorable and nearly hopeless condition would fully justify the effort for improvement.

Upon inquiry as to the hereditary disposition in the families of both parents, the mother made this statement: "My age is thirty-five years; my husband is about forty-five. We enjoy excellent health, and have always resided upon a farm. I am not aware of any member of my own or my husband's family being deformed. Our families can be traced back for two generations. About two weeks after becoming pregnant with this child, to the best of my judgment, I was thrown out of a sleigh into the snow, but did not sustain any injury to body or limbs, and consequently gave the circumstance but little consideration. Nothing else unusual occurred during my pregnancy. Previous to this great affliction I have had several children, all of whom enjoy excellent health, and are of natural form."

The malformation of the child cannot be readily comprehended without a careful examination of the engraving. The legs were flexed upon the thighs, and the thighs upon the abdomen, presenting the appearance of an extraordinary development of the thighs and nates, the retracted condition giving a spherical form to this portion of the child's body. The length of the apparent legs from the flexed thighs was three inches, the feet being of ordinary size, yet having but three toes on each. A line subtended from the centre of the tuber ischii to the posterior extremity of the os calcis measured four inches. A firm tissue beneath the skin retained the legs in an arched condition, and no appearance of a knee-joint existed. The limbs superficially presented a uniform semicircular curve from the abdomen to the feet, the latter being slightly adducted, the ankles having free motion.

On placing the child upon her back, the thighs seemed movable upon the pelvis, and the limbs could be separated laterally to a normal extent. By manipulation a very limited motion was also detected in what appeared to be the middle of the thigh, giving indications of the existence of a knee-joint.

The discovery of these yielding points induced us to sever subcutaneously what appeared to be the extended posterior muscles of the thigh, on both legs, two inches from their insertion into the os calcis, and then to apply extension, with moderate force, to the limbs. They slowly yielded, and developed imperfect knee-joints, at about what would be the lower third of normal thighs. The left limb extended more readily than the right, both limbs deficient of fibula, and, for want of the maleola, everted feet.

After three months' effort at extension, resulting in a very considerable yielding of the limbs, it was deemed advisable to sever the parts as before, but about an inch higher up, and to not more than half the depth of the first operation. This afforded most decided relief; and in about a month from that time the little patient commenced to walk with tolerable firmness, the right leg, however, being shorter than the left.

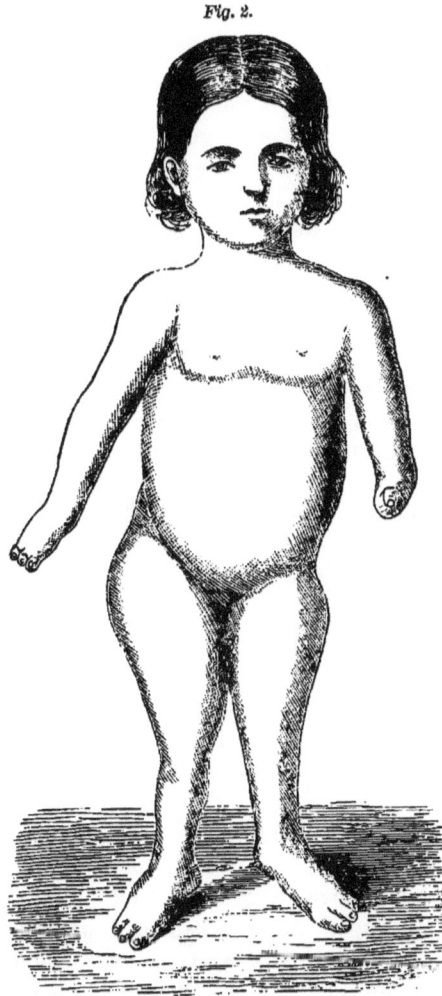

Fig. 2.

Within a year from the last operation, we severed a portion of the constricted parts in the right leg, about an inch above the insertion of the tendon upon the os calcis. This was attended with rather an alarming result; suppuration supervened, and it was with some difficulty that the child was relieved —the leg being only slightly improved as to length.

However, the child now walks with considerable freedom, and her present condition is fairly shown in the accompanying engraving (Fig. 2). It will be observed that there is a deficiency in the left forearm, there being only a stump of about two inches in length from the elbow-joint, and that covered by loose integument. From this stump a prominence proceeds, very similar to the projecting tip of an elephant's proboscis, and which can be extended at will — in fact, made so useful as to draw a pin from the clothing, or to convey food to the mouth. This arm is movable at the shoulder and also at the elbow-joint. Within the loose integument, there is a rudiment of the ulna, if not of the radius, of about two inches in length, upon which, by careful examination, there can be distinctly felt the first and second phalangeal bones of two fingers, and apparently a rudimentary portion of a third. These parts of fingers have the power of flexion and extension against the firm segment of an ulna, thus grasping and retaining small objects. The right arm remains extended from the shoulder, where there is normal mobility, the elbow being fixed and without any appearance of a joint. To this arm is attached an imperfectly developed hand, having carpal and metacarpal bones, terminated by three webbed phalanges, that can be but very slightly flexed at their extremities, being united by a dense integument extending to the distal joints, where there is only a partial separation. This segment of a hand is quite limited in motion, and it is only by approximating it to the stump of the left arm, that it can be used to convey food to the mouth, and only then by bending the neck and lowering the head obliquely forward.

Not having any desire to theorize upon the primordial development of the fœtus, we have simply noted the condition of children whom we have treated, and will merely recapitulate the several points having a bearing on the subject.

First: The parents were of normal development, in the enjoyment of good health, and void of hereditary predisposition to deformity.

Second: There is a presumable cause for the deformity of the child in the concussion received by the mother two weeks after conception.

Third: The result is a child of bright intellect, having a normally developed head, neck and body. Hence, there was no apparent deficiency in the cerebro-spinal development, and a uniformity of deficiency in having but three fingers upon the right hand and the rudiments of three fingers upon the left, terminating two inches below the elbow, and an arrest of development in the feet, three toes on each.

Fourth: Retraction of the legs upon the body; presenting what would appear to have been a very early retraction of the flexor muscles of the thighs, inserted far below the knee, and thus closely approximating the heels to the nates — the largely developed muscles covering the bones of the lower limbs so as to obscure the knee joints.

It is not improbable to suppose the right arm to have been held extended by a rigidity of the muscles at a very early period of gestation; and the elbow joint rendered thus fixed so firmly in the erect position by the condensing of the surrounding tissue — all tending to prove the unfavorable influence to have been of early date in the progress of development.

The accompanying illustration represents a case of arrest of development of great uniformity; the deficiency being confined to the feet and hands. The feet were flexed firmly upon the legs, the distal extremity of the os calcis pointing downward, and upon which the child could not have maintained an erect position, without assistance.

From this dependent condition the child was speedily relieved, having been brought for treatment to the "Institution for the Relief of the Ruptured and Crippled," when about four weeks old. The photograph was taken when the child was one year old.

Fig. 3.

The deformity in this case consists in a deficiency of the index and middle digits and the carpal and metacarpal bones of both hands; and of the middle and adjoining toes and corresponding tarsal and metatarsal bones of both feet. In this instance no hereditary predisposition can be traced to the parents, nor apparent lesion of the cerebro-spinal system of the child; nor is there record of any injury sustained by the mother during gestation. In both of these cases there has been arrest of development and muscular retraction. In the first one, of the flexors of both the lower limbs, to an extraordinary

18   ORTHOPÆDIA.

degree. In the second, of the flexors of the feet, producing talipes calcaneus. Each of the children possessed good health, and a normal condition of the mental faculties.

The third case we present is one that cannot be attributed to spasmodic contraction of the muscles during intra-uterine development, as it is not a distortion of the joints, but an actual angular bending of the lower third of the tibia and fibula, without any retraction of the muscles, other than that of a slight flexion of the

Fig. 4.

toes. This was relieved by opposing graduated pressure and counter-pressure applied to the limb, without restriction in exercise.

This child was first presented for treatment at the "Institution for the Relief of the Ruptured and Crippled," at the age of two years. (Figure 4.) The mother appeared delicate, but said that she enjoyed good health, as did also the father, who is a robust laboring man. The mother had no recollection of having received an injury during pregnancy with this child, and the deformity was discovered

DEFECTIVE PHYSICAL FORMATION. 19

by the midwife on its delivery. The patient was under treatment for two years, when the limb was restored to normal form. (See fig. 5.)

Fig. 5.

Another case of angular curvature of the tibia and fibula came under our notice for treatment where both legs were bent inward and nearly at a right angle. The tibia and fibula were bent at the lower third of the leg, the feet being perfect in form. The mother, a colored woman, strong and healthy, stated that when about three months advanced in pregnancy she was compelled to jump from a second-story window to the pavement below, but apparently sustained no other injury than that of a severe shock, which did not, however, prevent her, even for a day, from performing her ordinary labor, that of laundress. Treatment was commenced with this child when six months old, a similar course to that of the former case being pursued and continued for two years, ending in a complete restoration to normal form.

Case No. 6 is of a female child, *primipara*, delivered by a midwife, who stated that after the birth of the child she observed a band encircling its right leg below the knee, which she divided with the scissors, the band leaving a deep cincture, as represented in the engraving (fig. 6). The mother, a strong, healthy young woman about eighteen years of age, brought the child, when six months old, for treatment of varus of the impaired leg. Upon the dorsum of this foot the superficial integument was greatly hypertrophied, a condition that interfered very much with the restoration of the foot to its proper position. The drawing was made when the child was two years old, the foot being represented as in its original condition, in order to give a better comprehension of the case. At this age the cincture remained as at first observed, though the limb was equal to its fellow in size and strength. It was quite apparent, however, that a slight increase of the stricture made during fœtal existence would have arrested the circulation of the blood, and resulted in amputation of the limb.

Fig. 6.

This case tends to induce the belief that amputations *in utero* are made possible by encircling adventitious bands during gestation.

The only opposite view urged against the fact of such amputations taking place, is, that in cases of deficiency of a limb, the amputated portion has not been found on delivery of the child. But may not this be from the lack of careful examination of the placenta?

The amputation taking place at a very early period of gestation, when the member is very small, the missing portion may have become disintegrated. In this case, the most probable inference is that the adventitious band was formed when the child was quite advanced in gestation, as there was no apparent impairment of the blood-vessels, which had attained sufficient size and strength to maintain the requisite nourishment of the limb.

From the cases here presented it is reasonable to entertain the opinion, that an arrest of development is the cause of deficiency in a limb — as that of a fore-arm, and attached thereto a rudimentary terminal portion, or a deficiency in the number of digits, carpal and tarsal bones, as shown in the third engraving. Also, that the deficiency of a limb, having a regularly rounded terminus, may be the result of an amputation, from an encircling of the limb by an adventitious band, either at an early stage of development, or only a partial impression made at an advanced period, as in the case represented by the fifth engraving.

The seventh engraving represents a child having both limbs impressed with these cinctures, including one foot — the child now living, July 10, 1872:

Edward Molony (Fig. 7), aged four months, of Irish parentage, brought to the Dispensary department of the "Hospital for the Relief of the Ruptured and Crippled," requesting only the relief of the club-foot of the child, when, on examination, these cinctures in the limbs were observed.

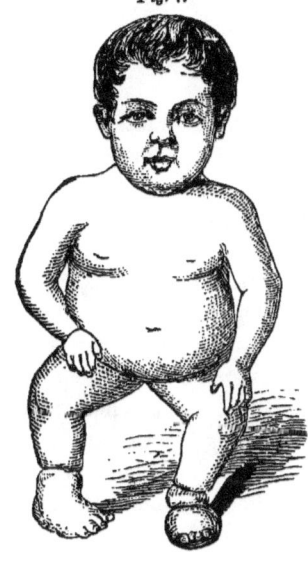
Fig. 7.

The parents of this child were both healthy. The mother, twenty-five years of age, had never complained of any uterine trouble — had three children at full term of gestation — two having perfect limbs. Forceps were used in the first labor; the two following labors were natural and easy. Subject of present history quickened at fourth month of uterine life, and was felt most in the right side. During the gestation, a girl, with marked talipes-equino varus, resided in same room with the mother, exciting her constant fear of a like deformity afflicting her expected babe. A street musician with drophand, frequently came by her door during this period of gestation. At the birth of this child no cords, either natural or adventitious, were discovered encircling the limbs; cinctures were found, as represented in the engraving, near the middle of the right leg extending to the bone. Above and below, the leg was of normal size, but flabby; the foot of this limb presents a case of talipes varus, having tense contraction of the plantar fascia.

On the left leg, the cincture is immediately above the malcoli, quite to the bone; another encircling the foot over the metatarsophalangeal articulation, shallow compared with the others. These two cinctures present the appearance of a cord having been passed round the leg, and crossed on the instep and round the foot. The stricture here has seriously affected the foot, which is greatly hypertrophied, and apparently contains misplaced metatarsal bones — all this portion of the foot presenting a ball-like appearance, having rudimentary toes projecting.

In reference to blemishes, their cause cannot be satisfactorily explained. They are presented in some children in most extraordinary forms and degree, such as simple discoloration, growth of hair on extraordinary parts, and of different color upon the same person or parts of the body. By persons of vivid imagination, the idea is often entertained that these blemishes on children are similar to or resemble some object that has seriously impressed the mother's mind during the period of gestation.

In the breeding of horses, it is a very common practice to place various markings of white paint upon the stallion, in order that the foal may exhibit a similar appearance, and it is said that the desired object has been obtained in this way. This device is of very ancient date, as we have before remarked.

Regarding the influence of these mental impressions, many mothers have been impressed with the idea that the remembrance or ideal of some unfavorable circumstance, occurring long before the birth of their children, would be indelibly stamped upon them, is a matter of popular and almost universal belief. That there was a true representation of the object thus impressing the *enciente* is not so apparent to others as to the mother; yet an abnormal appearance, and on the part designated by the mother previous to the birth of the child, has been in numerous instances verified.

CONGENITAL LUXATION.

To Guerin, of Paris, and Prof. John M. Carnochan, of New York, we are indebted for some very valuable information upon the subject of congenital luxation of the hip, particularly as the number of subjects thus afflicted has been so limited as to afford but little practical knowledge in the way of treatment; and their writings have done much toward attracting general attention to this phase of aberration, which certainly has an equal claim to the skill of the surgeon with the more commonly observed abnormal conditions of suffering humanity.

The joint most commonly affected is the hip, and cases are often presented in which both the hips are luxated, though the other joints are equally liable to congenital dislocation, arising from the same and generally admitted cause of these abnormalities, namely, muscular retraction.

When the tendency to congenital luxation arises at an early period of gestation, the articulating cavities, it is reasonable to suppose, are of imperfect formation; and the luxation may be either partial or complete. Hence, in some cases, the child's true condition remains undiscovered for two or three years after birth, when it is made known by its feeble attempts at locomotion. After this period, complete dislocation ensues, and the final condition of the person is as represented in the accompanying engraving (fig. 8), the copy of a photograph of a patient who, when about four years old, was placed under our treatment, and is now seventeen years of age.

The luxation in this case was first discovered by the mother, when the child was nine months old. She immediately conferred with her physician, who informed her that it would be impossible for the child ever to walk. Subsequently, the child was placed in a chair, which, by the propelling power of its legs and feet, it moved about the floor—remaining in this condition until six years of age. During this period the mother died, and the charge of the child reverted to an aunt who, shortly after, desired our opinion as to the feasibility of placing the little cripple on crutches.

Upon examination, we found the thighs flexed upon the body, with a disposition to yield at the hips; and the knees unimpaired.

After two years' treatment, the patient was enabled to walk, independently of support to body or limbs; and at fifteen years of age danced with as much ease as others of the same age. This facility in dancing has been noticed in others similarly affected.

The effect of the dislocation of the heads of both femurs, is to produce a very peculiar appearance in the locomotion of the individual. The want of a firm central axis permits alternate elevation and depression of the hips, bending the body alternately in a lateral direction at each step. The motion is not such as is observed in a bow-legged child, or in a sailor, who has some firmness when the step is taken, but the gait is one of a constantly yielding movement, quite peculiar, and determining the condition, at once, to an experienced observer.

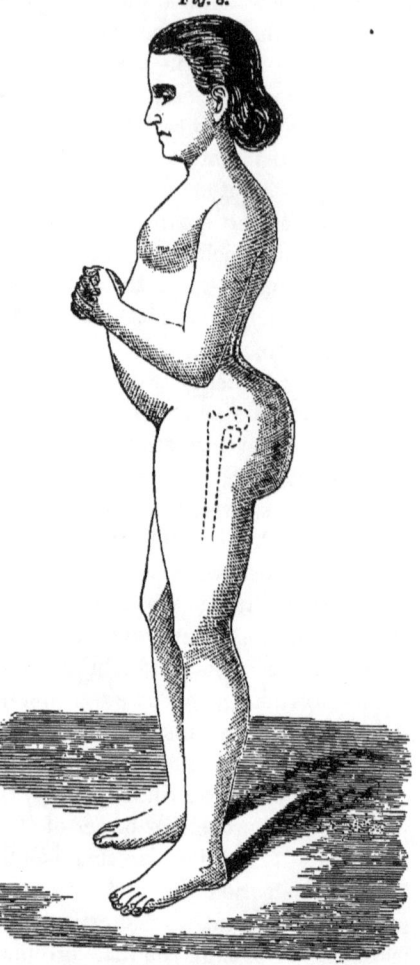

Fig. 8.

This malformation has been traced through many generations of individuals of the same descent. Dupuytren relates the history of a family of very aged persons, in which a woman of eighty; two maternal aunts, each seventy years of age; and a sister's child, were all thus afflicted. In this family, a woman who married a sound man had a daughter with this ailment. The daughter married a sound man, whose father had connate luxation of both femurs; and she gave birth to four children, in two of whom was represented the hereditary infirmity.

This hereditary disposition, that cannot be disputed, invites consideration as to the cause of the ailment; it is said to be muscular retraction. Is it not possible that there may be an arrest of development of the cotyloid cavity? For, in nearly all of the cases of this kind, the subjects have been in possession of more than ordinary good health, and devoid of any appearance of spasmodic disposition, or spastic contraction of muscles, the supposed cause, primarily, of intra-uterine ailments attributed to retraction of muscles.

ANCIENT TREATMENT OF DEFORMITY.

As we have stated, the treatment of deformities is evidently of very ancient date. Hippocrates taught the treatment of talipes 500 years before the Christian era; and, from his time to the sixteenth century, there is no other recorded dissertation to be found on this subject. It is not to be inferred, however, that in the interim there were no efforts made for the relief of these cases: The works of this great physician were preserved; and his instructions (though no doubt greatly modified) made available by his successors. That others treated deformities of the feet during the life-time of Hippocrates is certain; for, in his writings on the subject, he instructs that, after the bandage has been applied, the foot should be kept in a proper position by means of a leaden shoe made in the form of those of Chio; giving at the same time the model of boots to be applied if these should prove ineffectual.*

After the days of the father of medicine, Marcus Aurelius Severinus gave a treatise upon talipes; and following him, Ambrose Paré. We learn from the works of the latter, as translated from the Latin in 1665, the construction of an artificial arm, hand, and leg, by which extraordinary skill is displayed; and of apparatus for the relief of every variety of deformity of the human body — some of which were very ingeniously devised, and would compare very favorably with some of those in use at the present day.

In 1630, Fabricius Hildanus devised a method of treatment of club-foot; an apparatus for extending contracted fingers, and an extension splint for the relief of contracted knee joints.* This last is described as follows: "About the middle of the splint is a screw, to which is attached a ring, which encircles the knee. A pad is interposed between the knee and the patella; and by turning the screw, the joint is drawn backward toward its natural position."

---

* "Observationam et Curationum Medico Chirurgicarum."

This description is a tolerable representation of Amesbury's splint for a similar purpose, which is now in general use.

In 1651, Glisson treated deformity by extension. He states: "That the parts may the more be stretched, hang leaden shoes on the feet, and fasten weights to the body that parts may the more easily be extended to an equal length." This seems to be the earliest writer who recommends extension by weights and cords — now practiced as novel treatment, and in great repute; not only for the relief of deformity arising from contraction of muscles, but in the treatment of hip-disease, and fracture of the bones of the lower extremities; yet 217 years have elapsed since the introduction of this therapeutic agent.

For the treatment of deformed limbs and of the back, Scultetus, in 1656, described a mode of treatment by extension and counter-extension from the head and feet. Two years later, Circœus described shoes for the cure of talipes, very like in appearance and quite equal in principle to some of the apparatus in favor at the present day.

It would appear that in that day, as now, there were certain periods in which the medical profession became more than ordinarily interested in special subjects. Some twenty-nine years elapsed, when Isaac Mincius divided the sterno-cleido-mastoid muscle for the cure of torticollis, and Heister invented steel supports for the treatment of spinal curvature.*

Up to this period the treatment of deformity of the human body was altogether adventurous, when M. Andry's works appeared,† and the term *Orthopædia* was for the first time given as a proper designation for the treatment of deformity in children. Since then it has been applied to the treatment of deformity in general. This author states that he has "devised the term Orthopædia from two Greek words: *Orthos*, straight, and *pais*, a child." And, that "orthopædia" is to express in a single term, the plan of his work; which is to teach different modes of preventing and correcting the bodily deformities of children.

An interim of about fifty-five years now occurs, until the time of Venel, of the Canton of Berne, and Tiphaisne, of Paris; the latter a mechanic, and the former a professional adventurer, who, from

---

* Article in the "British Medical and Chirurgical Review," for October, 1861.
† "L' Orthopædiè," par M. Andry, 1741.

selfish motives, kept his discoveries in treatment as secluded as possible. To accomplish his unprofessional designs, he established an institution in which he attained much notoriety for his successful treatment of talipes. His enterprise was only of short duration, and terminated at his death. Dr. Ehremann, a patient of Venel, acquired a tolerable knowledge of his apparatus and treatment, which was subsequently practised by Brückner, of Gotha, and Naumberg, of Erfurt. The latter published the results obtained from Venel's treatment in 1796.*

In the preface of a translation of Scarpa's monograph "*Sulle Piedi Ferti*," into German, by Malfatti, a description is given of the proceedings of Scarpa, which led to the construction of his celebrated, and, with slight modifications by noted Orthopædists, the most approved shoe at the present day for the treatment of talipes varus and valgus. It is stated that, in 1781, Scarpa resided in Paris, and passing by Tiphaisne's abode, he noticed in the window drawings of deformed feet and of those that had been cured. He became immediately very desirous of obtaining information in regard to this successful treatment, and sought an acquaintance with Tiphaisne. The only information he derived thereby was from an incidental remark made by this skilful man in the course of general conversation, viz.: "Nature will not yield to violence, but only to gradual force." This excited still more the curiosity of this ingenious surgeon, and he determined to obtain the information at almost any peril—not for any sinister purpose, but that he might make known his discoveries for the benefit of his suffering fellow-man. It is stated that Scarpa availed himself of an opportunity, during the temporary absence of Tiphaisne, to enter the private room of the latter through favor of the housekeeper, where he found only a steel spring lying on a cushion.

Although his conduct in this special matter is not commendable, yet it does not detract from his indomitable energy and ingenuity, by which he gave to posterity so valuable an invention. From this very circumstance it is most reasonable to award the credit of the invention to Scarpa, the most eminent anatomist and surgeon of his day. He published his remarkable work about the beginning of the nineteenth century, after having made himself familiar with the ailment, and the condition of the deformity, which latter know-

---

* "Abhandlung über Verkrümmungen."

ledge no one had hitherto acquired. He discovered and asserted that the bones of the tarsus, in cases of talipes, are never luxated, but only partially separated from mutual contact and turned according to their smallest axis. In all probability it was the knowledge of this important and impressive truth — *that gradually increased elastic force is the most effectual means of redressing distortion, both of body and limbs, when resulting from contracted muscles* — that facilitated the designing and construction of his celebrated shoe, so simple in design, and yet complete and efficient in purpose, because of the continued leverage force from elastic springs, tending to oppose the contorted form of the yielding foot.

The application of elastic force is one of the most important practical principles in the construction of apparatus to overcome contraction of muscles or other elastic tissues. Fixed force will, if applied to contracted muscles, be resisted, to the exhaustion of the vital energy of the parts.

ANCIENT TREATMENT OF SPINAL DEFORMITIES.

The treatment of spinal deformity in the time of Hippocrates was — as we gather from his writings, in which he gives a description of the structure of the spine and the modes of treatment when contorted — accompanied with some very rude methods. One of these was to suspend the patient by the feet, to produce what he termed succussion. Another, was that of extending the patient, by means of bands under the arms and about the loins, and then forcing the spine inwards by the use of a lever.

This most dangerous treatment was also practiced by Paré in the sixteenth century.* Means were devised by this extraordinary physician for the treatment of all manner of deformity — even that of cleft palate, by the application of a well-fitting gold or silver plate to the defective palatine roof, by means of an ingenious method of retaining them *in situ;* and, to the present day, no improvement has been made upon his appliance. Recently, however, it has been superseded by a most ingenious device made of India rubber, enabling the individual thus afflicted to articulate distinctly, and to

---

* "Memoirs de l'Academie Royal de Chirurgie de Paris,"—Vol. IV.

swallow fluids with normal facility, by Norman W. Kingsley, M. D., a noted dentist of the city of New York.

Paré's metallic corset is said to have been the earliest piece of mechanism intended for the relief of the distorted spine; and, as shown by the drawings, does not differ very materially from an apparatus constructed by Dr. Abbe, of Boston, and, subsequently, by Dr. Brewster, of New York city, in 1850; being a perforated metallic casing of tolerable form, and in segments hinged together, into which the body tended to conform by continued pressure.

In 1768, M. Le Vasher devised an apparatus for supporting the head and gradually extending the spine in cases of curvature. A strong, well-fitting corset, applied to the body, sustained a movable iron rod fixed to a socket behind, and extending over the head in the form of an arch. From this arch was suspended a curved bar of polished steel, extending from ear to ear, and to which were attached chin and occipital straps. When adjusted, the head was held erect, and the weight of the same taken from the spine, a most desirable apparatus for the relief of caries of the cervical vertebræ.

In 1779, David, of Rouen, France, and Pott, of England, introduced, probably, the first rational ideas of the pathology of spinal disease, and placed the treatment on a scientific basis. Schmidt, of Marburg, in 1794, constructed a similar supporter, the base to be placed upon the costa ilii, with two metal bars to pass up on either side, terminating at the axilla, and having crutches which could be elevated to the extension of the body, and retained *in situ* by a band across the dorsum of the patient. This form of support is now in common use for the treatment of caries of the spine, both in this and the old countries. Mr. Sheldrake, a very ingenious mechanic, of London, and a man possessed of more than ordinary practical good sense, in or about the same year published a book, in which he described a variety of means for the treatment of deformity, and made some excellent suggestions as to the construction and use of apparatus, except in the case of caries of the spine. This ailment he treated by extension and confinement in the recumbent position, a treatment advised by Darwin in his Zoonomia (vol. 2), who gave the first intimation of the supposed advantage of keeping patients, laboring under spinal disease, in this position during treatment. This was in 1796, and it was subsequently highly extolled as a treatment of disease of the hips, and, as an auxiliary extension by means of attached weights and pullies to the foot or feet of the patient, is favorably considered at the present day by many eminent surgeons abroad and at home, while others condemn it as compromising the normal energies of the patient to the increase of impairment of health and aggravation of the abnormal condition. And we claim the verification of this in many cases thus treated.

SURGICAL MEANS OF REDRESSING DEFORMITY.

Section of the tendo-Achillis was favorably considered as a method of treatment for talipes for nearly a century before its introduction as a generally recognized practice, and is now so modified as to make it a safe and efficacious mode of treatment in the hands of skilful surgeons. It will be seen, by reference to the annals of medicine, that, as far back as 1769,[*] Hains, a surgeon of Dijon, with the view of establishing the fact, availed himself of a series of experiments, the result of which was favorable. He severed, both partially and wholly, the tendo-Achillis of cats and dogs, and, although the animals were left entirely to themselves, and no precaution taken to exclude the air, all the wounds became perfectly healed. Subsequently, and within a very limited period thereafter, Molinelli[†] opposed the generally received opinion of that time, and stated that wounds of the tendo-Achilles do not heal favorably; yet he made a statement in the "History of the Academy of Bologna," that four cases were noticed by him, in which the tendo-Achillis was divided transversely, and, being accidental, was complicated, yet healed kindly.

In relation to the division of tendons as a treatment for deformity of the human feet, the earliest information we have upon the subject is that Thelsenius, of Frankfort, proposed that the tendo-Achillis should be severed in the case of a young lady, seventeen years of age, laboring under talipes varus, and that, under his direction, the operation was performed March 26, 1784, by a surgeon named Lorenzi. This is reported as a successful operation, although the external integument was largely divided before dividing the tendon, when the heel descended two inches. The cure is said to have been complete on the 12th of May, in the same year, fifty-two

---
[*] "Journal de Medecin," Janv.
[†] "Comment. Academ. Scientaire: Bononiens," p. 189–196: 1773.

days being a comparatively limited period for the cure of club-foot, *i. e.*, to heal and perfectly straighten the foot after so severe an operation.

In the year 1806, on the 10th of May, the operation of Lorenzi was repeated by Saratorius in a case of talipes equinus, resulting from abscesses on the posterior of the leg, the patient being a boy thirteen years of age. The first step of procedure was to apply a tourniquet to the femoral artery. A longitudinal incision, four inches in length, was then made over the tendo-Achillis, the integument dissected, and the fascia divided on a director to the same extent. The tendon was severed transversely, but, the foot not yielding, the incision was extended to the os calcis; the cicatrices were severed and the tendon isolated, but the joint failed to yield. Saratorius informs us that he employed his whole strength, when, the joint giving way, such a noise and crackling ensued as though the whole of the bones had been broken. Symptomatic fever set in and suppuration followed, though not to any great extent. After nine weeks the wound cicatrized, complete anchylosis of the ankle having resulted, though the patient was able to walk easily. This is the first information we have upon the adventurous practice of overcoming adherent attachments about joints by violent force, which has been introduced as a novelty within a few years past in orthopædic practice, upwards of half a century having elapsed.

Michælus of Marburg, in Germany, is reported as having treated successfully several cases of talipes equinus, in 1809, by partially severing the tendo-Achillis, being enabled thereby to bring the feet at once into a natural position. It seems, though, that this would be impossible without complete section of the tendon.

Benjamin Bell, in remarking upon the generally received opinions on the treatment of distorted limbs, expressed himself, in 1801, as follows: "It has been a prevailing opinion among practitioners that little advantage is to be derived from any remedies that we can employ for distorted limbs, and they have seldom made any attempt to cure them, in consequence of which this branch of practice has been almost universally trusted to itinerants or to professed bonesetters." From this the notorious bone-setters may claim a respectable antiquity. "In this, however," says this eminent surgeon, "we are wrong. In saying so, I can speak with confidence founded on much experience. Having early in life observed the misery to which patients were reduced, I was resolved to make some attempt for the relief of such as might apply to me, however small the chance might be of succeeding; and, in various instances, I have had the satisfaction of relieving, and, in some cases, of curing completely patients who had been lame for several years, and where it was not expected that any thing could be done for them. Various machines have been invented for the removal of distortions of the spine by pressure. All of these, however, do harm, and ought not to be used. It must at once appear to him who is acquainted with the anatomy of the parts and the nature of this disease, that the displaced bone is never to be forcibly pushed into its situation; and if this cannot be done, it is evident that much harm may ensue therefrom. In all distortions of the spine, it is an object of the first importance to support the head and shoulders. If this is neglected, the weight of the head tends almost constantly to increase the curvature."*

This eminent surgeon could not have considered the advantages to be derived from lateral support applied to the body, as a basis of support to the suspension of the head by means of a steel bar and straps. It is true that pressure upon a projecting portion of a diseased spine is injurious, while that of lateral support is of inestimable value in the treatment; as it is a preventive to motion and attrition — the principal cause of absorption of the impaired tissues. Above all, a well-regulated régime is essential to an increase and diffusion of the vital forces that ensues when rest to the spine is obtained by lateral support sustaining the incumbent weight of the parts above the lesion, thus permitting the patient to take exercise in the open air. The support of the head and shoulders is all that is intimated by this eminent authority. The support of diseased joints by suitably-devised apparatus can only be considered as auxiliary to other therapeutic agencies in their treatment; and then with a favorable result only when the patient's health can be improved.

The remarks of Benjamin Bell in regard to the indifference of the educated practitioners of his day as to availing themselves of practical information tending to the relief of the deplorable condition of deformed persons, are equally applicable to the profession at the present day. It is only a very limited number of men, celebrated for their attainments in the medical profession, that have devoted a moiety of their attention to this most legitimate branch

---

* "A System of Surgery," 7th ed., vol. VII, p. 197.

of medical science. Hence, adventurous treatment has been tolerated by even the most eminent in the profession — as noted in the early history of medicine.

Certain practical discoveries were made many years before the accepted introduction of the subcutaneous division of muscles and tendons, that should have invited earlier attention to this invaluable practice. John Hunter, in his "Treatise on the Blood, Inflammation and Gun-shot Wounds," published in the year 1794, narrates certain physiological discoveries that should have prompted the subcutaneous operation for the relief of contracted muscles and tendons in cases of irreducible dislocations presenting in accidental injuries, if not for the cure of congenital deformities or the sequence either of infantile paralysis or chronic disease of the joints. It is patent that, if it had been made available in the one instance, it would certainly be equally applicable in others.

This eminent surgeon says: "The injuries done to the sound parts I shall divide into two descriptions, according to the effects of the accident. The first kind comprises those in which the injured parts do not communicate externally; as concussions of the whole body or of particular parts, strains, bruises, and simple fractures, either of the bone or tendon, which form a large proportion. The second consists of those which have an external communication, comprehending wounds of all kinds, and compound fractures.

"Bruises which have destroyed the life of the part may be considered as a third division, partaking at the beginning of the nature of the first, but finally terminating like the second.

"The injuries of the first division, in which the parts do not communicate externally, *are but seldom subject to inflammatory process; while those of the second, commonly, are both inflammatory and suppurative.*"

To Mr. Hunter must be awarded the merit of being the first to invite attention to the true pathological sequence of injuries, as well as ideas suggestive of subcutaneous surgery, as it is known that he performed successful operations upon tendons of dogs, which he severed subcutaneously. He, at different periods, killed the animals operated upon, finding in every case the tendons he had divided subcutaneously firmly united. These experiments were performed in the year 1767.

It is truly remarkable that forty-nine years should have elapsed from the time of Hunter's experiments upon quadrupeds before

Delpech performed the operation upon the tendo-Achillis of man, which, however, proved so unfavorable as to deter him from further effort. That he was impressed with the essentials to success must be conceded, as stated in his work "L'Orthomorphie," published some thirteen years after his unsuccessful operation of making a longitudinal incision of one inch on either side of the tendo-Achillis, thus leaving a portion of the skin over the tendon when severed. This procedure may be, considered as premising the subcutaneous severing of tendons for the relief of contorted muscles. This occurred in 1816. That the one failure should have deterred so noted a surgeon is extraordinary and almost incomprehensible. It should have induced sufficient courage and desire to prompt at least another effort; and, in the event of that proving successful (which it undoubtedly would have done), he would have been placed on the highest pinnacle of fame. But this was to be left for the eminent Strömeyer, of Hanover, who availed himself of the rules laid down by Delpech, which were truly worthy of consideration, and inaugurated a series of practical experiments which resulted in a decided success.

Strömeyer performed his first operation on the tendo-Achilles in the year 1831, and published an account of the subsequent results of treatment in the years 1833 and 1834, thus furnishing the initial of scientific tenotomy — the success in this new treatment being one of the greatest achievements in modern surgery, and so considered by all recognized authority in the medical profession.

There appears to have been for several years among eminent practitioners a reluctance to countenance or endorse the great discovery of Strömeyer, or rather a disposition to cast a doubt upon the verity of his success; but truth again prevailed, to the benefit of mankind, when the enterprising Dieffenbach, of Berlin, entered as a laborer in this field of new discovery, thereby inciting others to do the same.

In 1835, Duval, of Paris, was the first to sever, subcutaneously, the tendo-Achillis in that city; then Bauvier, and, about the same time, Pauli, in Germany, and Jules Guerin, of Paris — the latter being a most indefatigable investigator and daring operator. During this period, Bonnet, of Lyons, and Scoutetten, of Strasburg, entered with great energy into the scientific treatment of club-foot; and, in less than five years, several hundred cases of, not only club-feet, but of all superficial contracted muscles in all parts of the body and limbs producing deformity, were severed subcutaneously. Many of the deformed were thus restored to normal conditions, with scarcely an unfavorable case resulting from the practice, placing opposition at defiance; for the scrutiny of the most eminent in the profession, both at home and abroad, was brought to bear upon these adventurous practitioners, as they were considered; and, if there had been a permanent injury sustained by a patient, or any entirely unsuccessful case, they would have been noted and exposed, so strong was the opposition and so severe the criticism.

Mr. Whipple, of Plymouth, England, informed Mr. W. Adams, of London, that he severed the tendo-Achillis in May, 1836. This was the first operation of the kind known to have been performed in England. Dr. Little, whose account is exceedingly interesting, divided the tendo-Achillis in the city of London, February 20, 1837. The history is given by Mr. W. Adams in his Essay; to which was awarded the Jacksonian prize of 1864, by the Royal College of Surgeons, London.

In 1837 the first publication was given of the operation having been performed in this country by the subcutaneous severing of the tendo-Achillis and other tendons for the relief of club-foot. This was by William Detmold, M. D., since Professor of Orthopædic Surgery, who in this year arrived in New York city from Germany. He treated several cases of club-feet by the preparatory operation of severing the tendo-Achillis by the new method, of which he published the account in the "Medical Journal" of Philadelphia. Professor Nathan R. Smith, of the University Medical College of Maryland, states that he performed the operation previous to this date; but, as a report of the case was not published, the credit is due to Dr. Detmold. This enterprising surgeon and teacher stated, in a course of lectures, delivered in 1842, that he had operated in about four hundred cases of club-foot. Many other surgeons, during this period of five years, tested the new operation, but unfortunately failed, in extraordinary cases, to realize their expectations, because of their inexperience in the subsequent treatment. Hence, doubt and indifference has limited the practice to a few of the energetic and more enterprising practitioners, who, availing themselves of their experience in devising variously modified apparatus, have been more successful. This has, however, tended to each practitioner's lauding his own special method of treatment in the redressing of the deformity. The neglect, unfortunately, to divide the tendons of unyielding shortened muscles, tense fascia, and, in some cases, ligaments, when, if premised by careful perseverance in the application of various devices, manipulation, and the skilfully applied roller, they would have restored nearly all cases of this deformity to normal appearance.

And yet it is much to be lamented that many inexperienced in orthopædic practice attempt the treatment of club-foot, relying mainly on the operation, to the great discouragement of the patient — the tendons having, in numerous instances, been severed several times, and the case finally abandoned, after having inflicted much suffering upon the unfortunate cripple, who could have been easily cured by skilful treatment. These are about all the patients who are suffering from injudicious advice for this complaint after treatment, and not cured at the present day; except the very indigent, who could not heretofore avail themselves of the opportunities for treatment now existing; and those living in distant and sparsely inhabited parts of the country, where it is impossible for the patients to be supplied with proper appliances, and daily attendance from persons skilled in the manipulation and bandaging of their limbs; the latter auxiliaries being as essential to successful treatment as the severing of constricted tissues; and in many cases of slight deformity of the feet, skilful, persevering manipulation and bandaging will restore them to a normal condition.

CHAPTER II.

IMPAIRMENT OF TISSUES RESULTING IN CONTORTIONS.

Contortion of body or limb the result of impaired integrity of the fibrous tissues.— Paralysis of the extensor muscles.— Under cerebro-spinal irritation, the muscles not entirely under the control of the will.— VARIOUS FORMS OF CONTORTION OF THE FEET.— Talipes varus, talipes valgus, talipes equinus, and talipes calcaneus, their definition.— Non-congenital contortions exceed the congenital in the proportion of three to one.— Talipes varus, valgus and calcaneus generally take place during intra-uterine development.— CONFORMATION OF THE FOOT.— Difference of construction between the human foot and that of the bear.— The articulation and elasticity of the foot, and its relation to the leg.— The malleolus internus and externus.— Early diagnosis of tendency to deformity of the highest importance to patients.— A thorough knowledge of anatomy necessary to a correct diagnosis.— The outer contour of the foot in its normal condition.— ABNORMAL POSITION OF BONES, LIGAMENTS AND MUSCLES IN TALIPES VARUS.— Luxation invariably found.— The Astragalus least liable to deviation, especially in cases of dental paralysis.— Change of position of the cuboid bone.— Marked deviation of the os calcis, and the relative articulating surfaces.— Position of the cuneiform, metatarsal, and phalangeal bones.— Diminution of the plantar aponeurosis.— Change in the normal relation of the muscles.— In extreme cases, the peronei paralyzed.— Duverney's assertion that the loss of contractile power is a primary cause of talipes varus.— TENDENCIES TO ABERRATION OF FORM.— Yielding of the ligaments tends to produce varus.— Impairment of the normal conformation of the arch.— Diagnosis of congenital varus.— Difference between congenital and non-congenital varus.— Talipes valgus, its leading features.— Talipes equinus, its modifications and chief features.— Talipes calcaneus, differences between that variety and the preceding.— Talipes calcaneus occasionally attributable to careless treatment in the subdivision of the tendo-Achilles.— Views of William Adams on talipes equinus.— His diagnosis of the affection.—The ligaments subject to structural impairment.— THE CONDITION OF THE MUSCLES IN TALIPES VARUS.— If proper remedial measures are at once taken, a perfect and simultaneous cure results with the growth of the body.— Position of the tibiales posticus and anterior tendons in severe cases.— INFLUENCE OF LIGAMENTS IN TALIPES VARUS.— Method of obtaining accurate diagnosis of structure of ligaments.— Infantile paralysis generally the cause of non-congenital contortion of the feet.— Permanent deformity arising from compression of the foot, as in the Chinese.-- The plantar fascia.— Condition of the legs in cases of prolonged contortion.— CONTORTION ARISING FROM PARALYSIS.— With proper treatment, care and perseverance, restoration to normal power almost certain.— SURGICAL TREATMENT OF CONTORTED FEET.— Special study and experience on the part of the practitioner necessary to ensure a successful result in the treatment of contortions of the body or limbs.— PREPARATORY STEPS IN TREATMENT.— Description of dressings.— Preparation prior to operation— MEANS OF RESTORING FOOT TO NORMAL FORM.— Course of procedure after operation.— Treatment of patient after successive paralytic seizures.— Suppuration of tendon occasionally supervening in cases of strumous diathesis.— SEVERING THE PLANTAR FASCIA.— The treatment necessary in such cases.— R. M. Tamplin's testimony as to the danger of puncturing an artery in tenotomy.— Manipulation essential as an adjuvant.— Application of the Scarpa shoe beneficial.— Congenital club-foot curable by perseverance in proper treatment.— THIRD STAGE OF TALIPES VARUS.— The treatment.— On severing the posterior tibial tendon.— Mies and William Adams on severance of the tendon.— Continued extension.— TREATMENT OF PUNCTURED ARTERIES.— Subcutaneous severance of tendons.— Firm coagulation of the blood an incident in the diagnosis.— TENOTOMY KNIVES, description of.

CONTORTION of body or limb is the result of impaired integrity of the fibrous tissues. The ligaments are the primary sustaining media of the skeleton when in their normal condition; and by their elasticity are made subservient in a limited degree to the action of the muscles. Conjointly, they maintain all that pertains to the normal form essential to the movement of the body and limbs, constituting, with the cerebro-spinal functions, the locomotive condition of the animal economy. From this view of the animal system we will consider the diversions that tend to contortions.

Existing contortions, thus considered, would indicate a loss of integrity in the ligaments without regard to the inimical cause, as it is impossible for a permanent contortion of the skeleton to take place without a yielding of the ligaments to the unfavorable influence that tends to the final contortion. The yielding of ligaments will impair normal muscular antagonism to the extent of the loss of the essential equilibrium of action that sustains the form, and thus contortion is induced (from that of a slight impairment to that of an almost entire degeneration) from the increasing loss of normal integrity of the ligaments.

A paralytic seizure of the extensor muscles of a limb impairs their muscular tension, and thus permits the flexors to shorten and remain quiescent. Long-continued unfavorable position, from whatever cause, tends to a similar result. These inimical influences impair the muscular tissues primarily, and the ligaments secondarily and *vice versa*.

Cerebro-spinal irritation influences the normal motor-power of the muscles, producing the most intractable cases of congenital and non-congenital contortions. The muscles, under this abnormal influence, are not subject to the entire control of the will, but only partially so, and by a spasmodic movement, the contortion yielding for the time to slowly increasing force — in many instances to normal force. But from every effort of the will to progressive movement, the spasmodic disposition is exerted to a resistance terminating in contortion. These remarks as to the conditions attending the impairment of tissues resulting in contortion, though necessarily brief, are essentially requisite to the proper consideration of the several varieties of contortions of the body and limbs.

VARIOUS FORMS OF CONTORTION OF THE FEET.

Contortions of the feet appear to be the most numerous of all to which the animal conformation is subject; the non-congenital exceeding the congenital three to one, according to the statement of Mr. W. Adams, which statement agrees with our own observation.

The following four designations constitute the recognized varieties of contortions of the feet:

1st. *Talipes varus:* inversion of the anterior portion of the foot, with elevation of the heel.

2nd. *Talipes valgus:* eversion of the anterior portion of the foot. Of this we have two forms: *first*, a depression of the heel; *second*, an elevation of the heel.

3d. *Talipes equinus:* elevation of the heel, and extension of the foot.

4th. *Talipes calcaneus:* depression of the heel. Of this there are four intermediate forms: *first*, the foot is slightly inverted; *second*, the foot everted from slight yielding of the plantar tissue; *third*, an approximation of the heel to the anterior of the foot from slight contraction of the plantar tissues; *fourth*, a decided eversion and elevation of the anterior portion of the foot. All these may be included as varieties, though the classification is of but little advantage in the practical treatment. As such cases present, they will require to be treated according to the indications derived from a general knowledge of the subject.

The *first, second* and *fourth* of these varieties of contortion of the feet, are those generally recognized as having taken place during intra-uterine development, and some of them in the early formative stage. Of the latter, the bones have conformed in shape to the contorted condition; rendering it almost impossible to redress the deformity. Fortunately, there are but few such cases. Of this condition of contorted feet, we have only had to treat about eight extreme cases in thirty years, of orthopædic practice — the patients being much improved after persistent effort; the instep remaining prominent and the feet short and cramped in appearance. However long this continued course of perseverance in treatment may be required, if improvement is progression, the effort is remunerative in the relief afforded the patient in his rescue from a wearisome, dependent, and despondent existence, even though the foot may not be entirely restored to its normal form.

CONFORMATION OF THE FOOT.

In order to obtain a proper knowledge of the contortions of the feet, attention must be given to the conformation of the foot in a normal condition, and for this purpose we give the following wood-cut (Fig. 9), representing the skeleton of the human foot in ordinary form, and its relation to the leg, shown by an outline of the surrounding integument. The tibia and fibula, being separate, present more clearly the tibio-tarsal relation.

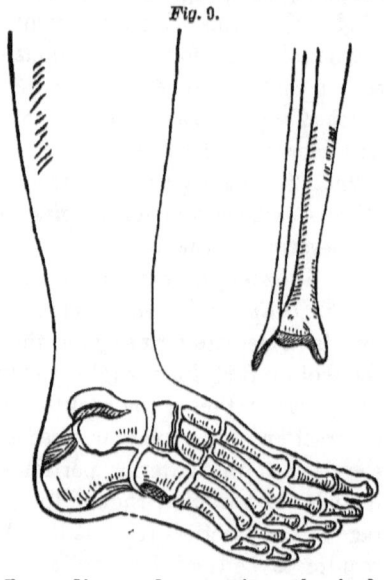

Fig. 9.

A line drawn from a central point of the distal portion of the os calcis curving to a central point of the great toe, bounding the outer side of the foot, nearly describes a semicircle, and both feet placed together completes the circle. In ordinary locomotion the body is sustained by a complete circular bearing of the outer margins of the feet, limited in breadth of base, being totally different from the plantigrade feet of the bear — the whole surface of his feet being planted at every step. In the foot of man the step is accomplished by the heel touching the ground first, the body advancing and concentrating the force upon the metatarsophalangeal bearing — the great toe mainly governing the regular movement, which is completed with elastic force. tending to facilitate the succeeding step. The plantar surface of the foot of man does not leave an impress upon the ground, as does that of the bear or lower grade of animals, which, unlike man, are not subject or liable to lateral deviation of the feet.

This invites our attention to the articulation and elasticity of the human foot, and its relation to the leg. The tibia rests upon the *rounded* surface of the astragalus, and is maintained laterally by the malleola, internus and externus, sustained by strong ligaments attached to the os calcis. The lower portion of the *malleolus internus*, in an ordinary sized man, is elevated from the ground about *three inches*, and the *externus* about *two and a half;* admitting of lateral deviation and stress upon the sustaining ligaments that are liable to elongation. In the foot of the bear the malleola are only about *an inch* above the plantar surface, and are of equal elevation; and this animal, being a quadruped, does not require the elasticity of the biped, man, to enable him to walk. Hence the difference between the two plantigrade animals, the latter not being subject to contortion of the feet.

Unfavorable impressions during gestation, as we have before stated, impair the articulatory relations of the bones in the foot to an abnormal condition of increasing tendency. The slightest yielding of any point gives direction to the subsequent condition, definable as a variety of contortion bearing a special designation. An early diagnosis of tendencies to deformity of the feet is of much importance to the patient, and a prompt decision is required from the surgeon. To obtain the requisite practical ability for this, a *thorough* knowledge of anatomy is required as well as careful clinical observation — the former, being a branch of surgical science, must be *well* studied in order to obtain a practical insight into the intricacies of orthopædic surgery. The outer contour of the foot in its normal condition is first to be carefully considered. The foot presents an elastic arch of unequal lengths of span, diminishing laterally. This is not so apparent as is really the case. On the inner face, the arch of the skeleton is clearly defined; on the outer, the yielding integument fills the shortened arch to a level with the bearing of the foot. This will be more clearly observed by reference to the skeleton of the foot as shown in Figures 10, 11, and 12.

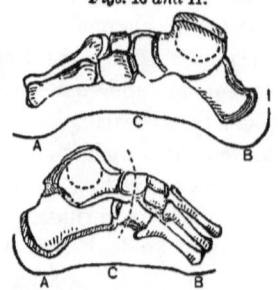

Figs. 10 and 11.

Figure 10 represents the inner face of the arch; A and B the points of bearing; C the outline of the integument of the decreased arch.

Figure 11 represents the outer face of the arch; A and B the points of bearing; C the integument sustaining the arch.

This diminishing of the span laterally presents a curvilinear face to that portion of plantar surface constituting a transverse as well as a longitudinal arch; see dotted line in fig. 11.

Fig. 12.

The outer face of the arch is more distinctly observed when the individual is balancing on the anterior portion of the foot.

Figure 12: A the posterior portion of the foot elevated; C the point pressed upon; B the defined arch.

ABNORMAL POSITION OF BONES, LIGAMENTS AND MUSCLES IN TALIPES VARUS.

Among all the articular changes that take place in the bones, in cases of varus, and of the most extreme forms, they are never found to be completely luxated, there being only a deviation, with partial separation of the articular surfaces from the yielding of ligaments upon the outer portion of the foot, and the fixed shortened condition on the incurvation, and, in some cases, adaptation of the bones in development to the abnormal conformation.

The *Astragalus*, of all the bones of the foot, deviates the least in cases of varus, other than its relation to the tibia, admitting of limited lateral deviation, and more especially in cases resulting from dental paralysis.

The anterior articulating surface, because of its normal position, presents a protuberance upon the dorsum of the foot, and is rendered thus prominent by the rotation of the navicular bone inward and obliquely across the extremity of the astragalus to a close contact with the internal malleolus.

The *cuboid* bone becomes changed from its normal position, being separated from the lesser apophysis of the os calcis and rotated upon its lesser axis. It is chiefly upon this bone that the weight of the body is borne when walking, and limited to the bursæ that is formed over this bone, as before described.

The os calcis presents the most decided deviation of any of the bones of the foot, when complete varus has supervened. It is so completely rotated upon its lesser axis that its posterior tuberosity is in contact with the internal malleolus, and drawn so forcibly upward by the tendo-Achillis that it is scarcely definable, the anterior articulating portion being equally depressed, thus making a very decided change in the relative articulating surfaces with that of the astragalus.

In conformity with these changes we find the three cuneiform, the metatarsal and the phalangeal bones all directed upward and inward, tending from a horizontal to a vertical position. The ligaments that sustain the bones in this position are greatly impaired, shortened or lengthened in conformity to the malposition of the bones. In varus, all those on the inner side of the foot are tense and firm, though on the outer side some are greatly elongated, especially at the articulation of the cuboid with the os calcis. The *plantar aponeurosis*, in extreme cases, unless severed, will be found diminished in breadth and exceedingly tense, being an almost insuperable obstacle in the redressing of the contortion of the foot.

As to the muscles, their normal relations are greatly changed. The *gastrocnemii solei* and *plantares* are much shortened, and the tendons partially separated in the common sheath near their insertion, their shortened condition impairing other muscles, from continued stress, and are the principal muscles implicated in producing several varieties of contortion of the foot. In extreme cases of varus, the two tibials, the flexor longus of the toes, and the adductos all contribute to the contortion. By the shortened condition of these muscles the peronei are so much impaired in tone as to nearly lose their power of contraction and to be in a paralytic condition. Duverney and others consider the loss of contractile power in the peronei muscles as a primary cause of varus. That varus has been the result of this derangement cannot be doubted, nor that a shortening of the flexor tendon has produced a similar result from a continual extension of those muscles. In either case, the muscular equilibrium has been impaired, and may have been so impaired, by the extraordinary yielding of the ligaments. Hence these abnormal conditions of muscles or ligaments tend to contortion of the feet.

TENDENCIES TO ABERRATION OF FORM.

It is the yielding of the ligaments which sustain this outer portion of the arch that tends to produce varus. The normal bearings are relieved by the settling down of the central portion of the outer margin of the foot, inclining the anterior portion inward.

Figure 13, the skeleton of the foot; A, the elongated ligaments that sustain the metatarsal and cuboid bones in normal relation, and also the ligament that sustains the os calcis and cuboid.

Fig. 13.

This yielding of the ligaments that sustain the outer arch of the skeleton of the foot impairs the normal conformation of the arch; not apparent, however, to a casual observer. The base of support is now concentrated upon the center of what constituted the arch of the outer margin of the foot, the two distant points of support having yielded, and gradually limited the base of support to a very small space.

Fig. 14.

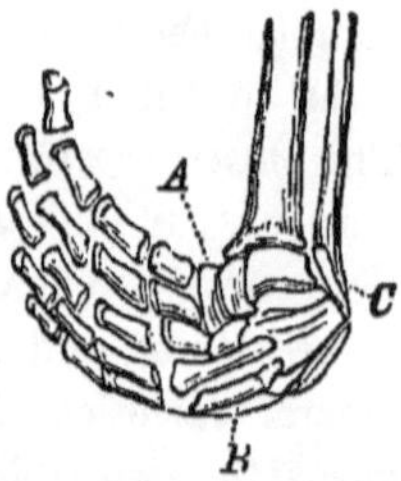

Figure 14. A and B represent the central line of separation; C the articulating surfaces of the astragalus and os calcis, changing position from the scaphoid and the cuboid.

The ligaments yield to an extraordinary degree, permitting certain muscles to shorten and become impaired in their integrity from the abnormal quiescent condition tending to the partial arrest of nutrition.

Congenital varus usually presents but a slight curving inward of the child's foot before it has walked. The curving inward of the anterior and posterior portions of the foot would seem to be the result of pressure from the weight of the body after the child has commenced walking upon the limited abnormal base.

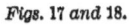
Figs. 15 and 16.

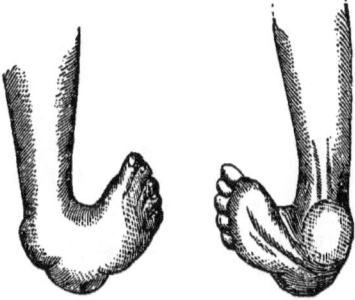

Figures 15 and 16 present the congenital varus before walking. Non-congenital talipes varus differs in the first stage from that of congenital, because of the weight upon the feet from the commencement of the difficulty. The anterior portion of the foot in either case inclines inward, and in walking, only one-third of the outer margin touches the ground. First, the heel being elevated, giving thereby to the patients an awkward appearance in walking; they are then said to be pigeon-toed, and finally club-footed, when deformed as seen in Figs. 17 and 18.

Figs. 17 and 18.

The eversion of the foot, termed talipes valgus, is attended with much less contortion than varus. Valgus arises from a yielding of the inner margin of the arch; tending only to a flattening and spreading of the parts laterally, as it is impossible for the outer margin to turn upward. The os calcis, in some cases, becomes considerably elevated from shortening of the muscles. Figures 19 and 20 represent this condition of the foot.

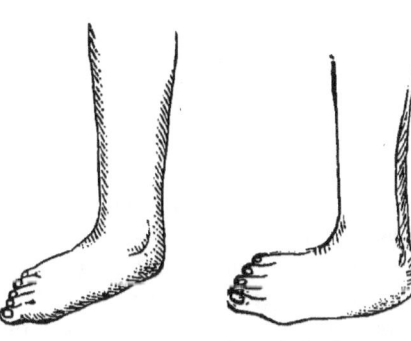
Figs. 19 and 20.

This distortion represents the flat or splay foot.

Of the contortions of the skeleton of the foot, varus and valgus represent the type of all the several varieties included under the head of club-foot.

In talipes equinus there is but a very limited change in the skeleton of the foot; the principal concentration of force being upon the metatarso-phalangeal bearing—allowing the toes to conform to the elevation of the posterior portion of the foot, as represented in Figures 21 and 22. A slight inclination inwards entitles the case to be designated equino-varus, or a reverse change, equino-valgus. Thus the several variations have obtained their designation.

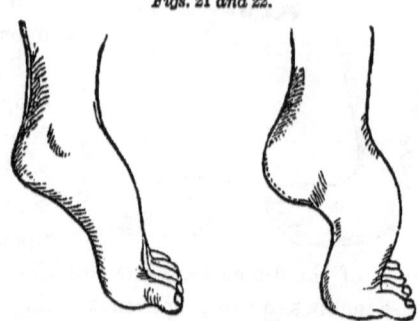

Figs. 21 and 22.

Talipes calcaneus presents a very decided change in the skeleton. The proximal end of the os calcis is elevated to the contracting of the longitudinal arch of the foot, the heel becoming the most determined point of bearing, and in some cases sustains or is constituted the only point of bearing. In this we have a shortening of the ligament and aponeurosis of the plantar surface of the foot as a consecutive result of a yielding mainly of the gastrocnemii plantar and soleus muscles. This is one of the abnormal conditions of the fœtus, and has resulted also in some cases from the subcutaneous division of the tendo-Achillis — a truly lamentable occurrence that can only be attributed to a careless subsequent treatment. Figure 23 fully represents this condition in a mild form. That the ligaments are subjected to unfavorable changes is generally admitted; that they may be impaired by posture and become permanently shortened, or, by loss of tenacity, lengthened; in this way causing an alteration in the form of the skeleton to the impairment of the equilibrium of normal force in the muscles is likewise generally admitted.

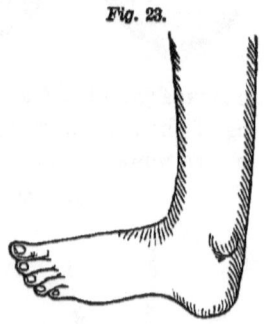

Fig. 23.

Mr. William Adams, on page 85 of his prize essay, referring more especially to talipes-equinus, says: "The ligamentous structures in front of the ankle-joint and on the dorsal aspect of the foot, especially the ligament between the astragalus and navicular bone, are found to be elongated in proportion to the degree and duration of the deformity; while those on the plantar aspect of the foot are contracted and shortened to a corresponding extent. The structural shortening and adaptation of the ligaments in the sole of the foot connecting the tarsal bones — especially the calcareo-cuboid and calcareo-scaphoid ligaments and the plantar fascia — always correspond to the severity and duration of the deformity, and offer the greatest resistance to its removal. The lateral ligaments of the ankle-joint are also elongated in their anterior portions, and contracted at their posterior borders. The posterior ligaments of the ankle-joint become structurally shortened in adaptation to the altered relations of the os calcis and tibia, which, in severe cases, are in contact at the ankle-joint."

In the same work the author remarks, at page 144: "The majority of writers have too little regarded the condition of the ligaments in cases of clubfoot, nor has any allusion been made to the subject in some of the principal works on Orthopædic surgery. In Dr. Little's work we have no description of any structural alterations in the ligaments either in infantile or adult cases of club-foot." * * * "The ligaments gradually adapt themselves in length and form to the altered form and position of the bones they naturally serve to connect, in proportion to the time the deformity has existed previous to birth, and also in proportion to the severity of the muscular contraction which determines the exterior form of the foot."

By this very eminent authority, it is admitted that the ligaments are subject to structural impairment, attributable to unfavorable muscular influence tending to this abnormal condition of the ligaments. Is it not reasonable to believe that this structural lesion of ligaments may be the primary cause of muscular shortening in some cases, and muscular shortening in that of others, impairing the integrity of the ligaments? The ligaments certainly do yield to continued elastic force other than that of muscular contraction; as in the case of acquired deformity of the foot from continued malposition in the favoring of a sensitive part, or unfavorable position from whatever cause; for instance, the continued resting upon a limb in playing the harp, which is known to result in contortion of the body. The unfavorable position may be sustained by the muscles; but the primary fixed condition is attained by ligamental adaptation — some shortened and others elongated — and in extraordinary congenital cases these ligaments are rigidly fixed with that of an abnormal adaptation of the bones to the incurvature of the foot

THE CONDITION OF THE MUSCLES IN TALIPES VARUS.

The muscles in congenital *talipes varus*, at an early period after the birth of the child, are usually of normal size; the legs fully developed, and the feet apparently, but slightly, varied from normal form, the inner margin of the feet having a vertical tendency. For a few weeks after birth, also, the contorted foot can be restored to the desired form by the hand, and without apparent pain to the child, that is, in ordinary cases; and if sustained with appliances, a cure is effected simultaneously with the growth of the child.

In this condition of the foot, the gastrocnemius soleus and plantaris may be considered as alone concerned in the production of the deformity; when, after a year has elapsed (the child walking), the tibiales posticus and anticus will be found tense and somewhat unyielding; also, that of the ligaments on the inner side of the foot and plantar aponeurosis; the tendency increasing with its age and physical strength.

The relative position of the tendon tibialis posticus, in severe cases, is to be found rather on the internal malleolus than behind, in its normal position, and is the result of the obliquity of the os calcis.

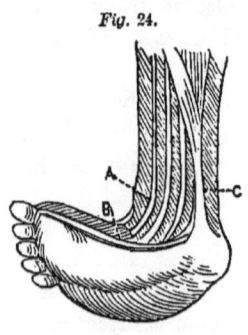

Fig. 24.

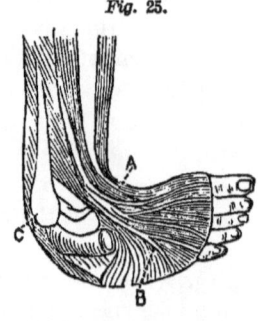

Fig. 25.

Figure 24. A Posticus tibial tendons; B, Plantar aponeurosis; C, Tendo-Achillis.

Figure 25 represents the anterior tendons. A, Anterior tibial tendon; B, Peronæus tertius; C, The os calcis, from its oblique position, is seen to be pressing the external malleolus, and thus gives convexity to the dorsum of the foot.

The three first-named muscles are the most implicated in talipes varus, the anterior tibial tendon and plantar aponeurosis being only implicated in severe cases, as represented in figures 17 and 18.

INFLUENCE OF LIGAMENTS IN TALIPES VARUS.

If congenital varus remains unrelieved during the growth of the foot, the ligaments present one of the most formidable obstacles to the restoration of the foot to its normal condition; and even in infants, when much deformity exists, tend to retard a speedy cure. This influence is determined by the rigidity of the foot; hence, a tolerably certain diagnosis may be obtained by grasping the foot and making an attempt to redress the deformity, when, if the ligaments are not very tense, the foot will assume an almost normal form. At the same time, however, care must be taken to observe the resistance of the child. The best time to examine the foot is when the child is asleep.

In non-congenital cases of contortion of the feet, which arise from infantile paralysis or unfavorable influence from various causes, this tendency to rigidity of the ligaments is not so prevalent a concomitant as in that of congenital contortion, being much more readily restored to normal form.

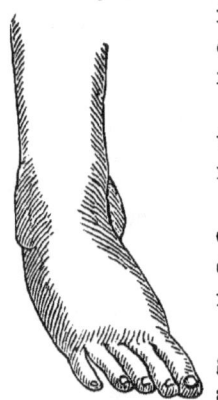

Fig. 26.

Fig. 27.

Figure 26 represents the first appearance of talipes varus from unfavorable position, or the result of partial paralysis.

Figure 27 is the representation of the result of compression, moulding the foot into permanent deformity, as in the foot of the Chinese lady, as represented in Mr. W. Adams' work on club-foot.

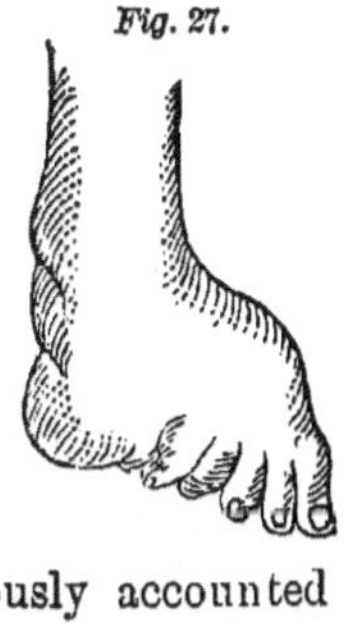

The plantar fascia, in some cases, tends greatly to maintain the deformity, and is often a serious obstacle in the treatment of contorted feet of adults from its shortened condition; in all cases, therefore, it should be carefully examined. In some it is only a tense band, and in others a body of considerable breadth. When divided, the foot slowly extends, all other parts being disposed to yield.

The legs of persons who have had contorted feet for a considerable period are generally remarkably attenuated, the cause being variously accounted for, but it is usually attributed to the quiescence of the muscles, resulting in firm contraction and consequent rigidity.

impeding the circulation essential to nutrition, the structure of the muscles having long remained in an abnormal condition. This occurs chiefly in confirmed congenital club-foot. The facing of the muscles by sub-cutaneous section tends to their improvement in development, but not to the extent of the normally-conditioned limb, nor to equal strength, yet they apparently recover their original strength, the patient frequently being able to walk for short distances without impediment or difficulty.

CONTORTION RESULTING FROM PARALYSIS.

In non-congenital cases of talipes some of the muscles become shortened, because of their tendency to recover from impairment in tone, while others remain flaccid; *first*, from the paralytic influence; and, *secondly*, from fixed extension. If not relieved fatty degeneration ensues at an early period, and continues until an equilibrium of action is established in all the muscles of the limb — a favorable result obtains from treatment in nearly all cases of infantile paralysis, and occasionally in children up to the age of sixteen. The sarcolemma of the muscular tissue maintains its integrity, and under a favorable condition reinstates muscular fibre to nearly the full development of the normal muscle.

Years of perseverance in treatment may be required in extraordinary cases, but when properly constructed apparatus is applied and kept in repair, and physiological treatment judiciously prescribed and enforced, the restoration of power is most certain.

SURGICAL TREATMENT OF CONTORTED FEET.

The treatment of contortion of the body and limbs requires special practical knowledge more than any other of the ordinary branches of surgical science. The mechanical and scientific acquirements necessary to insure successful practice consist in the ability to devise the therapeutical means for treatment that will be applicable to the apparent indications that present themselves, or may arise in the course of treatment. And it is quite true that this attainment of practical ability to construct surgical apparatus

IMPAIRMENT OF TISSUES RESULTING IN CONTORTIONS. 51

is within the province of the general practitioner. A scientific, skilful surgeon never hesitates to prescribe the apparatus required for an extraordinary fracture or lesion of the body or limbs when something of peculiar form is demanded. So in the treatment of contortions, an essential qualification is the ability not only to prescribe the necessary apparatus, but to devise such modifications as the various conditions may suggest; and we cannot conceive of any actual division of mechanical and surgical treatment for the cure of contortion of feet, limbs or body. If the subsequent treatment or devising of apparatus is submitted to the skill of the mechanic, it is a procedure, strongly indicative of a want of practical experience on the part of the surgeon.

PREPARATORY STEPS IN TREATMENT.

The primary steps in the treatment of contortion of the feet are necessarily the severing of tendons, commencing with that of the tendo-Achillis and its accompanying tendons conjoined and attached to the distal portion of the os calcis.

Before commencing the operation, the dressings should be in readiness, consisting of soft cotton fabric firmly rolled and of the following dimensions for each foot; For an adult, one of two and a half inches wide and two and a half yards long, with another of a yard in length. For infants, dressings of half the size and length. The additional requisites are: A strip of court plaster, a small pledget of soft woven fabric folded three or four layers thick, about two inches square, and wet with water; a sufficient quantity of lint to protect the points of pressure from the shoe, being at the inner side of the heel, the cuboid bone, and the outer edge of the anterior portion of the foot, and on the instep; the shoe to be applied, and a sponge and basin of water.

The patient is now placed on a table or bed, face downward, and with the feet projecting over the edge. Care must be taken to have the patient properly held by assistants, and especially the limb to be operated on. If an adult, have an assistant to maintain the foot firmly in position, and keep the tendon extended moderately, being attentive to direction, and relaxing the tension at the instant the knife has severed the tendon.

MEANS OF RESTORING THE FOOT TO NORMAL FORM.

The patient properly prepared for the operation, we carefully examine the tendon to determine its size and relative position, and depth beneath the external integument; then insert the knife, flat surface under the tendon, about one inch above the os calcis — passing it first obliquely downward and close to the tendon — and then depress the handle, to avoid penetrating too deeply, when under. The tendon now being fairly upon the blade of the knife, turn the cutting edge to the tendon, pressing the knife firmly, and, with a short sawing movement, sever the tendon slowly and carefully — to the protection of the other tissues — then turn the knife flatwise and withdraw it with care, so as to avoid enlarging the aperture, which must be closed instantly and covered with the court plaster. The folded pledget, being moistened, is then applied over the court plaster and secured by a few turns of the roller. The dry lint is next applied and secured by the roller, and when this is completed, the foot placed carefully in the shoe, and lightly secured by placing on the instep the pad, having tapes attached, which pass through an opening in the heel of the shoe, and are returned and tied on the instep pad. The anterior of the foot is then secured with the short roller, as will be described in another chapter. After the operation the nurse is directed to have the leg carefully supported upon a pillow or other soft material, and kept in a horizontal position, sitting or lying, as may be most agreeable to the patient, who is kept for two weeks confined to his room and precluded from walking or standing, and during that period, after which time he is permitted to gently press the foot upon the floor, using as a support the back of a chair which, when inclined to walk, he can push before him. This is continued for two weeks more, when he may be permitted to walk independent of support.

When the sequence of paralytic seizures are presented for treatment — the leg greatly attenuated and the foot distorted by a shortening of the tendo-Achillis — after severing the tendon in such cases, the foot should be extended and secured by a straight splint for ten days. This will insure the union of the tendon, though it will, for a comparatively limited time, retard the restoration of the foot to proper form. One of the most unfortunate conditions in which a patient can be placed is to have the tendo-Achillis ununited. His condition is pitiable indeed; compelled as he is to wear a heavy and expensive apparatus for the remainder of life, and even this giving only tolerable facility in locomotion. How damaging to the reputation of the surgeon are such sad catastrophes, occurring as they have, simply from a lack of attention.

We have never seen arteries punctured tending to any serious result; while Mr. Adams and others inform us it has unfortunately occurred; but ordinarily with little injury to the patient, hemorrhage having been arrested by pressure.

Suppuration of the tendon sometimes ensues, and this we have specially observed in patients of strumous diathesis, but never to the extent of arrest of the recuperative process. The period of rest enjoined was protracted, but the patient's condition readily improved by alterative tonics and liberal diet.

SEVERING THE PLANTAR FASCIA.

In cases where there is much shortening of the plantar fascia, together with shortening of the tendo-Achillis, we divide both at the same time. We have had cases, however, that required only, for the perfect restoration of the foot, the severing of the plantar fascia. The patient being placed in a favorable position, and properly supported, in the case of an adult, the plantar surface of the foot is supported and extended by an assistant—we carefully examine the tense mass of tissue, defining its breadth as nearly as possible. We now insert the tenotome (the same size used in the severing of the tendo-Achillis) about midway of the length of the fascia and close on the inner border pass to the supposed extent of contraction, closely engaging the mass to be divided; then turn the cutting edge to the fascia and gently maneuvre to the completion of the operation. The foot will elongate, but not to the extent that might be reasonably expected from the severing of a fascia, to all appearances strongly contracted. The longitudinal arch of the foot yields slowly even to a very firm pressure upon the instep, the bearings of the plantar surface being supported upon a metal plate, which affords the facility for making that pressure by means of the roller.

The section of the fascia being completed, care must be taken to close the wound, apply the court plaster and moistened pledget, and then the roller, with lint to protect the points of bearing upon the foot and guard against any injurious pressure. The shoe is next applied with some firmness, and if the tendo-Achillis has not been severed, the patient is allowed to bear weight upon the foot immediately, if so disposed; in fact, encouraged to do so.

We have never witnessed any injury proceeding from the division of this fascia, although well authenticated cases exist where an artery has been punctured and followed by most serious results. Mr. R. M. Tamplin informs us that he assisted in an operation where an artery was punctured, and that after the lapse of several weeks, a most difficult operation was necessitated and performed to arrest the bleeding of the wounded artery, well devised compression having been of no avail. Hence, the necessity for extreme care in performing this operation upon the plantar fascia, more especially in the case of very young subjects.

These are the only surgical conditions requiring tenotomy in the first and second stages of talipes varus. By the careful application of means for extension with that of manipulation, the latter being a very essential adjuvant to the treatment, a cure may be obtained in nearly every case. In congenital cases, where the foot can be straightened with the hand, we apply a modified Scarpa shoe (described in another chapter) at the age of six weeks. If, during the following five or six weeks of persistent effort, the perfect adjustment of the shoe be found impossible, the parents of the child are advised to suspend further treatment until it is six months old, when we sever the tendons and give special attention to the redressing of the deformity. That many cases of congenital clubfoot· can be cured by perseverance in well-directed manipulation, a doubt cannot be entertained. The failure may be attributed to the inattention of mothers and nurses who become discouraged from the slow progress made in these first efforts; and the great difficulty attending the securing the foot in a retaining shoe, and more often from indiscretion in bandaging the foot too tight for that purpose; and the result produced excoriations, subjecting the child to much suffering and delay in the progress of extension. Hence, many cases are treated by the preparatory step of tenotomy, which renders the foot less resistant to subsequent treatment.

THIRD STAGE OF TALIPES VARUS.

In the third stage of talipes varus, cases present a more complicated character; the foot, being decidedly more folded upon itself, does not yield readily to force. The tissues that tend to sustain this abnormal form are additional to those described as requiring tenotomy in the first and second stages. Now we have the posterior tibial, flexor longus digitorum and anterior tibial tendons to elongate. Many surgeons advise the severing of all the apparently tense tendons at one time in cases of only ordinary tenseness, which, they state, facilitates subsequent treatment. We, however, of late years seldom sever the three last-stated tendons, and are quite as successful in redressing the contorted foot as when we divided them, the rigidity of the foot being dependent upon the ligaments, and these cannot all be reached by the knife.

In the treatment of talipes (the generic term given by Menzel*), like that of practice in other practical procedures in surgery, intelligent, experienced practitioners do not confine themselves to any precedent that may be varied without adventure, to the imperilling the life of their patient, or that it is seemingly possible to make an improvement upon, in relieving the suffering from unnecessary pain, and facilitate cure by a more perfect coaptation of surgico-mechanical apparatus. Each, of course, will devise various means of arriving at the one desired object, the most speedy and perfect cure of the patient. Hence, it is well for the young aspirant in surgery to examine carefully the practice and ingenious devices of those most noted for their skill, determining which course he will pursue, and then make it available by his own ingenuity, even to that of an entire change in treatment and reconstruction in the appliances.

ON SEVERING THE POSTERIOR TIBIAL TENDON.

The sub-cutaneous section of the posterior tibial tendon, and that of the flexor longus digitorum are often included in the one operation, which we defer to a subsequent period. Our reasons for so

---

*"Dissertatio Indergeralis medica Talipedibus varis auctore."--- D. M. Menzel, Terbing, 1798.

doing are, that by dividing the tendo-Achillis and plantar fascia, the tense resistance is relieved to a considerable degree, and the contortion overcome by careful manipulation, together with suitable apparatus. We avoid any risk of wounding arteries from the severing of the posterior tibial tendon above the internal malleolus, inasmuch as a space presents, by this procedure, between the internal malleolus and the os naviculare, where the tendon may be severed with comparative safety, and without danger of wounding the artery or nerves. Again, if the tendo-Achillis and plantar fascia are not first divided and the foot everted by means of extension apparatus, it is impossible to sever the posterior tibial tendon at the point we have just described.

Mies, of Hanover, states that the scaphoid bone is displaced and firmly held in contact with the inner malleolus by the contraction of the posterior tibial muscle, the tendon of which, therefore, does not pass below and in front of the inner malleolus, as in the normally-conditioned limb.

Mr. William Adams says that: " In slight congenital, and in most of the non-congenital cases, this operation may be performed in the situation recommended by Mr. Syme (that of severing the tendon in front of the inner malleolus), but in such cases its division is seldom required."

It is very true that this operation is but seldom required for talipes varus, or that of the anterior tibial, which is readily severed near to its insertion where it crosses the ankle joint, as it is there quite prominent, except in very fleshy infants, and even then can be detected by bearing in mind that it is inclined to the inner side of the foot, from having become shortened, and, when detected, the finger should be kept upon the tendon, and the knife so inserted and guided as to have the point pass beneath or over the tendon *flatwise;* then turned, and the tendon severed.

By the sub-cutaneous section of the tendo-Achillis (or other tendons when necessary), the patient is not only more readily cured of contorted feet, but is relieved from a vast amount of suffering, both physical and mental, incident upon treatment for contortion without severing the tendons.

Continued extension, in attempts at the accomplishment of so desirable an object, subjects the patient to great pain continued through years; and is often attended with results unfavorable to the patient's health.

TREATMENT OF PUNCTURED ARTERIES.

We do not hesitate to sever tendons sub-cutaneously when persistent in maintaining a contortion, and in a measure free from complication with important blood-vessels and nerves. Arteries in the feet have been wounded, as that of the posterior tibial, and relieved by styptics and pressure. Mr. William Adams relates the case of an exceedingly interesting cure that occurred in his own practice: "On April 13, 1853, I wounded the posterior tibial artery in a child seven weeks old. The blunt-pointed knife was used, and I was not aware of the accident at the time of the operation. As neither the arterial jet nor the sudden blanching of the foot indicated the mischief, the artery was probably only wounded."

"After the lapse of ten days a deep pulsating tumor was discovered; and direct pressure by a graduated compress and bandage was applied. The pulsation diminished, but a very small slough formed in consequence of the pressure not being relieved for four days, from neglect of the mother to attend at the hospital, and a copious arterial hemorrhage took place. Pressure a little above the aneurism appeared to command the bleeding, and was therefore tried, but discontinued on the second day from the extension of the swelling to this part of the leg."

"This pressure having been removed, a second arterial bleeding followed. On May 12, 1853, I injected from five to ten drops of the concentrated solution of perchloride of iron, as recommended by M. Pravaz, of Lyons, into the centre of the aneurism, which was probably an inch in diameter. At the time of the injection clotted blood plugged the small cutaneous ulcer, through which the extremity of a long and finely pointed glass syringe, containing the styptic, was introduced and carried to some depth."

"Both before the injection and for five minutes afterward, M. Lonsdale compressed the femoral artery, so as to insure the blood acted upon being, as nearly as possible, in a stagnant condition — a most essential point. The first effect observed was that the loosely-clotted blood, filling the cutaneous ulcer, became firmer, and that from ten to twenty minims of straw-colored serum oozed through the ulcer; affording conclusive physiological evidence of the firm coagulation of the blood, which was also indicated by a general feeling of hardness over the sac. A piece of lint and a light bandage was applied."

"The cutaneous ulcer showed itself the next day to be contracted, and plugged with a firm, black clot. The surrounding skin, which previous to the operation had presented a tense, shiny, swollen and slightly reddened appearance, was now pale and less tumefied. The aspect of the limb was remarkably changed, and a process of shrinking and contraction appeared to have commenced, so that no inflammatory results were apprehended. Progressive improvement took place; the ulcer healed in a week and the shrinking and contracting advanced. On May 25th a deep puckered cicatrix, and a little deeper seated induration alone indicated the former seat of the aneurism."

We have given the particulars of the case as practical information, to enable the surgeon who may be so unfortunate as to require more than the ordinary means of arresting the hemorrhage, resulting from an accidental puncture of an artery; or, as in the case of Mr. Adams, having punctured the artery, meet with the subsequent aneurism resulting in hemorrhage. We have accidentally wounded the posterior tibial artery, but never failed in arresting the hemorrhage by graduated compression, carefully applied for about twelve hours; then relieved and only moderately supported by compress and roller. Special care must be practiced in the arresting of hemorrhage, from an artery thus wounded, by applying a small compost compress as soon as possible, even to a small jetting stream, when moderate pressure will arrest the bleeding. If permitted to flow for any duration of time, the difficulty of arresting it will be in proportion; and, if secured from flowing externally will become infiltrated from the increased impetus, and finally form, as in Mr. Adams' case, a pulsating tumor.

TENOTOMY KNIVES.

Orthopædists differ considerably in regard to the form of the knives to be used in their practice; some having blades of various shapes, while others are satisfied with one or two forms, with which they perform all the operations pertaining to orthopædic surgery — all agreeing, however, that the blade must be of such a form and size as will make the smallest possible aperture in the external integument; and hence they have rounded shanks and short cut-

IMPAIRMENT OF TISSUES RESULTING IN CONTORTIONS. 59

ting edges; some straight and spear pointed, some curved, and others of a blunt pattern. See Fig. 28.

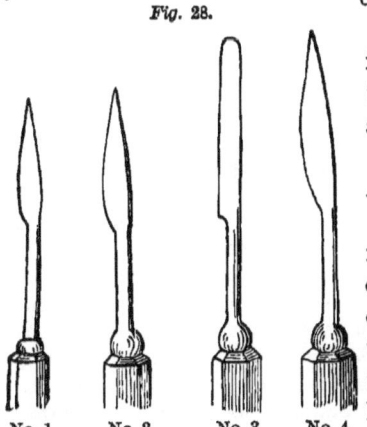

Fig. 28.

No. 1. No. 2. No. 3. No. 4.

The No. 1 is suitable for operating on infants, and on tendons on the anterior aspect of the feet of adults.

No. 2, for severing the tendo-Achillis in adults in ordinary cases.

No. 3, for severing tendons or muscles where special care is required to avoid wounding an artery or nerve, incision being made through the integument to the tendon or muscle, then the blunt knife inserted so as to engage only that which is to be divided.

No. 4 is simply a large sized tenotome.—at times required for severing a muscle or broad fascia. The forms of tenotomes claim some consideration. The smallest knife in use is three-fourths of an inch in length of cutting convex edge; the back nearly straight, having a cutting edge of about one-eighth of an inch, tending to a point, the remainder of the back being obtusely angular. The breadth of the blade is full three-sixteenths of an inch in the middle. The knife should be thick enough to ensure good strength, and increased in size for special purposes, as that of dividing broad fasciæ, muscles, and tendons of adults. The shanks should be rounded, and of about three-fourths of an inch to the blade of the tenotome. The handles are from three to four inches in length, flat or octagonal, as may be desired.

CHAPTER III.

GENERAL REMARKS ON THE TREATMENT OF TALIPES.

Mr. William Adams' treatment.— His modification of Scarpa's shoe.— Illustrative cases.— Detailed description of modern treatment.— Transverse tarsal joint a chief point to be considered in treatment.— India-rubber cords unreliable.— MEANS OF CURE IN FIRST STAGE OF TALIPES VARUS.— The author's modification of Scarpa's shoe.— Treatment of adults in talipes varus.— Extraordinary case of cure.— The modified Scarpa shoe applicable to nearly all varieties of club-foot.— SECOND STAGE OF TALIPES VARUS, AND TREATMENT.— Prognosis as to cure.— Treatment in cases of excoriation.— Fallacy of Prof. Syme's statements.— THIRD STAGE OF TALIPES VARUS.— Description of condition of patient.— Mode of treatment.— Treatment occasionally deferred.— Causes of valgus, and condition of the bones.—Treatment of talipes valgus.— The tendo-Achillis must be severed in severe cases.— Talipes equino-varus the result of unskilful treatment.— Description of apparatus for first stage of talipes valgus —Treatment of extreme cases.— Extension apparatus.— Method of application of the modified Scarpa shoe.— Treatment of extreme cases of talipes valgus.— First stage of treatment.— Second stage of treatment.— Faradization of much service, when applied to the paralyzed muscles.— TALIPES EQUINUS.— Distinguishing features in this ailment.— Authorities differ as to origin.— Congenital talipes equinus extremely rare.— CONTRACTION OF THE PLANTAR APONEUROSIS.— Objectionable points in Scarpa's shoe.— Professor Mütter's shoe.— Causes of talipes equinus.— Most frequently the result of paralysis, induced during dentition.— Other causes.— Prognosis as to cure.— Treatment.— Action of the apparatus.— Talipes equinus from spastic influence.—TALIPES CALCANEUS.—Diagnosis.—The Hospital for the Relief of the Ruptured and Crippled, a priceless boon to the suffering poor.—. Cause of congenital talipes calcaneus a disputed point.—Seldom, if ever, met with in the adult.—Treatment of congenital talipes calcaneus.—Severance of tendons exceptional.—Treatment of non-congenital talipes calcaneus.—Tenotomy seldom required in its incipient stages.—Description of extension apparatus.

BEFORE describing our treatment of the several varieties of talipes, a brief quotation from the writings of one of the most eminent orthopædic practitioners in London, in relation to the primary treatment, subject, however, to modification, may not be amiss.

Mr. William Adams, in his work published in London in 1866, page 106, says: " When any lateral inclination of the foot exists, usually inversion, or an inclination to varus, I make use of a cog-wheel placed in the sole of the Scarpa shoe, which is divided transversely, as represented in figure 14; the mechanical centre of motion is thus made to correspond to the transverse tarsal joint in the foot, which is the anatomical centre of motion in this portion of the deformity.

"It is an error to attempt to overcome the inversion or eversion of the foot (or rather of its anterior portion in front of the transverse tarsal joint, which can only be involved in any such movement,) by a cog-wheel acting laterally, and placed opposite the ankle-joint, as in the ordinary Scarpa shoe. * * * For the sake of adjustment in very slight cases this may be used; but if any decided degree of inversion or eversion has to be overcome, I never use the so-called *double cog-wheel* action at the ankle-joint. I prefer acting more directly upon the transverse tarsal-joint by means of the instrument represented in figure 14."

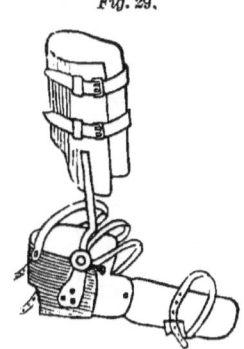

Fig. 29.

Figure 29 represents Mr. Adams' modified Scarpa shoe, alluded to as figure 14 in his book, page 106, which he terms the transverse tarsal arch.

Figure 29: Scarpa shoe, modified by Mr. Adams, with a transverse division in the sole plate corresponding to the transverse tarsal-joint in the foot, used to control inversion or eversion of the foot. In Mr. Adams' book, page 118, it is stated that the transverse division in the sole plate is moved by means of a cog-wheel, a means of fixed force which we consider objectionable.

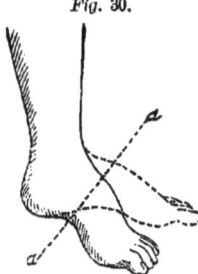

Fig. 30.

Figure 30. "Case of Talipes Equinus in a boy. The deformity depending chiefly upon the foot being bent upon itself from the transverse tarsal joint indicated by the line *a, a*."

Mr. Adams attaches much importance in the treatment of club-foot, of whatever variety, to a knowledge of the transverse tarsal joint, and to the modified Scarpa shoe as invented by himself; the sole-plate in this consisting of two portions that admit of lateral movement, and controlled by a double cogwheel to the redressing of the deformity of the foot.

Toward the accomplishment of the same object (avoiding, however, any fixed force where an elastic one can be applied), it will be observed in the subsequent pages that we apply the roller so as to invert or evert the foot upon the *undivided* sole-plate of the Scarpa shoe. This is done by reversing the turns of the roller — passing it after a few turns over the foot, then in a reverse direction under the plate, drawing the contorted foot upon the plate, or an approach thereto, limiting the force to the endurance of the patient. A few turns are then made over the foot and plate to secure the position; the heel having first been firmly fixed by means of the instep pad and tapes. A curved elastic spring is made to oppose the inward tendency of the foot and ankle, having its centre of force upon the transverse tarsal joint and the single elastic spring tending to unfold the foot, and also to limit extension and admit of flexion by the limited joint.

The transverse tarsal joint is to be carefully considered in the treatment as the axis of motion or point of yielding; and that the ankle joint is of minor importance, as it sustains to a great degree its abnormal relations from flexibility, whilst the anterior of the foot is contorted in all the varieties of lateral deviation from abnormal stress upon the ligament. That the transverse tarsal joint described by Mr. W. Adams is the centre of motion may be admitted as well as that of rotation in varus, inward and upward, of the anterior and posterior extremities of the foot; inclining the feet when thus contorted to overlap each other in walking, from the incurving of the anterior portion of the foot — hence the primary treatment is to unfold or extend the contorted foot which will, in nearly every case, restore the leg and foot to a normal condition. And is most successfully accomplished, as the celebrated Scarpa determined in his day, by elastic force from a steel spring; which is a *lateral* force, whatever form the shoe may have. Much depends, however, upon the practical skill of the surgeon in the manipulation of the contorted foot, and in the utmost vigilance to avoid undue pressure upon salient parts. Abrasions of parts of the foot arising from this cause will often retard progress in treatment, even to the extent of permanent injury.

We consider India rubber cords unreliable and cumbersome in their application, and hence, have abandoned their use.

In the first condition of talipes varus in children or adults, the sole of the foot rests upon the ground in nearly normal form; the

GENERAL REMARKS ON THE TREATMENT OF TALIPES. 63

outer margin presenting an increasing curve, and the plantar bearing, limited to the centre of the curve; the external malleolus becoming more prominent from pressure of the proximal end of the os calcis, which is, at the early stage of deformity, tending outward, and the anterior of the foot presenting little or no change in appearance, other than inclining inward and upward, as seen in the drawings in the previous figures.

This condition of a tendency to an inward curving of the feet, presents in the new-born babe, and in all stages of life even to that of old age; induced from various causes affecting the ligaments, muscles and tendons. In children, most frequently, from paralysis, contusions, and ulcerations of the feet; in adults, from caries, indolent ulcers, accidental injuries of the legs, careless walking (on the outer edge of the feet) and continued from habit, and in aged persons from a weakened condition of the muscles on the anterior aspect of the limb; but most frequently from a contraction of the plantar fascia. Persons thus afflicted become unsteady as to maintaining the centre of gravity; hence, walk awkwardly with the toes inverted, and at every step describe the segment of a circle with the feet. The acquired club-foot of adults but seldom, if ever, increases to that of the second stage of varus.

MEANS OF CURE IN THE FIRST STAGE OF TALIPES VARUS.

The condition of the patient in this stage of talipes varus, in children, is, to ensure careful, judicious treatment, quite favorable for success. The division of the tendo-Achillis is advisable, in cases of adults, as an auxiliary to the subsequent treatment. The wearing of the SCARPA SHOE (*modified*) will suffice for children in most cases if properly applied. Constant attention must be given to the parts of the foot most exposed to pressure, and fresh lint applied from time to time, as the parts present indications of inflammation. This care will facilitate the cure; as there will then be no interruption to the extension of the foot while the new formation of tendon is yet in a yielding condition, as well as when extension alone is being relied upon.

When the foot is restored to normal form, the modified Scarpa shoe should be worn night and day for a year or eighteen months. During the day an ordinary laced boot can be worn over the shoe—the laced boot having a broad, low heel, and fitting closely to the foot. At all times the foot should be nicely secured in the steel shoe by tying the heel down firmly in the heel-cup of the metal shoe, with tapes attached to a pad placed on the instep at its junction with the leg. This is most readily accomplished by passing the tapes through an open space left in the heel of the shoe.

THE AUTHOR'S MODIFICATION OF SCARPA'S SHOE, AND MODE OF APPLYING IT IN THE TREATMENT OF TALIPES VARUS.

Fig. 31, A, the encircling band for the leg, and vertical spring attached to the shoe having a limited movement by means of an extended point D, striking a small projecting stud, thus keeping the sole of the shoe at a right angle with the upright elastic curved spring. The encircling heel band of sheet steel, forms a cup for the heel, and is set above the sole about a quarter of an inch, admitting straps to pass through beneath this band, to be returned and tied on a cushion placed on the instep of the foot. At C the heel band is extended so as to give ample support to the outer margin of the foot.

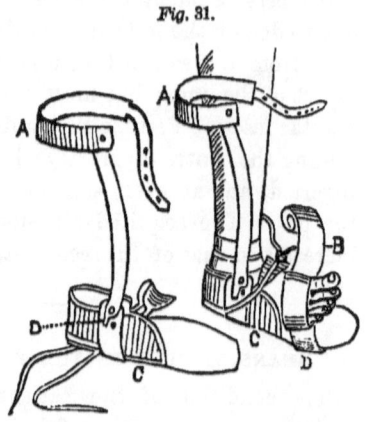

Fig. 31.

The shoe being applied to the foot in a case of talipes varus, the heel is secured in the cup, and the band for encircling the leg, A, left outside the leg until the foot is secured by the roller B. The roller is seen to first encircle the foot by a few turns, and then at D, reversed and passed under the sole; thus a purchase is obtained for everting the distal portion of the foot, and secured by a few turns of the roller over the foot and plate, the foot being well protected with some soft material intervening between the plate C and the foot. The foot being now secured, the band A is to be passed round the leg.

By the elastic force of the vertical spring the tibio-tarsal normal relation is eventually restored, and the meta-tarsal by the reversing of the roller on the foot under the sole plate. And the extension

GENERAL REMARKS ON THE TREATMENT OF TALIPES. 65

of the tendo-Achillis, by means of the instep strap and the limited joint, admitting of motion to the tibio-tarsal joint. The advancing of the body concentrates the weight upon the front of the foot, and extends the gastrocnemius muscles, one of the most important steps in the treatment.

In the treatment of adults laboring under this stage of talipes varus, with the tendo-Achillis in a tense condition, a cure can only be accomplished by the section of the tense tendon; and failures to cure, in most instances, may be attributed to the neglect of this necessary preparatory step in the treatment. Even in aged and corpulent persons, the severing of the tendo-Achilles is admissible; and is, in fact, the only means of affording them relief from the painful contraction of that tendon.

By severing the tendon, aged persons have been relieved from a decrepit condition. Mr. William Adams, in the appendix to his treatise on club-foot (page 368), gives the history of a remarkable case of a gentleman fifty-four years of age, and weighing two hundred and ninety-four pounds, who had been under the treatment of some of the most eminent surgeons of Europe without being benefited by their treatment, and who was subsequently cured by him through the severing of the tendo-Achillis of both feet.

The modified Scarpa shoe is the best appliance in these cases, as it can be firmly affixed to the foot immediately after the operation, and, by means of the elastic, curved, upright spring, normal lateral relation of the ankle joint is restored, and affords a continuous support with elastic force tending to overcome the excessively contracted tissues of the opposite aspect of the limb. The posterior muscles of the leg are limited in their contractile efforts by the limited joint of the shoe at the ankle, which, while it admits of motion in the flexors, at the same time limits the contraction of the extensor muscles of the foot. The flexors are thus relieved from the continuous extension that has partially exhausted their power of contraction, and the firmly secured heel becomes a fixed point for the accomplishment of the desirable object of flexing the foot upon the leg, while a firm support is given to the outer margin of the foot.

This extended surface of support is of inestimable advantage in furnishing *an extended point of bearing*, as when limited to a small space, is a sure means for the destruction of the integument pressed upon.

The general applicability of the modified Scarpa shoe to nearly all varieties of club-foot constitutes it one of the most useful of surgico-mechanical appliances, requiring only experience and dexterity in the application of the roller to have it meet nearly all the indications that may present in the treatment of contortions of the feet.

SECOND STAGE OF TALIPES VARUS AND TREATMENT.

In the second stage of talipes varus the foot rests almost entirely upon the outer edge, the os cuboid bearing part of the pressure. The inner margin of the foot is now more inclined to curve inward, and the dorsum of the foot is nearly vertical, the external malleolus prominent, from the increased obliquity of the os calcis, and the heel more retracted than in the first stage, the motions of the joint more limited and the whole foot less pliable.

This form of talipes varus is more often the result of paralytic seizures, or of acquired habit resulting in deformity. Congenital cases frequently present for treatment, but never after two years of age; as their contorted feet have then assumed the condition belonging to the third stage.

Prognosis, as to cure in the second stage of talipes varus, may be considered quite certain under judicious treatment.

In this stage of the ailment we have the gastrocnemius, plantar, and soleus muscles considerably shortened; requiring for successful treatment the severing of the tense tendons of one or more of these muscles. This preparatory step of course saves the patient much suffering, which would otherwise result from their continued extension.

The subsequent treatment requires similar appliances to that of the first stage of talipes varus; only more force is required, and an extended duration of time for the cure. The points most exposed to pressure are very liable to excoriation, and must be defended from pressure with lint, carefully prepared and used in ample quantity; never permitting cotton fibre in an unmanufactured condition to be used. Cotton fabric that has been long worn and frequently washed, if carefully and evenly folded to the thickness of a quarter of an inch, and every day opened up and refolded, will effectually protect the parts from concentrated pressure.

In cases where excoriation has ensued the pressure must be diminished and the foot, if possible, kept in the brace. A dressing of mutton tallow applied once a day to the excoriated parts affords relief in nearly all cases. If more than ordinary inflammation and suppuration should supervene, ordinary surgical treatment must be persevered in to the relief of the patient, when the former treatment must be again commenced, and if the tendons, which have become tense during that period of non-extension, are not again severed, it is only prolonging the time of treatment and hazarding a failure in the perfect restoration of the foot. Even under more favorable circumstances the repetition of the operation facilitates treatment, and tends to a more perfect cure than a reliance upon continued extension.

When deformity of a foot has been reduced, all has not been accomplished that is required to ensure a continuance of the normal form, and more especially if the patient is young and the ailment congenital. The tendency to contortion remains and, if not carefully controlled for a year or two, a relapse is sure to ensue to the detriment of the surgeon's reputation, as application will then be made to some other person to complete the cure.

In the treatment of varus the "Scarpa" shoe, in its modified form, should be worn day and night during the period above stated. As it is worn *within* a shoe but little inconvenience attends its use. The construction of the outer shoe, however, is of importance; as it should fit closely over the metal shoe, and be made to lace firmly only over the malleola — lacing above impedes the circulation.

We have been thus minute in describing treatment of this ailment, and advising long-continued vigilance, as it is the most common contortion of the feet and apparently to the inexperienced practitioner, the most readily cured; being so advised by the first authority in general surgery, and by what Mr. Adams denominates "erroneous doctrine inculcated by eminent authority;" as that of Prof. Syme, who states that contortion of the feet is curable by the severing of the tendons alone, and *without adjunct mechanical treatment.*

The experience of nearly forty years' practice in the treatment of these ailments has impressed us with the most indubitable evidence of the fallacy of Professor Syme's statement, and of the existing necessity for the perseverance and extreme care we have so repeatedly advised, to ensure permanent relief in these cases of contorted feet.

It is much to be regretted that such inconsistent statements should be made by so eminent authority; tending, as they do, to impair confidence in the treatment, because of the disappointment on the part of both patient and surgeon. Not only have practitioners of eminence, from their reliance upon so high an authority, been disappointed and discouraged from the failures resulting from this, but their reputations have been seriously compromised as, in many instances, careful subsequent treatment at other hands has in these same cases effected a cure; and by persons of usually limited attainment in the medical profession, to the encouragement of adventurous treatment.

That eminent surgeons, in the general practice of surgery, have sufficient practical experience, or are disposed to devote their valuable time for *months* to the treatment of a patient laboring under contortion of body or limbs, is not to be expected, when a capital operation requires only two or three weeks' time, and results in relief from much suffering, and often a valuable life saved. Hence this is, imperatively, a special department of surgery, too frequently assumed by persons of very limited knowledge in their profession, because of the required patience and labor essential to insure successful treatment which they can well afford.

This is to be regretted, as orthopædic surgery, to insure success in treatment, requires absolutely scientific attainment in anatomy, physiology and surgery; the comparatively simple operation of applying counteracting force to the cure of deformity requires an *intimate* knowledge of anatomy, and the improvement of the abnormal conditions of the muscles demands a more than superficial knowledge of physiology, while the arresting of any inflammatory condition that may be induced from pressure under ordinary circumstances properly applied, upon parts deficient in vitality, requires, in some cases, even more than ordinary skill in surgery. Therefore inexperienced surgeons who will assume the treatment of this class of ailments, performing only the initiatory step of severing the tense tendons and delegating the subsequent treatment to instrument makers, inexperienced students, or nurses, are seriously accountable to their patients for dereliction of duty.

THIRD STAGE OF TALIPES VARUS.

The condition of the foot in this stage, from continued use in an unfavorable position, becomes greatly contorted. The anterior portion presents the inner margin, upward and inclining inward, the plantar aspect vertical, and having deep fissures. The dorsum is rounded and somewhat irregular, because of the partial displacement of the tarsal bones. About the position of the internal malleolus appear slight corrugations of the skin, presenting a seeming outline of the inner margin of the heel. The posterior is round and tense, from which arises an acute prominence of the tendo-Achillis extending upward. The base of support is represented by large bursæ with hardened surface, an apparent provision of nature to prevent injury to the bones and ligaments subjected to this undue pressure, the result of the abnormal condition. However, this is disposed to inflammation, thereby subjecting the patient to much suffering.

Both feet being affected, the patient, in walking, carries the anterior of each foot alternately over the other with a semi-circular motion, and, from the limited base of support, is, of necessity, compelled to keep constantly in motion when attempting to sustain the body in an erect position, exacting a more extraordinary demand upon the muscles of the thigh for sustaining force than in the normal condition of the feet, tending to an increase of strength. The leg below the knee is attenuated to mere skin and bones, presenting the appearance of a very low degree of vitality, subject to abrasions of, at times, alarming aspect, and difficult to heal. The great stress and tension upon the muscles thus compressing the blood-vessels, nutrition is impeded in the soft tissues. The vessels, being less impeded, sustain the growth of bone, as it is but seldom that we find a club-footed limb shorter than its fellow in a normal condition.

TREATMENT OF THIRD STAGE OF TALIPES VARUS.

In this stage of deformity of the foot, many ligaments and several tendons are at fault, as well as the plantar aponeurosis, and to obtain a tolerable prognosis, requires careful examination as to their influence in retaining the foot in its contorted condition, when it may be found that the severing of one or more of the shorter tendons may suffice in redressing the deformity. The tense condition of the tendo-Achillis will indicate the necessity of its section, and, in some cases, will be all that will be required in the way of tenotomy for a tolerable restoration of the contorted foot, *i. e.*, to enable the patient to place a portion of the plantar surface to the ground, a very great relief when afforded.

There are cases, and not a few, where treatment has been deferred to a late period (and, if congenital, most unfavorable cases for treatment). From dental paralysis and constant exercise upon the impaired limb or limbs, has resulted in the *third* stage of talipes varus. The most serious impediment to the redressing of the ailment is the shortened plantar fascia, though this obstacle may be readily detected by careful examination, when an apparent thickened mass of tissues, limited in breadth at the middle of the concave arch, and quite tense, is found to extend from the base of the os calcis to the distal extremity of the metatarsal bones. The section of this tense substance, the tendo-Achillis being previously severed, will be the only requirement for the interference of the knife, to insure a sufficient preparatory condition for relief by judicious subsequent treatment, in nine-tenths of the cases that will come into the hands of the surgeon.

In extraordinary cases, the division of both the tibialis posticus and anticus tendons may be indicated, to the relief of the impediment of the contorted condition of the foot; but such cases seldom result in a perfect cure; the feet have become so fixed in their abnormal condition, from the adaptation of the bones, ligaments and aponeurotic shortening, that the division of the tendons will not, in all cases, be sufficient for the restoration of the normal form, the ligaments being primarily at fault and so deeply situated as to preclude the possibility of severing them with safety. Much improvement, however, may be obtained from perseverance in manipulation and suitably constructed appliances.

CAUSES OF VALGUS, AND CONDITION OF THE BONES.

Valgus is the result of some local impairment of the sustaining tissues, in nearly all cases that present for treatment, though an exceptional case of congenital valgus will at times occur. It is generally the effect of some continued extending force upon the plantar aponeurosis, as that of exercising too freely in dancing when young, or high heeled, thin soled shoes, when worn by delicate children or feeble adults; sprains and injuries resulting from jumping, or falling on the feet from a great height; partial paralysis, preternatural laxity of the ligaments of the foot and ankle, as in scrofulous or rickety children.

The ligaments and muscles implicated in valgus have been stated. The astragalus is rotated on its *long* axis inward, and there is a separation of the head from the cavity of the naviculare, the latter being depressed, and its prominence most marked, appearing as the key of the arch displaced to the depression of the instep. The os calcis is rotated on its *short* axis and obliquely outward, and its posterior extremity extremely elevated. The cuneiform bones are closely pressed, and are nearly *in situ ;* the phalanges assume a more or less vertical position as the contortion increases, inducing a most painful condition, limiting the patient's ability to take walking exercise to a much greater degree than that of talipes varus. It is the next most common variety of talipes, and known by the term *valgus*, and characterized by the contortion tending outward, the feet having become flattened from the yielding of the plantar fascia; the bearing being upon the entire plantar surface inclining to the inner edge, and tending to an increasing depression in front of the internal malleolus, which becomes quite prominent, with a roundness increased from the advanced position of the naviculare. This is the *first stage* of valgus.

TREATMENT OF TALIPES VALGUS.

For the relief of talipes valgus, tense tendons present that require to be severed. In cases of moderate contortion, it will suffice to sever the peronei and extensor longus tendons; but in some more severe cases, the tendo-Achillis will require severing as well as that of the tibialis anticus and extensor pollicis.

The operation can be performed by determining the location of the extensor longus in front of the external malleolus, and close in to the ankle joint where it will be found prominent. Care must be taken in inserting the knife flat beneath the tendons to be severed, and that it reaches only to the breadth of the tendon to be divided; by this precaution, the anterior tibial artery can be avoided.

Arteries and nerves are but seldom wounded when proper precaution is taken; *first*, to apply resisting force to the tense tendon, rendering it more prominent, *then* placing the forefinger of one hand upon the tendon and carefully inserting the knife beneath, limiting the cutting edge to the breadth of the tendon. This precaution will apply to the severing of nearly all tense tendons. There is an exception, circumstantial, however, in the case of section of the biceps tendon when shortened at the knee. The perineal nerve has been severed from indiscretion or want of practical knowledge. In cases of long standing, the nerve is shortened as well as the tendon, and cannot be avoided, even when the tendon is cautiously severed; but from the shortened condition of the nerve also, it presents a tense margin, and the operator, supposing that he has not severed the entire tendon, again inserts the knife and *severs the nerve*. The result of this is a paralysis of the peroneus and extensor muscles, terminating in talipes equino-varus.

The first stage of valgus is readily relieved by a steel spring having an elevation to correspond with the normal width of the foot, and secured in the shoe at the heel by two or three screws; the anterior portion left free to glide upon the inner sole of the shoe. The spring is constructed of thin sheet steel, and the arch raised with the hammer. The form, when cut out of sheet steel, is as represented in Figure 32.

No. 1 is the form when cut out of the sheet of steel, and No. 2 represents the arch, 3 showing the outer edge, which has been drawn into form by the raising of the arch 2.

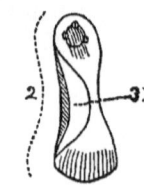

Fig. 32.

No. 1.   No. 2.

This spring should be worn for eighteen months or two years. This has been our common treatment for over twenty years, and when carefully persevered in, has never failed to cure valgus in the incipient stage and even in advanced conditions — sometimes where other treatment had been objected to, as the spring can be worn in the ordinary shoe. If in the first stage of valgus, the tendency is not arrested, greater exaggeration ensues; the bearing of the foot inclines to the anterior and inner edge, the heel in some cases being elevated outward from a shortening of the gastrocnemius and soleus

GENERAL REMARKS ON THE TREATMENT OF TALIPES. 73

muscles, the plantar aspect becomes convex, the internal malleolus and the scaphoid quite prominent, while the dorsum presents a complete depression of the arch. At this stage of the ailment, the patient suffers much pain in attempting to walk, and has but a very limited control over all efforts in that direction.

TREATMENT OF EXTREME CASES OF TALIPES VALGUS.

Cases that have advanced to an extraordinary degree of contortion have the following tendons shortened:

The extensor longus and peronei will be found prominent and tense in front of the malleolus externus, and to facilitate the treatment should be severed before any effort at extension, by means of apparatus, is made; as the use of the latter without the former preparatory step will only encourage the patient to prolong the treatment by persisting in wearing the extension apparatus in hopes of a cure resulting without submitting to tenotomy — an operation greatly dreaded.

In most cases the tendo-Achillis will be found so tense as to likewise require division. However, this may be deferred until after the section of the above mentioned tendons which in many cases, will be found all-sufficient for relief of the foot.

The extension apparatus is that of the modified "Scarpa" shoe used in cases where there has been a shortening of the tendo-Achilles. The ankle joint, in the apparatus, being limited precisely as in the

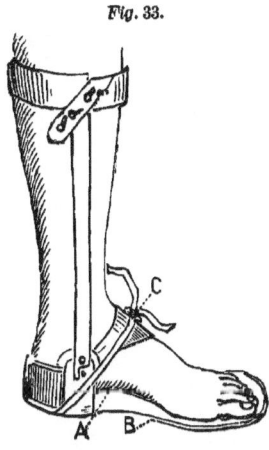

Fig. 33.

treatment of talipes varus; the only change being that of having the upright spring applied to the *inner* side of the leg, tending the foot inward and sustaining it at a right angle with the leg, as well as permitting a free movement of the foot upward. [See Fig. 33, a varus shoe having the spring reversed.] A, the inner margin of the foot; B, the metal sole plate; C, the foot secured by the instep pad and tapes.] This movement you should instruct the patient to perform by placing the front of the foot on some fixed object which would allow the heel to be depressed. This will afford an exercise of great value in all cases where the gastrocnemius, soleus and plantaris muscles are in a shortened condition, and even in some cases, obviates the necessity of severing the tendo-Achillis.

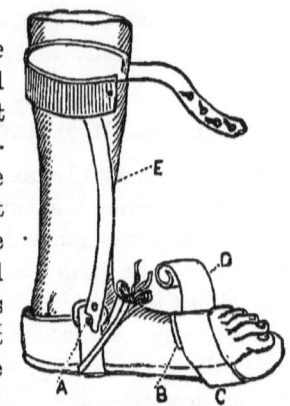

Fig. 34.

The method of application of the modified "Scarpa" shoe is represented in Fig. 34. In this engraving the front of the foot is limited in its upward tendency, as in nearly all cases of valgus the extensors of the foot are paralyzed. It will be observed that the joint A can be readily reversed from that constructed for extending the gastrocnemius, soleus and plantar muscles. The limited joint sustains the foot at a right angle with the leg, the heel being fixed.

To apply the roller to the foot, a few turns should be taken over and under (B), and then pass under the plate in a reversed direction (C), by which means the everted foot will be drawn on the plate, there being a sufficiency of the roller at D to secure the foot in its place.

This having been completed, the curved elastic spring (E) is to be adjusted by passing the attached broad padded band around the leg and the strap buttoned, which completes the dressing — the foot having previously been protected at the points of pressure by means of lint and the rollers.

This is the first dressing, which is to be continued during the reparative process of the several tendons, or, for about three months, when it will be found that the foot is relieved from the everted condition, and presents only the depressed instep.

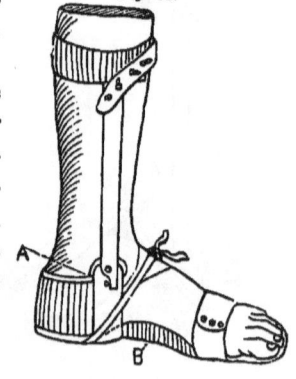

Fig. 35.

For the elevation of the instep, we have a piece of cork properly shaped for that purpose and attached upon the soleplate of the "Scarpa" shoe, as first applied. (See Figure 35.) A well-padded cushion (A) is secured by a plate on the inner side of the joint at the ankle. The cork is attached on the foot plate (B), and a roller secures the ball of the foot firmly to the plate — the heel being firmly held by the tapes tied on the instep.

The patient, in the second stage of the treatment is now advised to wear over the "Scarpa" shoe, a laced boot with a low heel. and to continue wearing the brace, day and night, for at least eighteen months; then be relieved to the wearing of a shoe prepared with the elastic spring, as worn in the treatment of talipes valgus in its incipient stage and represented in Fig. 32.

There are cases of talipes valgus, where the tendo-Achilles and soleus muscles are so paralyzed as to tend to a depression of the heel, and properly termed calcaneo-valgus. (See Fig. 36.)

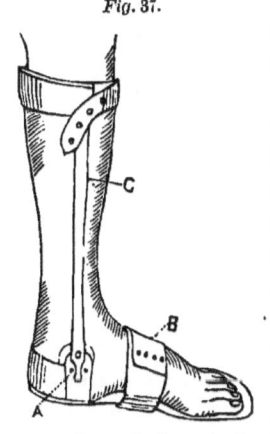

Fig. 36.

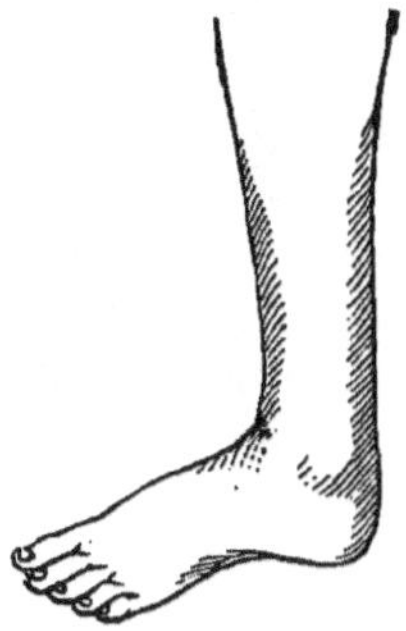

The foot, in this condition, is attended with much inconvenience to the individual when walking. When an effort is made to take a step the anterior of the foot is involuntarily extended, and to avoid tripping the foot has to be raised higher than ordinary, and the inner side of the foot advanced — the whole limb being more or less rotated outward; the heel striking the ground first and the front of the foot coming down mechanically when the attempt is made to advance the other foot.

For the relief of this condition of the foot the modified "Scarpa" shoe is most applicable, having the joint so limited as to admit of elevating or depressing the extremity of the anterior of the foot only two inches as shown in Fig. 37.

Fig. 37.

The construction of the limited joint is seen at A, and the foot secured to the plate by means of a roller at B. The vertical spring C is of only sufficient strength to support the foot in normal position, permitting lateral motion in the ankle joint from its elasticity.

The modified "Scarpa" shoe should be worn day and night, and during the day within an ordinary shoe. At stated periods of about twelve hours, the foot ought to be stripped and the tendo-Achillis and soleus carefully manipulated for ten or fifteen minutes; then have strychnine ointment applied to these muscles, twelve grains to the ounce of simple cerate, care being taken to cleanse the surface with warm water and soap, or an alkali, previous to each application of the ointment.

Faradization is of much service applied to the paralyzed muscles, when the foot is thus supported, and decidedly injurious when not supported, as its tendency is to shorten the flexor muscles and increase the contortion, since it is impossible to localize the electrical influence sufficiently to obviate this tendency. The application of all modifications of electricity is objectionable under the above stated circumstances, because of its increasing the contractile power of the muscular tissue, and it cannot be limited in its influence to one or more paralyzed muscles.

All the muscles of the limb, if not of the whole body and limbs, are influenced by induction, when electricity is applied to any part of the system. When applied to the foot or hand of a delicate female, it has been known to induce the menstrual flow. The concentrated form of electricity produces a powerful stimulant effect analogous to that of acupuncture, and is then diffused through the medium of the circulatory fluids, they being decided conductors of electricity.

TALIPES EQUINUS.

Of the various contortions of the foot, those within the category of *talipes equinus* are the most simple, the first indication being an inability to flex the anterior of the foot to a right angle with the leg, relief for which is afforded by the wearing of a high-heeled shoe, indicating a shortened condition of the extensor muscles of the foot. (See Fig. 38.)

Fig. 38.

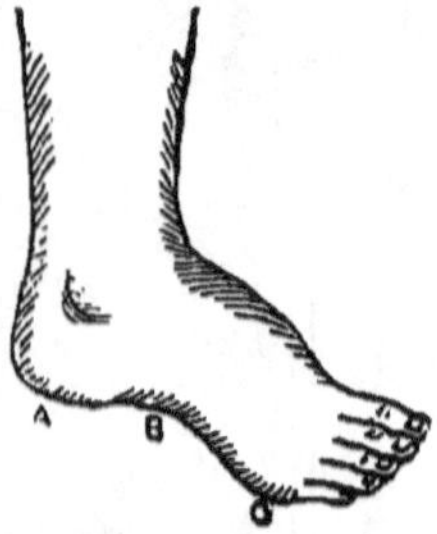

When the patient is not relieved of this restricted movement of the foot, it increases to a most extraordinary condition of contortion; assuming more or less of the two forms of contortion just treated of, and in partaking of one or the other, is denominated equino-varus or valgus.

The most marked character of talipes equinus is the elevation of the heel and tendency to vertical position of the metatarsus, concentrating the entire bearing upon the extremities of the metatarsal bones, which leads to an increase of breadth in this part of the foot, and very decided contraction of the plantar fascia.

There are extraordinary cases of this ailment, the result of paralysis, limited to the flexors of the foot.

The individual thus afflicted walks upon the dorsum of the toes and ends of the metatarsal bones, and from long continued stress advances the bearing on the dorsum to the tarsus. (See Figs. 39 and 40, for representation of this condition.)

Figs. 39 and 40.

The distinguishing feature in talipes equinus resulting from paralysis of the flexors, is, that there is not the tense and prominent elevation of the tendo-Achillis, as in that of other influences tending to this contortion; hence, it will be observed that the heel is round, full and prominent; indeed, the os calcis is in a horizontal position, or very nearly so.

Authorities differ much as to the origin of this ailment, all agreeing, however, that cases of congenital talipes equinus, compared with the non-congenital, are extremely rare. Mr. Tamplin, with all his large experience, expresses a doubt as to the congenital origin of the ailment. Dr. Little, of London, is positive in having seen two cases out of the many that he has treated. Mr. Broadhurst believes he has seen one or two in his extensive orthopædic practice. Mr. William Adams believes, from report, which he considers worthy of reliance, in three cases, which he has classed as congenital talipes equinus, and these three are out of *three thousand* cases of different kinds of deformities that were treated by the late Mr. Lonsdale and himself. Truly a limited number, and not very positive proof that they were congenital!

Mr. Adams, in treating of the pathology of this ailment, says:

"Talipes equinus is, with very rare exceptions, a non-congenital affection. It is often claimed to be of congenital origin by parents, but upon close inquiry we generally learn that the symptoms were not observed until the child had began to put its feet to the ground."

This agrees with our experience, as *we have never had positive assurance of a single case of congenital talipes equinus during a practice in orthopœdic surgery of nearly thirty years.*

CONTRACTION OF THE PLANTAR APONEUROSIS.

The condition of the foot when contorted by the plantar aponeurosis, having, from long duration, become inordinately shortened, is such that treatment by means of the improved "Scarpa" shoe, would be very slow indeed, the shoe, in some cases, requiring to be worn for years to accomplish a perfect cure. This is because of the fixed or unyielding means of extension thus employed, and which is never so effectual as elastic force, or motion in opposition to the resistance. The foot placed upon the plantar plate of the "Scarpa" shoe, which is unyielding, and an effort made to extend the plantar fascial tendons and ligaments, by pressure made upon the instep, the resistance is, as in that of all fixed apparatus for the redressing of contortions of the body or limbs, to be compared to force made by means of screws and cog-wheels. This fixed power induces pain, the result of which is an irritable, unyielding condition of the muscles in which it is desirable to effect permanent extension to a limited degree, and yet not impair their normal functions. Elastic force or limited motion has not the tendency, as has fixed extension, even of a moderate degree, to excite an irritable condition of the extended muscles or other tissues.

The extension of muscles and ligaments by elastic force is not a new discovery, if we may credit what is stated of Scarpa's visit to Tiphaisne's house in Paris, in 1781, where he made the discovery that elastic steel springs were used by that then noted man, for the cure of club-foot, and which had been kept a secret from the public, except in the single instance of his remark to Scarpa that "nature will not yield to violence, but only to gradual force." From this intimation, coupled with his discovery of the steel springs, Scarpa is said to have designed his club-foot shoe.

The ordinary "Scarpa" shoe is not all of the apparatus we have

GENERAL REMARKS ON THE TREATMENT OF TALIPES. 79

relied upon for the redressing of contortions of the feet; nor did the inventor rely solely upon it. He constructed a shoe admirably adapted to the elongating of the plantar aponeurosis by means of elastic force, a modification of which we have made for this purpose as represented in Fig. 41.

Fig. 41.

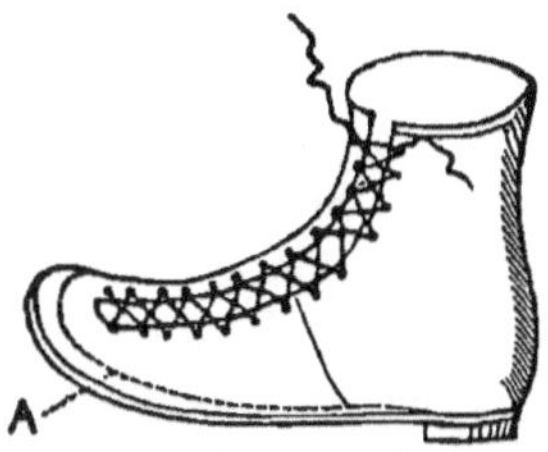

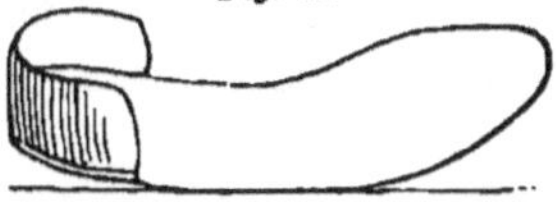

Fig. 42. The spring as applied to the foot, and secured with the roller, being intended to be worn within the shoe, Fig. 41.

Fig. 42 represents the plate before it is applied to the foot — the anterior portion curved upward; but, being very elastic and nicely tempered, it yields to the pressure of the foot.

Fig. 42.

Scarpa's apparatus for extending the plantar aponeurosis consisted of a plantar plate and cup for the heel, with an elevating spring attached on the under side of the plate about one-third the distance from its anterior extremity. This made the apparatus objectionable since, because of its cumbrous proportion, it could not be worn within a shoe.

The late Prof. Mütter, of Philadelphia, devised a shoe somewhat similar to this (Scarpa's) by having the plantar plate rivited into the heel of a common laced boot, open to the toe, and attaching under the anterior portion of the plantar plate, an elevating spring that rested upon the inner surface of the sole of the shoe, with both ends curved so as to glide easily there; the centre of the spring being elevated to the plantar plate. This ingeniously constructed shoe was, however, intended by this most excellent orthopædist* to produce extension of the extensor muscles by elasticity in the anterior portion of the foot. That this shoe made a very decided impression upon the plantar-fascia there cannot be a doubt entertained. The appliance as we have modified it is worn with more comparative comfort to the patient, being much less cumbrous in the shoe, and by no means so objectionable in appearance as Dr. Mütter's appliance. It is of great importance in the treatment of a patient, to have appliances, when completeness will admit, of the construction to meet the indications that present; to have them con-

---

* Lectures published by Thos. D. Mütter, 1839, p. 59.

cealed from observation. The patient will be more reconciled to their use, and to the more favorable progress in treatment.

CAUSES OF TALIPES EQUINUS.

First: Long experience has confirmed the opinion that talipes equinus is most frequently the result of paralysis induced *during dentition*, and limited, mainly, to the flexor muscles. The extensor muscles being thus liberated from their normal tension shorten upon themselves to a quiescent or persistent state that results in a pathological derangement of tonic fixedness, limiting the demand for nutrition, as well as the recuperative tendency in the flexors, by keeping them in their extended position. The position of the foot during sleep, together with the weight of the bed-clothes upon the extended foot, during the night, of many hours duration, contributes largely to the aggravation of the ailment.

Second: There is but a comparatively limited number of cases attributable to spastic contractions, the supposed result of convulsions, occurring before or immediately after birth, and affecting nearly all the muscles of the body, indicating cerebral implication. The locomotion, for the want of co-ordination between mind and muscle, in such cases is performed with apparent difficulty and violent effort, because of inability to control the desired movement of feet or hands; walking upon the anterior portion of the feet with an irregular effort of both to maintain the erect position — a staff being of no use because of an inability to make it available by the use of the hands.

Third: *Scrofulous ailments result in talipes equinus.* Inflammation resulting in abscesses is another cause of the gastrocnemius muscles terminating in structural impairment, and in some cases of the joint and the muscles becoming atrophied.

Fourth: *Unfavorable position of the foot.* To relieve pain from any cause is liable to become a habit resulting in contortion, and most frequently that of talipes equinus. By persistent position atrophy is induced in the muscular and ligamentous tissues; and in addition to the shortened condition of the *triceps suræ* is that of the plantar aponeurosis. Wounds and bruises are a common cause of this ailment, and may be classed as traumatic.

GENERAL REMARKS ON THE TREATMENT OF TALIPES. 81

These several causes produce similar pathological effects; structural changes in the affected tissues, which are of less serious import in regard to treatment than the three first described, with the exception of the complete impairment of a nerve or great extent of injury to the muscles.

The prognosis in regard to the cure of talipes equinus may be considered as favorable in all cases where anchylosis has not taken place, as in complication of synovitis with that of shortened muscles. Infiltration of the areolar tissue immediately surrounding the joint having resulted in structural lesion, will somewhat retard progress, but never presents an insurmountable difficulty.

TREATMENT OF TALIPES EQUINUS.

The primary treatment of the four known causes of talipes equinus has been the one preparatory step that we have considered the most favorable to the patient in regard to the mitigation of suffering induced by efforts to overcome the contortion. And the one cause, in more than a majority of cases, has been the result of infantile paralysis (as described in the first division). This step has been *to sever the tendo-Achillis,* the subsequent treatment consisting in that of placing the limbs in a supporting frame, as represented in Fig. 41.

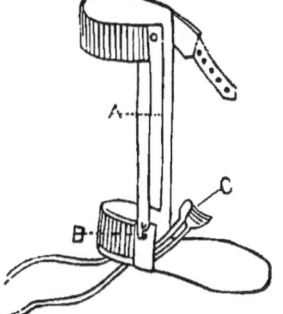

Fig. 43.

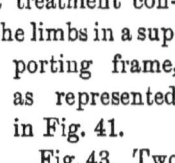

Fig. 44.

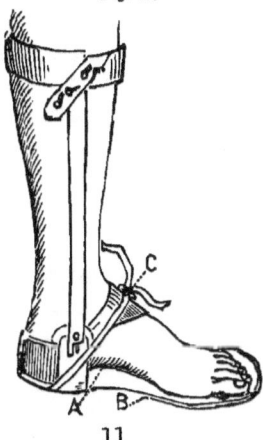

Fig. 43. Two upright bars (A), attached to the sandal by means of a limited joint (B). At C the instep pad and tapes are represented, by means of which extension of shortened muscles can be made.

Figure 44 represents the foot in the supporting frame A, the heel depressed; the anterior of the foot resting upon the sole-plate (B), and the foot secured by the instep-pad and tapes as shown at C.

It will be observed that the heel is

gradually brought down by means of the instep-pad and tapes— care being taken to protect all parts that may be exposed to pressure, and more especially in these paralytic cases, as vitality is very limited, and consequently, the parts infringed upon disposed to serious ulceration, tending to an erysipelatous condition of the foot and leg. A moderate extension is all that is required after the section of the tendo-Achillis in cases the result of paralysis of the flexor muscles, or the severing of the plantar aponeurosis alone (the extensors not being greatly implicated), although of several years duration. With gentle force gradually applied, the foot will settle down to the plantar plate, and when this is accomplished, careful attention must be exercised in regard to the patient's physical condition. If in ordinary good health, we permit our patient to walk with the support of a chair, resting on the back and pushing it before them; children, at the expiration of two or three weeks from the time the tendon has been severed, and adults in four or five weeks. We have seen patients walk in less time than one week after the operation, having only a supporting frame upon the foot, and no unfavorable circumstance to occur; but this has not been our practice; and, beside, we have been greatly favored in not having cases of ununited tendons, and this good fortune we attribute solely to these precautionary measures.

Having restored the contorted foot to normal form, the case now demands careful supervision, in having the supporting apparatus kept in perfect order, and worn night and day as long as there is apparent weakness in the flexors of the foot, or shortening of the plantar aponeurosis. It must, however, be borne in mind that the patient is most unfavorably conditioned when in bed at night without a support to sustain the foot at right angles with the leg; the weight of the bed clothes contributes largely to continued extension of the foot, as we have previously stated, and thus impairs the recuperative tendency in the paralyzed muscles.

In the treatment of talipes equinus resulting from other causes than that of infantile paralysis, greater circumspection only is required as to the resisting influences maintained by tense tissues, which tissues should be severed after having carefully and effectually tested manipulation and extension.

TALIPES EQUINUS FROM VARIOUS CAUSES.

Cases resulting from spastic influence are, in many instances, relieved by the simple severing of the tendo-Achillis, requiring only to be kept at rest during the restoration of the severed tendon. The operation often results in arresting the involuntary movements of the feet and legs. We have witnessed the most salutary results from this treatment, and do not hesitate to sever the tendon in such cases, and but seldom know of a failure to improve the condition of the patient, in enabling them to place the entire sole of the foot to the floor. This ability being accomplished, tends greatly to their improvement in walking, thus relieving them from a most dependent condition.

Talipes equinus, when the result of scrofulous ulceration about the joints, is entitled to the most serious consideration, as the severing the tendons does not always fulfill expectations; but the further treatment is not to be abandoned because of this — manipulation, as carefully and perseveringly practiced, with efforts at extension, will, in many cases, afford relief even to the degree of freeing the patient from apparent anchylosis.

Many cases of talipes equinus resulting from unfavorable position are, in their incipient stages, relieved by the extension support. This should be tried first, and if not successful in due time, or if the patient suffers from the pressure, the tendons should be severed to facilitate the cure. Traumatic cases having resulted in great shortening of the plantar aponeurosis, may not require the severing of the tendo-Achillis, but that of the aponeurosis, which will be sufficient to accomplish the cure, even when great deformity exists; requiring, however, in all cases, careful perseverance in manipulation and use of appliances so constructed as to yield to the pressure of the step of the patient, the anterior portion of the foot being kept constantly in a state of extension by elastic force. This is most readily accomplished by an elastic steel plate so designed as to be worn in the shoe, and, when neatly constructed, presents no external appearance objectionable to the patient — a point gained of valuable consideration, because of the appliance being worn more constantly and to his more immediate relief.

TALIPES CALCANEUS.

Congenital and non-congenital talipes calcaneus differ very materially in appearance and condition. In the former the dorsum of the foot is brought very nearly in contact with the leg, and is held rigidly in this position, the tuberosity of the os calcis presenting the only point of bearing on the plantar surface. As there is no contraction of the plantar fascia, the anterior portion of the foot is simply elevated, the bones maintaining their normal relation, with an excessive elongation of the extensor muscles.

Fig. 45.

Figure 45 shows a depression of the heel and elevation of the anterior portion of the foot flexed upon the leg.

The non-congenital condition of this contortion of the foot presents a depression of the heel and slight contraction of the plantar fascia, which retains the anterior portion of the foot in a horizontal position, the os calcis, in severe cases, being almost perpendicular.

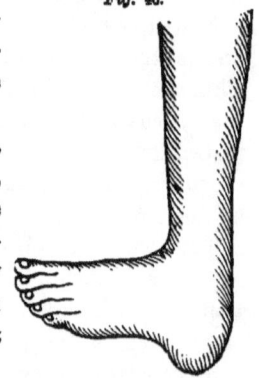

Fig. 46.

The patient being thus conditioned, walks as though on stilts (see figure 46), the anterior extremity of the foot projecting horizontally, and elevated some distance above the abnormal base of the heel.

This condition of the foot does not remain. The plantar fascia continues to shorten, doubling the foot upon its plantar face until it assumes the appearance and condition of a Chinese lady's foot previously represented, and if compressed in iron shoes could have been reduced to that condition.

Figure 47 shows the anterior portion of the foot depressed, and a sharp doubling upon the os calcis. When in this condition, the patient is seriously impaired in ability to walk, stumping as it were with a wooden leg, and readily fatigued from the slightest effort to use their feet, and if long continued from actual necessity, painful inflammation in the salient position of the os calcis and extremities of the metatarsal bones, after restraining the patient from walking for several weeks at a time, is not an unfrequent occurrence, a most serious embarassment to the indigent, often subjecting them to compulsory pauperism, successful treatment not being attainable through elemosynary institutions, and the want of means to purchase expensive apparatus. Happily they are now furnished (and have been for the past ten years) by the Hospital for the Relief of the Ruptured and Crippled, in the city of New York, being the first institution in the country which gratuitously supplied orthopædic appliances and kept them in repair, besides securing experienced and skilful surgical treatment, as long as may be required by either indoor or outdoor patients, over five thousand cases having been treated in the short period of twelve months.

Fig. 47.

CAUSE OF TALIPES CALCANEUS.

The cause of congenital talipes calcaneus is one of the mooted subjects; however, some of the supporters of the dynamic theory, as Mr. William Adams and Mr. Lonsdale, admit that probably it is the result of malposition of the fœtus, and pressure *in utero*. Mr Lonsdale observes: "I have seen four cases of congenital calcaneus where it could be distinctly traced to position *in utero*, they being breech presentations, with the legs extended upward, the feet being doubled upward and pressed against the tibia in front."*

Mr. Tamplin observes: "I have never yet met with *congenital talipes equinus*." As before stated, congenital calcaneus differs in the simplicity of the contortion, and the prognosis is proportionately favorable.

TREATMENT OF CONGENITAL TALIPES CALCANEUS.

The cases that require the severing of tendons are exceptional, and when such cases do present, the tendons for division are those of the tibialis anticus, extensor proprius pollicis, extensor longus

---

* Lancet, Sept. 1, 1855.

digitorum, and peroneus tertius. The apparatus for redressing the contortion is represented in Fig. 48. A sole plate, and cup for the reception of the heel; a single upright, elastic spring to pass up the back of the leg to an encircling band for the leg; an adjustable cup for the dorsum of the foot, having leather straps to fasten on the heel cup, and tapes passing down through slits in the sole-plate to be tied underneath.

Fig. 48.

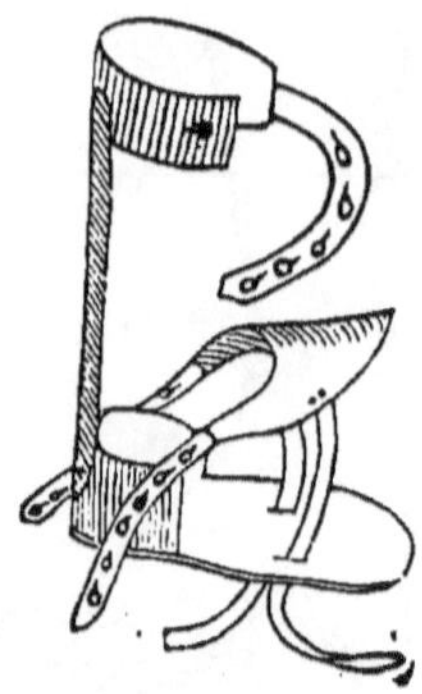

With this simple apparatus which we devised some twenty years ago, we have, in most cases, reduced the foot to normal form in a few weeks in cases not produced by paralysis; but in order to effect a permanent cure we have advised the continued application of the apparatus for five or six months.

We advise, at the outset of the treatment, gentle force to be made by the tapes; the brace to be removed twice a day and the foot examined and manipulated with the hand. If tender points present, protect them with lint. This formula is to be continued for a month or more after the foot is reduced to the sole-plate.

TREATMENT OF NON-CONGENITAL TALIPES CALCANEUS.

Non-congenital talipes calcaneus presents a most complicated anatomical condition for treatment. In more than a majority of cases of this contortion, it is the sequel of paralysis of the extensor muscles of the leg. The anterior portion of the foot is usually slightly everted in its flexed condition, being elevated above the dependent heel and nearly at right angles with the leg. To meet these indications with suitable therapeutic means has greatly taxed the orthopædist's ingenuity.

Fig. 49.

In this stage of the ailment, tenotomy is but seldom required as a preparatory step in the treatment. Extension apparatus [Fig. 49] must be relied on, so constructed as to support the os calcis and metatarso-phalangeal point of bearing when brought down to the plantar plate; the plate being held by means of a limited joint at right angles with the upright bars that pass perpendicularly up the leg, and secured by an encircling band. A broad strap passes from bar to bar about an inch above the instep, intended to restore the tibio-tarsal relation; pressure being made upon the dorsum of the foot by means of a roller.

The pressure upon the instep by means of the roller, and the tibial supporter in a fixed position tends to a restoration of the os calcis to a normal position, which is the principal indication to be met.

This apparatus, originally intended for the redressing of non-congenital talipes calcaneus, we devised some twenty years since, and in no instance have we found it necessary to modify or alter the original construction, other than to omit the use of the tibial band — which in most cases is not required.

In nearly all cases of non-congenital calcaneus, because of the partial paralysis of the extensor muscles, the apparatus must be worn for two or more years; subjecting the patient, however, to but comparatively little inconvenience—a laced boot being worn over the apparatus. But, in this, as in all other cases of contortion of the feet, the appliances must be *worn during the night.* The metal shoe should in all cases be worn independent of the ordinary shoe; as the foot can be much more readily controlled, and the expense of a night shoe saved.

CHAPTER IV.

INFANTILE PARALYSIS.

Contortions the sequence of infantile paralysis.— Illustrative cases.— Paralysed limbs susceptible to a restoration of power.— Proper and careful treatment essential.— Views of Ancients as to infantile paralysis.— Galen on palsy. — Pott, Baillie and Abercrombie on paraplegia.— Infantile paralysis arising from irritation of the primæ viæ.— Prognosis in infantile paralysis.— Special training in the department of surgical science necessary to the successful treatment of the disease.— The sequence of unrelieved paralysis.— The London "Lancet" on origin and treatment of infantile paralysis.— Our treatment of infantile paralysis.— Exposure of infant to cold or damp, chief cause.— Congestion of the peripheral vessels most commonly present.— Vesication of the spine, and irritants applied to the ankles and wrists generally successful in restoring the impaired functions.— In advanced cases an asthenic condition of the implicated portion of the spinal cord has ensued.— Mechanical and physiological treatment of infantile paralysis.— Patients usually in enjoyment of good bodily health.— Congested appearance of cuticle.— Atrophied condition of the wasted limb.— Essential paralysis.— Laborde's definition thereof.— Degeneration of the muscles.— Extraordinary cure of restoration of muscular power in a patient fifty-four years of age, suffering from equino-varus of both feet.— Another case of a lady thirty-nine years old.— Prognosis for treatment. — SPECIAL TREATMENT OF CONTORTION, THE SEQUENCE OF INFANTILE PARALYSIS.— If unequal tension of muscles observable, use of electricity inadmissible.— Method of obviating tendency to talipes calcaneus.— Cases demanding Tenotomy.— Mr. Wm. Adams' opinion as to severance of tendons in cases of infantile paralysis.— Method of treatment initiated by Richard Barwell.— Preference of author for severance of contracted tendons as a remedy in this class of ailments.— Personal efforts in primary treatment.— Electricity a powerful remedial agent.— Objection to the Swedish movement.— Recovery frequently ensues where no system of treatment has been attempted.— PARALYSIS the result of caries of the last cervical and first dorsal vertebræ.— Paralysis the result of adventurous treatment.— Case in illustration.— POSTERIOR GENU FLEXUM.— Leading features of this deformity.— In childhood, nearly every case curable, while neglect gradually transforms it into a permanent deformity.— Treatment of infantile patients.— Supporting apparatus.— Support to be continued until muscular power is restored.— Time required for cure of deformity.— Electricity and manipulation, when skilfully applied, valuable auxiliaries.— Contraction of iliacus and psoas magnus muscles.— INFLUENCE OF OCCUPATION TENDING TO PARALYSIS.— Writers and Painters specially subject to paralysis of the hand and wrist.— Frequently met with among shoe-makers, glass-cutters, seamstresses and compositors.— Paralytic premonitions.— Drop-wrist. — Treatment.— Apparatus.— Static electricity.— Effect on the patient, and results in ameliorating his condition.

CONTORTIONS THE SEQUENCE OF INFANTILE PARALYSIS.

The ancient and some modern writers have considered infantile paralysis as the sequence of apoplexy. Stall and Portal mention, under the caption of "Apoplexia Infantum," * what they consider to be the most probable causes, such as being subject to worms, dentition, and the improper administration of opiates.

M. Portal states: "I have seen children of a plethoric constitution afflicted during dentition with a true coma ending in palsy of the arms or legs." We have treated children laboring under coma and stertorous breathing, that, when relieved of the latter, the sequence hemiplegia, and others that resulted in paraplegia. The citation of a case or two will render the subject more comprehensible: One, the patient a little girl four years of age, of vigorous growth and rotundity of form, labored under coma and stertorous breathing. The eye balls were so contorted laterally that only about one-third of the dilated pupil could be seen. An emetic was administered, when a quantity of undigested pith of cabbage-stalk was ejected and the child restored at once to consciousness and, within an hour, to the ability of articulating. Paralysis of the right arm and leg was the sequence.

The other, a little girl three years of age, equally well developed, and giving a history of previous good health, labored under coma and stertorous breathing with constant agitation of the right side, and eyeballs divergent. An emetic was given, and failed to act for fifteen minutes. The patient was then placed in a tepid bath and ice applied to the head. After a few minutes there was ejected a quantity of unripe gooseberries. The sequence in this case, paraplegia. These children, being carefully treated, recovered from their paralytic condition in about eighteen months. Of younger children laboring under coma, stertorous breathing, and convulsions during dentition, we have treated many cases; and the result has been in nearly every instance partial paralysis of the limbs.

---

* 1 Portal, p. 62.

Another not uncommon cause of infantile paralysis is exposure from sitting or lying on cold, damp ground. Cases have been presented by intelligent parents where the mothers had been compelled to leave their babes with the elder children, who placed them upon the grass or damp ground while they engaged in play. From this exposure ensued the paralysis in one or more limbs. This is so common an occurrence that a reasonable doubt cannot be entertained as to its being a cause of infantile paralysis.

Many cases occur that are represented as the sequence of measles, scarlet fever and cholera infantum, and others, in which no reasonable cause for the paralytic condition can be given. The mother's or nurse's statement is, that the child previously enjoyed good health, when, on a stated evening, it had a slight fever, was rather restless during the night, and on the following morning was found to have one or more limbs paralyzed. In some cases, in a month or two, the arm has been quite relieved, the leg remaining powerless, or but slightly improved, with a tendency to contortion. In others, the arm remained powerless at the shoulder whilst the leg had recovered. Thus, it will be observed, that the limbs are variously affected, and in nearly all cases susceptible to a restoration of power.

The tendency, in these cases of infantile paralysis, is to contortion of the limbs affected, if not properly supported in normal form, and this contortion is the result of the recuperative tendency of some muscles, or the shortening of those least impaired. Mr. Wm. Adams states: "I believe all these deformities may be prevented by judicious treatment; in fact, that there never need be another example of deformity in these cases if the liability to their occurrence and the mode of their production were generally understood."

A careful diagnosis is greatly to the advantage of the patient, but practical information on the treatment is as essential to their relief, tending to the proper construction of appliances, and their modification as relief progresses. The normal increasing ability of motion in the muscles must be carefully considered, and the first required support diminished in ratio to the improvement, or a serious impediment to progress in strength of muscles will result from limiting the necessity for vital energy by continued support.

The ancients were evidently familiar with some of the now well known causes of paralysis in children, but they make no especial mention of the various inimical influences tending to a peculiar condition, as in that of paralytic seizures in infants. Galen makes the following remarks upon palsies:

"If a nerve becomes thickened and harder than natural, the propagation of its power is hindered; or if it be compressed by some hard body, it cannot afford a free passage to the power. If compression be made on nerves by cords or by the hand, by phlegmonous or schirrous tumors of neighboring parts, or by luxated or fractured bones, the nerves become first torpid, and afterwards entirely lose sense and motion."*

Cold, he states, often injures a single muscle, especially that situated upon the superficies of the seat, as when a person sits upon a cold stone, or remains too long in the cold. (This applies equally to children that are paralyzed from sitting on cold, damp ground.) Forestus treats of palsy in the hands from cold and moisture applied to the neck, and of palsy of the bladder from a similar application to the back.†

Mr. Pott states that a great proportion of cases of paraplegia, especially in infants, depends upon constitutional affection of the spine.‡ And he further states that adults are by no means exempt from caries of the spine terminating in paraplegia, but that he never saw a case of it at an age beyond forty.

Drs. Baillie and Abercrombie attribute all cases of paraplegia to lesions of the brain, and claim that children are subject to the affection.§

Mr. Charles Bell appears to be the first to invite attention to paralysis in children arising from irritation of the primæ viæ. He states that there is a class of local palsies which are neither preceded by pain, inflammation, nor disorder of the brain, but arise from irritation of the bowels, and that the loss of motion from this cause occurs often in children, "and that among those wretched Irishwomen who apply as out-patients to our hospital, and with a child at the breast with its arm hanging down, entirely without muscular power. In such cases I have invariably found, on questioning the mother, that there has been disorder in the bowels, with a passing of green stools, griping and spasms previous to the paralytic seizure of the limb."

---

* "Galen de locis affectus," lib. 4.
† Forestus, obs., p. 34.
‡ Pott, p. 20.
§ Abercrombie on Diseases of the Brain, p. 33.

92   ORTHOPÆDIA.

This is a true representation of many cases thus conditioned, and presented for treatment, except that children having green stools and spasms, have pain. Yet this cause afflicts but a very limited number with infantile paralysis, compared with that of the several causes previously stated, and paralysis the sequence of inimical impressions made upon the digestive organs and reflected upon the nervous centre, tending to a partial arrest of nervous influence, and limited to the impairment of only certain classes of muscles, and in some instances to that of a single muscle, or the flexor muscles of one limb and the extensor of its fellow, both limbs being thus so affected as to incapacitate the child to stand alone. Upon examination, however, certain muscles will be found to be in active condition, tending to shorten, the equilibrium of normal antagonizing force being interrupted. Contortion is the inevitable sequence under such circumstances, if not arrested by the aid of acquired knowledge in the treatment of such cases.

PROGNOSIS IN INFANTILE PARALYSIS.

The recuperative tendency in all cases of infantile paralysis exists largely, and, in many instances, even without skilled interference, the normal status is recovered. Inexperienced practitioners, however, are liable to place too much dependence upon this tendency, and to defer the simple treatment until the more complicated is required; and then, to obtain a suitable therapeutic agent, depend upon the instrument-maker's judgment. This sad practice is daily resorted to, even by those who make themselves proficient in the treatment of this class of ailments. Anxious parents, prompted by the love they bear for their afflicted children, make repeated efforts for relief, bewailing their mental sufferings, and expenditures they can but ill afford, having availed themselves of the services of those who are reputed to be the most eminent in the medical profession, and all without obtaining an arrest of the increasing contortion. Cases thus deplorably conditioned have been cured — patients that have suffered from eighteen months to twenty years of age.

The practical attainment of knowledge in this department of surgical science determines a favorable prognosis in nearly all cases of infantile paralysis, the ailment having now been brought within the rational influence of pathological and therapeutical science, and requiring only assiduity from the skilled practitioner, together with invariable obedience from patients and attendants, to ensure relief in nearly every case of infantile paralysis.

THE SEQUENCE OF UNRELIEVED PARALYSIS.

The result of inefficient or no treatment tends to atrophy and persistent contortion of the limb or limbs, which are much influenced by continued unfavorable position, as in the instance of permitting a child to daily remain in a sitting posture in a small chair, which is made an available means of locomotion by muscular effort of the arms and body. In this dependent condition many children are allowed to remain for life, because of the opinion of the family practitioners in medicine that such cases are incurable, and only because of their inability to relieve the patient, which, as must be admitted, is almost an impossibility for the general practitioner. Clinical instruction in this class of ailments is most difficult to obtain, and suitable apparatus to meet the various indications that present, impossible without experienced knowledge to devise the therapeutic agent to be constructed by the skilled mechanician.

As this work is only intended as a practical treatise on the several ailments included therein, we refer the reader for more general information to the most excellent work of Mr. Wm. Adams on Club-foot, and the invaluable works of Heine : " Practical Treatise on the Diseases of Children," translated by P. H. Bird, London, 1854; West, "Lectures on the Diseases of Infancy and Childhood," 3d edition, London, 1854, and many more recent works of merit on the subject at home and abroad too numerous to mention, but of great value in affording practical information upon the subject.

The impression made upon the nerve centres have not, as yet, been satisfactorily determined—subject-matter tending to careful and satisfactory examination not having been afforded, because of limited opportunity. Barthez, Riellet, Dr. Fliess, Mr. Wm. Adams, and recently Dr. J. Russell Reynolds, physician to the University College, London, have each and all come to the conclusion that the brain is but seldom, if ever, implicated in the primary impression that results in infantile paralysis. The latter authority states, in a letter published in the London "Lancet" of July, 1868, p. 35, the following :

"The muscles and bones are small, and the former often lose all of their characteristic structure; and until recently it has been affirmed that beyond these changes of the limbs nothing abnormal was to be found. You may read descriptions of post mortem examinations in which it is stated that the brain, spinal cord and nerves were found to be in their natural state; and it is only of late that by a process of more minute investigation, the real malady has been discovered. M. Laborde has shown that there is a distinct change in certain portions of the spinal cord; that the interior columns are more translucent than natural, and present a very appreciable greyish rose tint to the naked eye, and that a similar change may be observed, though to a less degree, in the lateral columns. The consistence of these tracts of nerve tissue is diminished; and upon microscopic examination there may be observed a marked proliferation of the elements of the connective tissues, cells and nuclei being dispersed in the midst of a finely granulated substance in which there are fibrils of extreme tenuity. In the parts which are most affected, tubules are either lost altogether or they present a varicose appearance, while the other portions of the spinal column preserve a perfect integrity."

In the statements of this authority we feel disposed to place much reliance, it having been sustained by practical results.

It has been our practice for many years, when applied to for treatment in the early stage of seizure of infantile paralysis to question the attendants as to the condition of the child at the time of the attack, or immediately preceding it, and if informed that the child was feverish or restless on the day or night previous — that being the only symptom of impairment of health — we endeavor to ascertain, by examination of the mouth, the condition of the gums; and, if no irritation is found there, inquire into the diet and condition of the bowels, or as to what exposure the little patient has been subjected; sitting on cold, stone-pavement, damp ground, wet floor or in excessive drafts of air. These are the most common causes of the slight fever that tends to infantile paralysis; and the treatment required is, with some regard to the exciting cause; at the same time applying rubefacients to the extremity of the paralyzed limb, and wet cloths at a temperature of about 40° Fahr. to the spine; to be continued for forty-eight hours, if relief is not sooner obtained.

This method has relieved patients in less time than twenty-four hours, and without the least tendency to relapse.

The patient, not being relieved by judicious, ordinary treatment, may be considered as laboring under a very decided impression upon some portion of the cerebro-spinal centres, most commonly congestion of the peripheral vessels. A similar condition is often the sequence of protracted disease; as that of scarlatina, rubeola, pertussis, and cholera infantum, and it is the exception to meet with a case of paraplegia from any other causes, where caries of the spine is not a concomitant.

In every instance where contortion of the muscles is not apparent in cases of infantile paralysis, and the seizure is of less than a month's duration, we vesicate the spine and obtain as large a drain of serum as possible by the frequent change of emollient poultices and the repetition of vesication to a reasonable degree, and if improvement is not apparent, irritants are applied to the ankles and wrists of the paralyzed limbs, and in this stage of the ailment this treatment has been most frequently attended with decided success. The drain of serum relieves the congested condition of the peripheral vessels of the spinal cord, and the irritants to the extremities tend to the restoration of the normal status of the impaired functions of the implicated nerves.

In the more advanced cases, or those of longer duration, when contortion has commenced, we then consider that an asthenic condition of the implicated portion of the spinal cord has ensued, and that irritant plasters constantly applied to the spinal column are indicated, with that of other therapeutical agents, to restore and maintain the normal form of the contorted limbs.

This stage of the ailment brings us to a consideration of what is usually denominated the *mechanical* and physiological treatment of infantile paralysis, that has resulted in contortion, and a degenerate condition of the muscles from a partial arrest of nutrition.

Patients in this advanced stage of the paralysis usually enjoy good bodily health and, as Dr. Reynolds remarks, it is rare to find on them bed-sores or other evidences of localized mal-nutrition of the skin. Normal warmth, however, is deficient and maintained with much difficulty by artificial means. The want of vital impetus in the circulation is apparent, from the congested appearance of the skin of the diseased limb, which not only fails to maintain its muscular development, but in some instances diminishes in circumference rapidly, yet it is but rarely that the bones fail to keep pace in growth with those in the corresponding but healthy limb.

The loss of equilibrium in muscular force induces an inability to control the limb, resulting in an apparent shortening of the leg when sustaining the weight of the body.

The sensibility of the paralyzed limbs is usually found to be in a normal condition, or nearly so, when carefully tested, and, as heretofore stated, it is only certain groups of muscles that are paralyzed. Thus it will be observed in some cases that an obvious wasting of the gluteal region of the affected limb has taken place, and no decrease of rotundity of the thigh, or an atrophied condition of the thigh and a tolerable development of the leg maintained. A similar condition will be found in the arm and shoulder, the mal-nutrition being most observable in the deltoid muscle, supra and infra spinatus and scapula, whilst the pectorales, major and minor, are seemingly unimpaired; or, all of these muscles may be paralyzed, and the muscles of the arm and hand active. Under this condition, the head of the humerus may be seen distinctly, and if the arm is not supported by sustaining the elbow from the opposite shoulder, a separation of two inches will eventually take place in the joints, rendering the restoration of the limb to usefulness impossible. Patients in this deplorable condition have applied for treatment, stating that they had been advised by their physician not to have the arm supported by a sustaining bandage, as it would deprive the limb of motion (actually an essential to the only reliable means of cure). It is true, exercise is essential to cure, but it cannot be taken with the arm in its pendent condition; the muscles should be manipulated about the shoulder, and the arm supported from the elbow, and exercised as constantly as possible.

As this ailment appears to be strictly local, it has been termed "essential paralysis," and noted as peculiar to young children. Laborde, in his treatise "De la Paralysie dite essentielle de l'Enfance," limits the susceptible period to from nine months to four years of age. This we consider a reasonable limit, but there are exceptional cases, both older and younger, presented to the orthopædist for treatment, as that of ordinary cases laboring under the sequence of paralytic seizure, contortion most commonly of the foot, with atrophy of the leg, a comparatively limited number having the muscles of the thigh impaired as well as those of the hip, the knee being advanced, bringing the leg to an obtuse angle with the thigh, while others have the sustaining ligaments of the knee flaccid, permitting it to fall backward. In cases where both limbs are affected, we usually have the advanced knee with the foot presenting a case of talipes equinus or equino-varus, and in the other, the knee inclining backward, and the foot presenting a case of valgus or equino-valgus.

In these cases, when of some duration, the muscles degenerate from a normal condition of development to that of containing deposits of flat globules; but the sarcolemma areolar membrane and connective tissues maintain their integrity to an ability of again, under favorable circumstances, resuscitating the impaired limbs. A similar condition exists in the nerve tissue, the nerve centre being excited into an approaching normal condition, all of which has been accomplished by judicious treatment in patients whose limbs have been extremely atrophied and powerless for twelve or more years. It is true that there are various degrees of improvement in the paralyzed limbs, some never being fully restored to an actual ordinary fullness and strength, though of decided usefulness to the patient, when so far restored as to sustain a perpendicular bearing upon the limbs of the superincumbent weight of the body. This approach to a normal condition tends largely to the future increase of strength in the impaired muscles at almost any age, which fact we have before alluded to — the statement of Mr. Wm. Adams, that he restored a gentleman of fifty-four years of age, who was afflicted with equino-varus of both feet, and compelled to use crutches in attempting to walk. Previous to the treatment of Mr. Adams he had consulted Sir A. Cooper, Mr. Keat, Sir B. Brodie, Sir Charles Bell, Magendie, Dupuytren and Frauquier; and not one of these eminent surgeons had given him relief, because of their want of practical information in regard to the severing of tendons and properly devised apparatus, the only reliable means of affording relief in such cases.

Another interesting case is spoken of by Mr. Adams: that of a lady thirty-nine years of age who was relieved after having labored under contortion of the feet, the sequence of infantile paralysis, from the age of five years — being so decrepit at the time of treatment as to render it necessary for her to be carried from room to room by a servant. The severing of tendons restored to her the use of her limbs and the ability to walk without support.

The prognosis for treatment are clearly defined in the results of the above cases; the lady being relieved of contortion of the feet, the sequence of infantile paralysis, and of thirty-four years' duration; and the gentleman being cured at the age of fifty-four years.

SPECIAL TREATMENT OF CONTORTION THE SEQUENCE OF INFANTILE PARALYSIS.

Contortions of the feet — the result of infantile paralysis in its incipiency — may be relieved in many instances by elastic supporting force; but what is better, when opportunity is afforded the practitioner, is by sustaining the affected limb or limbs, making available, during the early period, all the therapeutic means of restoring energy in the impaired limb. For local treatment, manipulation and electricity in its various forms should be carefully applied and varied, if one form fails, but in its most limited force, for ten days, when it may be increased moderately if improvement is manifested, without contraction of one or more muscles tending to contortion. Whenever an unequal tension of the muscles is observed electricity is inadmissible, and should be discontinued; instead, suitable means should be applied to sustain the limb in normal form; resisting the tendency of certain muscles to shorten, yet permitting the movement of the limb by limited motion, that the enfeebled muscles may be exercised.

If there is a tendency of the gastrocnemius to shorten, the form of apparatus as shown in Fig. 50, having a limited joint at the ankle, must be used to meet the indication. By taking hold of the sole-plate at the toe the foot can be flexed upon the leg, and the patient being unable to walk, passive exercise can be thus given to tolerable advantage. In this condition of the patient, the external application of strychnia has, in our practice, answered an admirable purpose when applied to the enfeebled muscles daily, for some months, carefully cleansing the surface of the skin with warm water and soap before each application —

℞ Strychniæ sulphatis, grs. xii.
Adipis ℨ i.
M. ft. ung.

and, as an adjuvant in the treatment we apply a narrow plaster of Burgundy pitch to the lower third of the spine, in cases of paralysis of the lower limbs. The patient's general health must of course be carefully considered, as they are not always in a favorable condition when laboring under this local derangement; and moderate exercise, inviting motion in the impaired limb or limbs, should also be made available.

In cases of paraplegia, where contraction of certain muscles has not ensued, or the paraplegic condition is not induced by caries of the last cervical and first dorsal vertebræ, steel supports to the hips and even to the axillæ are a desirable auxiliary to the restoration of power to the paralyzed limbs. The patient being placed in an erect position for a reasonable duration of time soon desires further privileges, and is apt to concentrate his will in an effort at locomotion which contributes largely to the accumulation of strength, however feeble the influence may be at the first attempt; for as long as vitality exists in the muscles they are susceptible of being influenced by the will to their improvement, and, in a large majority of cases of infantile paralysis, to perfect restoration when assisted by skilfully contrived mechanical support. This remark applies with equal force to the treatment of adults.

Fig. 50.

The accompanying engraving presents the form of apparatus we have applied to paralytic patients during the past thirty years.

The springs must extend from the axillæ to the feet; the feet supported by a soleplate and a cup for the heel, and, in some cases, a limited joint at the ankle, keeping the foot at a right angle with the leg. The other joints are so constructed as to be perfectly fixed by means of a strong steel slide as at A, intended to move easily over the end of the bar extended beyond the axis of the joint — the joint being now represented as free to permit of the patient's sitting down, while at B the slide is upon the extended bar, and sustains the limb in the erect position. A similar arrangement will be observed at the hip. The hips are encircled by a broad steel band from which ascends a bar to each axilla, and two more to the support of the back; being connected at proper distances by a steel band that crosses the back and extends through each axilla. This frame is maintained by some strong textile fabric; being laced in front and kept extended by whalebones.

Fig. 51 gives a back view of the apparatus. A is a buckle and strap attached to a broad webbing strap (B), which crosses the hips over the joints, and when the joints are locked assists in supporting that portion of the patient which otherwise would project backward.

Cases present for treatment, where there is such a relaxed condition of the muscles about the hip, and extension of the ligamentum teres, that the head of the thigh bone, if not sustained, will escape from the cotyloid cavity. To sustain the normal relation of the joint, the following apparatus has been constructed by us, and applied for the past twenty years, with great advantage to the patients.

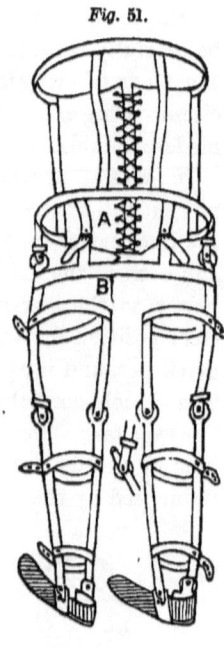

Fig. 51.

A represents a piece of sheet steel attached to the upright thigh bar below the thigh joint, and curved in the form of a hook, catching only the posterior portion of the bone, so as to sustain the head of the os femoris in the cotyloid cavity. The attenuated condition of the integument permits the bone to be grasped, and, by means of the belt about the hip, is perfectly sustained. B represents a limited joint at the knee when there is only an impairment in the strength of the sustaining integuments of the knee, permitting posterior flexion. C represents the joint that may be applied in case of necessity. D simply shows that gaiters can be worn over the apparatus when applied to the feet, which apparatus, being separate from the shoes or gaiters, is not objectionable in being

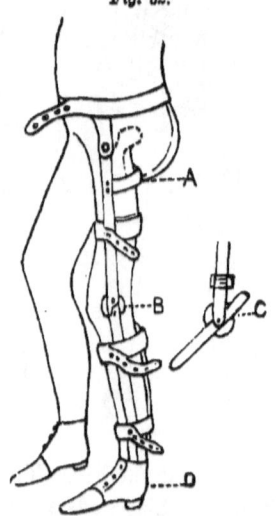

Fig. 52.

worn by the patient when in bed, an essential necessary to ensure a cure.

In some cases there presents an increasing tendency to elevation of the instep, in both children and adults, which, if not arrested in time, tends to talipes calcaneus, an actual elongation of the tendo-Achillis. This may be relieved by the daily use of a shoe made after the following plan: an elastic steel plate is curved upward and fitted to the inside of a shoe, the shoe being made to lace upon the foot from the toe. The plate is curved upward from about the anterior two-thirds of the length of the sole of the shoe, the elevation at the toe being about an inch and a half. The heel portion of the spring plate is firmly riveted to the inside of the

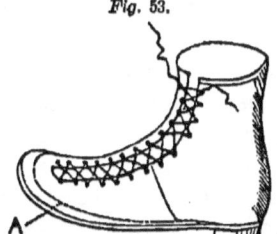

Fig. 53.

heel of the shoe, leaving the anterior portion free, so that in walking the spring will necessarily be borne down by the front of the foot, and thus, from resistance, tend to relieve the contracted tissues.

By Fig. 53, A points to the spring plate within the shoe as elevated at the toe.

Well regulated elastic extension will relieve incipient cases of contraction, if carefully persevered in; but after some duration of this contracted condition, it is with greatly increased difficulty

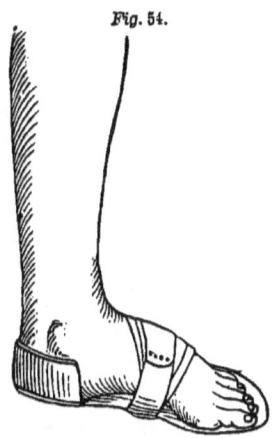

Fig. 54.

that it is overcome. The suffering caused by concentrated pressure is often unendurable, and in those confirmed cases of contracted feet relief cannot be afforded by means of the best advised elastic force, but division of the shortened tissue, even to ligaments, is demanded as an auxiliary aid. By this preparatory, or additional treatment, nearly every patient laboring under contorted limbs, the sequence of infantile paralysis, or other causes affecting adults, can be greatly relieved, and many cases of contorted and atrophied limbs restored to normal usefulness. When the plantar fascia has been divided, the apparatus represented is applied. (See Fig. 54.)

A steel sole-plate and cup for the heel. And when the roller is applied over the instep, a shoe can be worn over the whole.

CASES DEMANDING TENOTOMY.

A persistent condition of the shortened muscles tends not only to the arrest of development, but finally, to degeneration of the muscular tissue; depriving it of its contractile power, and thus arresting motion in the joints, and inducing and maintaining the contorted position. Not only does it deprive the patient of the benefit of healthful exercise, but in the case of an adult seriously affects him with despondency, and an almost constantly irritable state of mind, to the utter exclusion of any desire for society or pleasant intercourse. For the relief of this pitiable condition of the afflicted, we have a reasonable excuse for an adventurous surgical interference, if it may be so considered, when practicing tenotomy for their relief, and at almost any age of the patient. Entertaining this consideration, we would advise orthopædists not to delay making the effort to relieve extreme cases of contortion. When patient perseverance in well devised means of extension has failed, sever the shortened tendons and tense ligaments, and by all means avoid torturing your patient by fruitless efforts at long continued application of steel braces, however well devised for the purpose of extension; as the resisting force and points of bearing are unavoidably painful, even in cases that eventually yield to the treatment.

The chief objection that has been advanced, to the severing of tendons for the relief of contortion, the sequence of infantile paralysis is, that it deprives the limb of all self-sustaining power. This statement is not sustained in actual practice. That it has occurred is possible, when inexperienced practitioners have severed tendons, and failed to devise proper appliances to sustain the severed tissue and limb in normal position, or the foot in relation to the leg until the tendon was reunited. But we have never had a case in which the tendons have failed to unite, after having been severed, even in the most attenuated limbs, although the limb remained feeble for years, requiring support to enable the patient to walk — a privilege they had before been deprived of — and in nearly every instance improvement has been realized to the gradual accession of the ability to walk without a support. This has been the result of an experience of nearly thirty-five years daily devotion to the treatment of deformity, the result of various causes, a large proportion of the patients laboring under this condition of ailment.

For the preparatory stage of treatment, requiring the tendons to be severed and apparatus applied, in this class of contortions of the feet and limbs, we refer the reader to the description given in the previous chapter on congenital contortion of the feet; slight variances, we consider, may be left to the intelligent practitioner in surgery.

Mr. William Adams, in his most excellent treatise on club-feet, published in 1868, pp. 41 and 42, makes the following statement in regard to the severing of tendons in cases, the sequence of infantile paralysis:

" Hence," he says, " we can hardly be surprised to see a book issuing from the press with a leading chapter on '*the impropriety of tendon cutting and its evil results*,' and with another chapter on '*my new method of treatment ;*' but it is with no small degree of astonishment that we find the same author, a hospital surgeon, speaking of tenotomy thus: ' The operation of cutting tendons or muscles had been haunting the domains of surgery for about one hundred and fifty years, and had been gradually becoming a less adventurous proceeding, when, in 1832, Strömeyer demonstrated a method of its performance without danger, or at least, with very little danger of producing suppuration and sloughing.'*

" In the so-called method brought forward by Mr. Barwell, the feet are placed in an improved position and so retained by means of tin splints placed lengthwise down the leg, and a series of elastic india-rubber cords attached to the tin splints and passing in different directions corresponding to the paralyzed muscles. The foot and leg are previously covered with adhesive plaster which doubtless adds to the general support afforded by the apparatus, and to some extent prevents excoriations and blisters from excessive pressure, which, however, is still described as occurring in some cases. By the use of these elastic cords, the author speaks of supplying an *anterior tibial muscle* in one case, and a *posterior tibial muscle* in another, or a *tendo-Achillis ;* and in the same way many other muscles are supplied, so that the foot and leg are rigged like a ship. And no doubt.

---

*"On the Cure of Club-foot without Cutting Tendons," by Rich. Barwell. London, 1863.

a very useful compensation for paralyzed muscles is often thus afforded; but that this can in any way be regarded as a *curative means* for the paralytic affections which Mr. Barwell describes as being the 'head and front of the offending,' has yet to be demonstrated. If it should prove to be a means of cure for paralysis, it would be a valuable addition to those at present so frequently employed with but very limited success, but if this cannot be proved the new method seems to be as little deserving the unbounded praise bestowed upon it by Mr. Barwell as it can be shown to have any claim for novelty.

"On the latter point, I can state that the plan was certainly brought under my notice about ten years ago, by Mr. Bigg, of Leicester Square, who had a great variety of ingeniously constructed instruments upon this plan, with vulcanized india-rubber cords attached by hooks and passing in various directions according to the deformity for which the apparatus was made; all distortions of the feet, knee-joint instruments, spinal instruments, etc. Some of these instruments for the treatment of clubfoot were adapted by Mr. Bigg to cases in St. George's Hospital, and he told me the chief difficulty was to regulate the pressure without producing sores. The constant pressure from the elastic force could not be borne by the patient, and the plan of treatment was given up, only to reappear as '*my new method*' ten years later."

The cases adduced by Mr. Barwell in illustration of the success of the treatment belong almost exclusively to the class of non-congenital distortions associated with paralysis, or arising from debility and ligamentous relaxation, such as the ordinary flat-foot, etc.

This extended quotation has been made from Mr. Adams' book, as reliable information upon the credit that is to be given to treatment by means of elastic cords to redress contortions of the feet. Our faith in the "new method" as a curative means is weak; as all that could be desired in the treatment of this class of ailments has been attained by severing the contracted tendons, *i. e.*, where the foot could not be restored to normal form by the hand, and so maintained by means of a supporting brace without pain. Excoriation and pain indicate the necessity of severing the resisting tissue.

PERSONAL EFFORTS IN PRIMARY TREATMENT.

Our primary treatment consists chiefly in efforts tending to restore the normal form and relative position of the limbs to the body, as well as the maintenance of a perpendicular bearing, and this mainly by the aid of artificial support. The patient is to be encouraged to make efforts at locomotion, although at first considering it impossible. The attempt, being an effort of the will, imparts a favorable influence, and the erect position determines a recuperative tendency that, in a few days, is most apparent in nearly all cases that present for treatment, even those of long standing. The feeble circulation and consequent diminution of temperature so apparent in the purplish color of the cuticle indicates an enfeebled capillary circulation in the impaired limbs. They soon assume, even to a casual observer, a more natural appearance, and that too before the patient has improved in ability to move the limbs. Auxiliary means may now be made available and to some advantage, as that of manipulation by gentle pressure of the fingers, commencing at the extremity of the limb, and advancing to the body slowly, twice or thrice a day; flannel wrapped about the limb and covered with chamois neatly fitted; the limbs washed every twenty-four hours with warm water and soap, and the application, by means of a suitable covering, of the dry vapor of burning spirits. An India-rubber sack is a convenient means for this purpose, and also serves another purpose, that of applying an air pump securely to one end, and exhausting the air; by this means the veins become turgid and the capillary circulation improved.

Electricity, as described in another chapter, may be used, and varied to meet certain indications that present. If it tends to impair the increasing power of the limbs, which it will in some cases, or increases the contractile power of the shortened muscles, it must be discontinued. Passive exercise by means of ingeniously-contrived and complicated fixtures may be used, a practice known as the Swedish movement cure. This means of passive exercise we do not approve of as a curative means, as its general reputation would warrant. Long experience has impressed us most favorably with the treatment we have described, and we believe that the close observer of the practice we have adopted will not be disappointed in the result when patiently and skilfully applied. As all excesses in the treatment must be carefully avoided by close observation upon the patient's condition under the same, which, to insure success, demands most careful modification, as in that of all other efforts to restore energy to the enfeebled muscular tissue.

It must be borne in mind that nearly all cases of infantile paralysis, although terminating in contortion and impairment of the normal development of the muscular tissue tend, even without curative interference, to recovery, inviting, as it were, skilful assistance to the resumption of a normal condition, slow in extraordinary cases, and greatly retarded when unskilfully treated (as great injury may be sustained from excessive treatment).

In fact, recovery often ensues in cases where no treatment has been attempted, and in nearly all cases of long standing we are informed of improvement having taken place, as that of a recovered arm or leg, or one side, or the recovery of the ability to walk, both legs having been paralyzed.

PARALYSIS THE RESULT OF CARIES OF THE LAST CERVICAL AND FIRST DORSAL VERTEBRÆ.

A very common cause of paraplegia in children is caries of the spine when affecting the last cervical and first dorsal vertebræ. Nearly every patient thus afflicted, if timely support is not given to sustain the superincumbent weight of the head, will be found in a paralyzed condition — the paraplegia being more or less complete. These cases are relieved by the application of vertical support to the head, when in the incipient stage of the ailment. Perfect restoration to the ability to walk has been afforded within a month from the time of the application of the support. From this ready means of relief we are disposed to attribute the cause of paraplegia to a congested condition of the tissues in that region, making an impression upon the nervous centres or spinal cord in a remarkable manner not heretofore considered by pathologists, as the paraplegic condition ensues before angular curvature presents, that would tend to mechanical pressure upon the spinal cord; and relief afforded as before stated by extension of the spinal column. What is most remarkable in this pathological invasion is because of the paralysis only affecting the lower extremities and in exceptional cases the pelvic viscera as evinced in incontinence of urine, and more rarely impairing the retentive power of the sphincter ani; patients thus afflicted having the perfect use of them, and no functional impairment of the heart, lungs or stomach.

PARALYSIS THE RESULT OF ADVENTUROUS TREATMENT.

Traumatic injuries of the dorsal and lumbar vertebræ tend, if severe, in most instances, to paraplegia, or clonic spasm, which is but seldom the sequence of caries of the spine, unless in cases where violent effort has been made, with the intention of redressing the angular curvature, after it has existed for a considerable time. Such cases we have witnessed after adventurous treatment, and subsequently have restored the patients by re-establishing and maintaining the projection by means of mechanical appliances. This is an important fact to be remembered in the treatment of caries of the spine. Parents having children thus afflicted are exceedingly anxious to have the distortion of the spine, even on a venture, restored to normal form, and for that purpose resort to persons who make such pretensions. A case in point may more fully impress the reader with the importance of avoiding such treatment. A reverend gentleman of eminence in this city, brought his little son, a boy seven years of age, for treatment, the fifth, sixth and seventh dorsal vertebræ being diseased. An examination revealed the projecting spine in the dorsal region, and a course of treatment was improvised. A brace was applied, consisting of lateral support from the axillæ to the crest of the ilium, but avoiding pressure upon the projecting bones. Under this treatment the child maintained a tolerable condition of health, giving promise of an arrest of the disease. The father very frequently expressed himself as desirous that the projecting bones should be depressed. To this we gave no other attention than to advise perseverance in the use of the support to the child's body—which would prevent motion and consequent attrition—at the same time insisting upon the necessity of a generous diet and moderate exercise in the open air. From this treatment the little patient derived much benefit. The father finally ceasing to bring the ailing child for advice, as had been his custom, every two or three weeks. Inquiry was made, and information obtained, that our little patient had been placed under the treatment of a "doctor who had straightened his spine," to the great delight of the parents, and friends who had advised the change of treatment. We heard no more of the case for some years, when we were informed, by a member of this gentleman's congregation, that the aforesaid patient was completely paralyzed from the attempt made to straighten his spine.

In this condition he remained for several years, when his father again made application to have his son placed under our treatment. On visiting the (now) young man, we found him in a most deplorable condition; complete paraplegia had ensued, and to so great a degree as to require his knees to be tied together, and a strap placed about his body to retain him in the chair constructed for his comfort.

The first step in the treatment was to apply a suitable support to his body, and thus sustain the projecting spine in a quiescent condition, and to have an ointment of strychnia applied to his paralyzed limbs daily. The limbs were also manipulated and an irritating plaster applied to the lumbar region. This treatment enabled the young man, in the space of nine months, to walk, by taking hold of his father's and brother's arms, for support; and in less than eighteen months he could go about without assistance; firm ossification having resulted from the support applied to the spinal column.

This is only one case out of the many injured patients, whom we have treated to the entire restoration of power to their paralyzed limbs, from the result of this dangerous attempt to redress the angular projection by force, thus interrupting the restorative process of ossification by separating the approaching points of contact. To whatever extent the angular projection has advanced, it should not be interfered with, other than to render it support, and to the arrest of motion, that would interrupt the process of ossification.

POSTERIOR GENU FLEXUM.

This deformity of the leg consists in a relaxation of the crucial ligaments of the knee, and impaired tone of the muscles; allowing the head of the tibia on its anterior surface to approximate to the anterior surface of the condyles of the femur. The posterior surface being separated to a greater extent than is natural, the leg curves posteriorly, and the popliteal space is changed to an actually prominent curve, uniform with that of the leg when supporting the body.

Patients laboring under this condition of the leg suffer much inconvenience from the unsteady support of the knee in walking. The weakened limb when carried forward yields with a jerk, owing to the relaxed condition of the ligaments and muscles. The patient appears to sink when taking a step, and the very act of walking tends to increase the deformity. (See Fig. 55.)

Fig. 55.

This condition of the leg, or legs, is common also in cases of infantile paralysis, and the impairment usually confined to one leg, and in a limited number of cases the knee joints of both limbs are implicated in the weakness. And, as patients advance in age, in most cases, the limbs gain strength in their deformed condition and enable the individual to walk, though at the same time it becomes a permanent deformity. In childhood nearly every case is curable.

The treatment consists in sustaining the limbs in natural form. In cases of infantile paralysis, the support is usually required to extend from the pelvis to the sole of the foot, — the joints of the apparatus being so constructed as to limit the foot to a right angle with the leg, and the knee to a vertical position. The support should be worn night and day, and only removed to manipulate the limb, which should be performed twice in every twenty-four hours, while the patient is encouraged to walk, which exercise tends greatly (if not excessive) to increase the strength of the enfeebled muscles, when well supported and the feet permitted free action.

Fig. 56 represents the supporting apparatus: Two steel bars to extend up the leg nearly to the body, curved out laterally at the knee and attached at the upper ends to a broad steel band intended to half encircle the limbs and admit of adaptation to their contour, having a free attachment to the bars. When covered, attached straps complete the encircling of the limbs, studs being riveted into the steel band as a more desirable means of fastening than that of buckles; the lower portion consisting of joints at the ankles, heel-cup, sole-plate, instep pad, and straps to secure the feet to the apparatus. In such cases, a soft roller should be applied over the bars and back of the knee-joint, bringing the leg to a vertical position, then over the knee and two or three inches above and below it. The support should be continued until muscular power is restored sufficiently to admit of taking a few steps every day without the support of the braces, and then reapplying them again for more general exercise, being very careful not to leave off the supports too soon. For posterior curvature of the leg the reversed application of the roller to that required for protruded knee is to be made, and the brace worn within a laced gaiter.

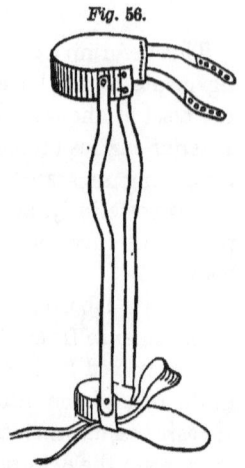

Fig. 56.

The time required for the cure of this deformity will be from five to ten years in many cases; the limbs having become attenuated, and fatty degeneration of muscles ensued, as in other cases of infantile paralysis. Yet these cases denote possibility of restoration, for as long as the sarcolemma maintains its integrity, we will have muscular development under favorable circumstances. This is an encouragement to persevere in the treatment of all cases of infantile paralysis. Patients having failed to be relieved from infancy to the age of eighteen years, can be much benefited, if not perfectly restored; also, relaxation of ligaments and muscles, from other causes. If the general health is unimpaired, the ailment is the more readily cured from properly constructed surgico-mechanical appliances. These remarks apply equally to the treatment of all similar ailments; electricity and manipulation being suitable auxiliaries when skilfully applied.

Contraction of the iliacus and psoas magnus muscles present, in some cases, most obstinate resistance to extension of the thigh, thus disabling the patient from attaining an erect bearing upon the limb to walk. This is often the greatest obstacle in treatment, because of the difficulty of applying suitable extension. The extension frame we have recently invented is a most successful means for the extension of those muscles.

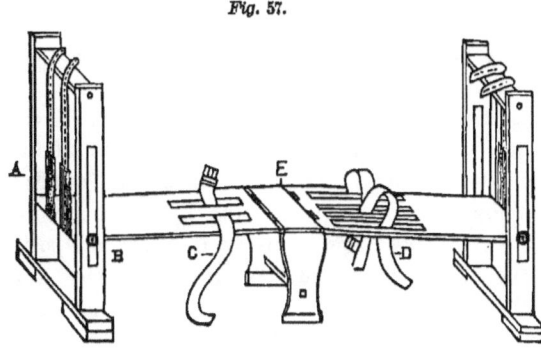

Fig. 57.

A, The end of the frame; the uprights having slots and screw bolts to secure the distal end of half of the bed plate, retained by hinges to a centre support, E. The bolt-head B, which when unscrewed leaves the bed-plate suspended by India-rubber webbing, and leather straps with holes so as to be attached to studs on the cross piece at the head of the upright frame. Thus the bed-plates can be elevated or lowered by an elastic support—each end being so constructed. Slots are seen in the bed-plates through which pass webbing belts to secure the body and limbs. C, The strap for the body. D, That for the legs, one or both.

By means of this apparatus, the muscles are made subject to effectual extension, and by elastic force, the hips resting upon the central fixed portion, upon which a pillow or folded blanket can be placed for protection from the hard surface, the body and legs being suspended by means of the elastic supporting straps, graduated as may be desired or borne by the patient. From fifteen to twenty minutes is about the extent of the time that the patient can endure the extension of the contracted muscles. Immediately on the relinquishment of the extension apparatus, the body brace and extension spring should be applied to maintain the limb in its improved condition.

INFLUENCE OF OCCUPATION TENDING TO PARALYSIS, AND THE TREATMENT.

In that peculiar variety of paralysis, the cause of which may be referred to local exhaustion from long continued position or exposure to cold water, as in some of the trades, is limited to the wrist and hand of adults. The first indication of the approaching impairment is indicated by cramp, and spasmodic twitching of the flexor muscles of the fore-arm and hand. This is found chiefly among writers and painters, as they are more subject to long continued strain upon the hand and wrist, though it is often met with among shoemakers, glass-cutters, seamstresses and compositors. These cramps and twitchings of peculiar muscles are but the premonition of an approaching paralysis, which finally ends in what is familiarly known as drop wrist or drop hand, an inability to raise the hand or extend the fingers. Relief from the malposition tends largely to the cure, by permitting the extensor muscles to recuperate in power when relieved from continued extension, and for which we have devised the following apparatus:

Fig. 58. A metallic casing partly encircling the fore-arm, and sustaining a steel spring that extends over the wrist to the fingers; the hand and fingers being partly covered with a glove, including the thumb, to which tapes are attached and passed through holes made in the crosspiece on the end of the spring, also from the thumb to the short piece, there being holes for that purpose. These tapes being drawn up and tied, determine the necessary relief to the impaired muscles, and fits the hand again for service; the flexor muscles being apparently unimpaired, or improved by the extending force.

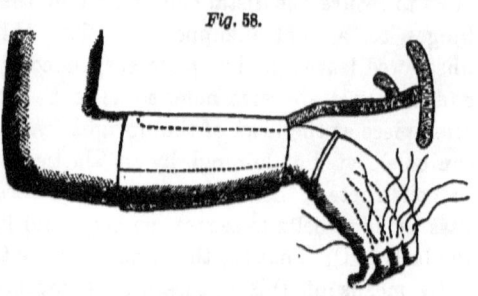

Fig. 58.

This relief to the extensors is a most potent curative means when kept applied day and night. Light shocks of static electricity, as obtained by friction upon glass, applied daily, have afforded restorative power after the failure of Faradization and the moxa. This mode of applying electricity is represented in Fig. 58.

MACHINE FOR ACCUMULATING STATIC ELECTRICITY.

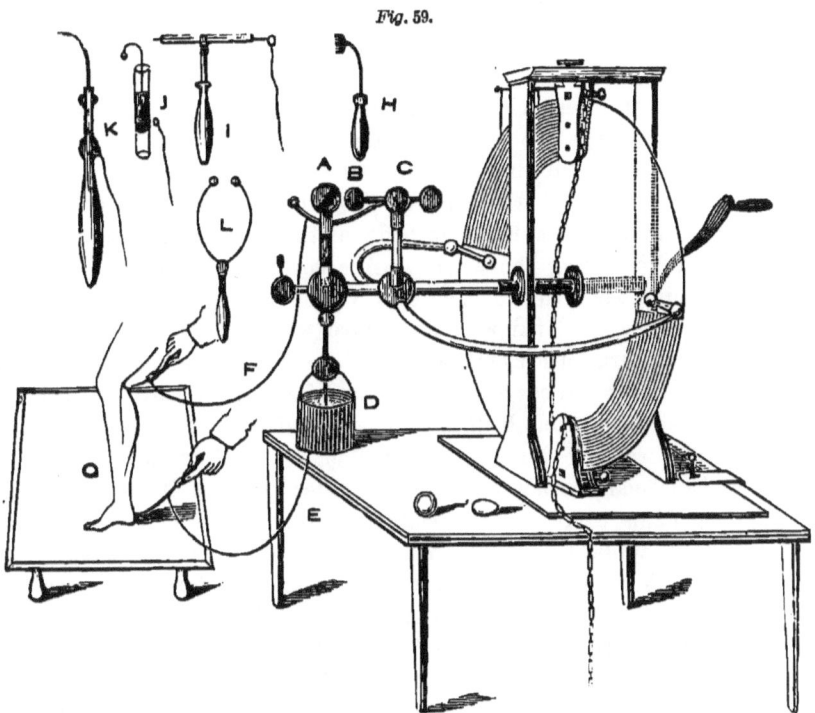

Fig. 59.

This diagram represents the electrical machine and apparatus. A and B do not quite touch, but have a space between them, which space can be increased or decreased by drawing the rod connecting B and C through the central ball. By increasing the space between A and B, the shock can be intensified to almost any desired degree, from a small Leyden jar; but if a more diffusive and mild shock is desired, it can be obtained by substituting the glass tube J. To insulate the patients, they can be placed in a chair on the table G, which is supported by glass legs. By conducting the electricity through the metallic cord F, alone, and exciting the electrical machine by revolving the glass plate, the patient becomes charged with electricity. At the left hand corner of the diagram are representations of several instruments used in the varied modes of applying electricity to patients. H is a metallic brush that will draw numerous sparks when held at a slight distance from the insulated patient, when fully charged with electricity. I is a glass tube having a wire passing through, and supported by a handle for applying electricity to internal parts. K is a pointed wire inserted into a handle, and serves to draw off electricity silently from the insulated patient, tending to produce absorption of enlarged glands and relieve spasmodic tension in muscles. L serves to discharge the Leyden jar by applying one ball to the body of the jar and the other ball to the ball on the stem entering the jar. Of the two balls on the table of the machine, one is round and covered with cloth, and can be attached to the wire K. The body of this ball is of wood, and prepared by having the whole surface covered with varnish and metal filings, and, when dry, covered with cloth. This is used upon patients when insulated and surcharged with electricity, and is an invaluable therapeutic agent to excite the cuticle, or indolent ulceration, and other indications of a similar character, by covering the ulcer first with a cloth, this being immediately effective, invariably improves the tone of the capillary circulation locally, and very beneficially when applied to limited parts of the body or limbs. Pains caused by rheumatism, neuralgia and other causes, are, by the application of this ball, in many instances, permanently relieved. The other ball is of wood, and of an ovate form; the base, or blunt end, to be attached to the wire K. This is used, as is the other ball, by the operator placing his foot upon the end of the metallic cord attached to the handle of the implement K, and holding the pointed end of the ovate ball at the distance of an inch or two from the inflamed part of the insulated patient, when it will, if repeatedly applied, in many instances relieve chronic inflammations, and is a valuable remedial agent. The pointed wire K, is to be used in a similar manner, by placing the foot upon the cord, thus conveying the electric current to the earth from the insulated and surcharged patient, through the insulated metallic cord, which, if attached to a gas-pipe, is much more efficient than when placed under the foot on the floor, in dissipating the electricity. Electricity in other forms can be similarly applied.

CHAPTER V.

ELECTRICITY AS A THERAPEUTIC AGENT IN THE TREATMENT OF PARALYSIS.

Friction upon amber and glass, or Static Electricity — Galvanism — Magnetic Induction, in the form of Electro Magnetism — Magnetism and Galvanic Induction combined — Difference between Static and Dynamic Electricity — Static Electricity more extensively useful as a therapeutic agent — Electric concussion, how accomplished — Electric sparks — Electric shock — Electrical Induction — Electrical sedative influence — Electrical Rubefacient — What constitutes Tetanus — The "Electro Tonic" State — Electrical Diagnosis — Duchenne on the interrupted current — Remak, an advocate of the continuous current — Thorough knowledge of diagnosis essential to the operator — Vitalizing power of Electricity — Reactive power — Matteucci's definition — Alterative action — Sedative action — Promotion of Nutrition — Promotion of Secretion — Promotion of Absorption — Capillary Circulation — Digestion and Menstruation — Use in Inflammation — Use in Congestion — Counter-irritation, Revulsion — Organic Contraction — Muscular Exercises — It is imperative that the Practitioner should understand the direction of the Electrical Current — How to use the Galvanic Battery — Results of Static Electricity — Dr. Golding Bird's valuable and interesting Classification — Dr. Todd's evidence as to its efficacy — Static Electricity as a remedial agent for Nervous Diseases — The Electrical Bath — Dropped Hands — Illustrative Cases of Paralysis from Lead Poisoning treated by Electricity — Effect in Rheumatic Paralysis, illustrated by Diagnosis of four patients — Twelve cases of Paralysis from various cases cured — Case of Hemiplegia — Irrecoverable injury results from injudicious treatment either by static or dynamic electricity — Case communicated by M. Bemond of Bordeaux — Static Electricity a remedy for Sciatica — Chorea relieved by Static Electricity — Testimony of Dr. Hughes, of Guy's Hospital, London — Amenorrhœa — Tumors — Aneurisms successfully treated by electric puncture — Opinion of Dr. Maurice H. Collis, of this treatment — Mode of operation — Tabulated results — Method of applying Electricity — Dr. Duchenne, the first to apply the principle of localization — Faradization — Electricity in its two places, as an excitant and a deobstruent — Points of application — Contraction of the muscles a certain signal of their electric excitation — Muscular sensitiveness, where most apparent — Points in the muscular system which should be avoided in electric excitation — Galvanism — Its influence on asthmatic affections — Galvanic influence greater than Faradic upon the muscular tissues — The continuous current, mode of application — The interrupted current — Faradization — Primary and Secondary batteries — Electricity only useful in a certain class of ailments — The nervous condition of the patient a vitally important element in diagnosis — Partial insensibility not readily detected by loss of tractile sense of the power of appreciating heat or cold — Increased articular sensibility a diagnostic symptom of paraplegia — Electricity available in arresting morbid, nervous or muscular sensibility — The circulatory system subject to the favorable influence of electricity — Extreme muscular or nervous excitement controllable by electricity — Faradization a potent remedial agent — Electricity inadmissible in the early stages of paralysis — In progressive paralysis, electricity may be used freely — Contractility of the muscle the only benefit derivable from dynamic electricity — Continuous current, effect on the nerve — Necessity of caution in the use of electricity as a therapeutic agent — The duration of the application an important consideration — Abnormal conditions of the cuticle, muscle, or nerve, not appreciably relieved by electrical influence — Injurious effects arising from a reckless use of electricity — Static electricity the most reliable as a method of relief — "Essential" or spinal paralysis in children, peculiar insensibility to electrical influence — Electricity not admissible when persistent contraction has taken place — After restoration to normal power, electricity an effective adjunct — Paralysis affecting the vocal organs immediately relieved by static electricity — Influence of static electricity in sudden and severe strains of the arm — Progressive atrophy, sciatica supervening, relieved by Fowler's solution — Treatment of slight injuries — GENERATORS OF GALVANIC ELECTRICITY — Galvano-Faradic Company of New York City — PORTABLE GALVANIC BATTERIES.

ELECTRICITY AS A THERAPEUTIC AGENT.

We will here introduce the subject of electricity as a therapeutic agent, the means of accumulating and various modes of applying it; and a consideration of its peculiar action upon the animal system, as well as the various means by which it is developed or excited — there being four different methods of obtaining electricity.

The *first* mode of obtaining electricity described, is by friction upon amber, then upon glass; and accumulating it from the atmosphere upon an insulated metallic cylinder, requiring, as we have before remarked, various appliances to the purposes for which it is intended as a therapeutic agent. The *second* is evolved by contact, and chemical reaction in the form of galvanism. The *third*, by magnetic induction, in the form of electro-magnetism, and *fourth*, by combination of magnetism and galvanic induction.

The first form is denominated static electricity, and is said to be stationary, or not active; while in the form of galvanism, it is said to be dynamic, as being essentially in movement and exercising power. A very great difference exists between the phenomena exhibited by these two forms of electricity: the *static* having, in a high degree, the properties of attraction and repulsion, and when brought into movement, exercising great mechanical power; the *dynamic* exhibiting its energy more in developing heat and producing chemical change. It is supposed that this difference depends, not on any essential diversity of character, but on the different states of the electricity developed in the two methods ; that excited by friction having little quantity, but great tension or intensity, by which it is able to overcome resistance, while that set in movement by contact and chemical reaction has feeble tension but large quantity. These terms, however, are rather conventional, and intended to represent certain qualities in convenient language rather than to be expressive of the fact — for it is by no means universally admitted that electricity is a distinct substance to which the term, "quantity," is at all applicable — unless as a figure of speech.

Static electricity is controlled by many forms of application, differing from the dynamic form, and, as we believe, more extensively useful as a therapeutic agent; producing two distinct influences upon the animal system, an *excitant* and a decidedly *sedative*. The dynamic is merely an excitant, not available in diminishing excessive nervous energy, without impairing vitality, as in the case of insulating patients, and exciting a powerful current of static electricity — surcharging them — and then drawing it off with a sharp metallic point connected with the earth. The influence is so great in this method as to induce syncope in delicate persons, as we have frequently witnessed; the influence is only temporary, however, tending to a normal restoration of nervous influence.

Silent current is the passage of a current of static electricity *through* the person, or a local part, by means of the insulated hand-conductors, as seen in Fig. 59, applied to the leg — the patient sitting in a chair or standing on the floor, and not on the insulated table. The hand-conductor F, having the cord attached to the prime conductor of the electrical machine, and also to a metallic point in the end of the handle, is applied to the desired part; the other hand-conductor E, having a cord also which should be connected with the gas-pipe, or other means of conducting the electricity to the earth. The points of the hand-conductors, or electrodes, as they are now termed, can then be applied to either side of a finger, and the electricity passed silently through the joint. This influence promotes absorption, and is made applicable to the treatment of indolent swellings and ulcers; it may be considered a gentle excitant of extraordinary influence.

Electric Concussion is accomplished by the patient standing or sitting, their person connected with the prime conductor of the electrical machine, which, when charged, imparts to the patient an excessive quantity of electricity; when, if connected with the earth, a silent current will pass, as before described, but, by separating the balls (A and B), a very light concussion will be felt throughout the system and can be increased to a painful degree by increasing the space between the balls. This is a very potent means of relief in a torpid condition of the system, or parts, as that of a limb, or part of a limb, or in tumors — tending to their dispersion. Persons of a sedentary habit, and dyspeptic, are much benefited by this mode of having static electricity applied.

Electricity taken from the prime conductor to the lower border of the ensiform cartilage, and the point of the other conductor applied to the region of the anus — the metallic cord attached to this conductor passing to the earth and a very decided concussion induced, has relieved most obstinate constipation of the bowels, and, in some cases, has given permanent relief, the bowels not being impaired, as by the use of drastic purgatives, but improved in tone. This application affords much relief in light paralytic seizures.

Electric Sparks. — This means of applying static electricity is accomplished by connecting the patient with the earth — (an important precaution in the use of static electricity, as the floor being, in a measure, very dry, insulates the patients and the effect of the electric application is rendered uncertain) — the necessary preparations being thus made and the electrical machine briskly excited, electricity is passed from the prime conductor through the cord attached to a metallic ball on the end of a hand conductor or electrode, and by approaching a part of the body or limbs with this charged ball, sparks of electricity strike the part with force and pungency. This can be increased or diminished in severity to the degree of cauterizing the part, or of exciting a large extent of surface, to the relief of neuralgia and rheumatism or other painful conditions, usually relieved by rubefacients, but much more readily than by means of the more ordinary methods.

Electric Shock.— By means of the Leyden jar the most powerful concussion is given, even to the degree of destroying life. When the ball of this jar is put in contact with the prime conductor of the electrical machine, and electricity accumulated, the interior of the jar becomes plus or positively charged, the outer coat being minus or in an opposite condition. If, now, any part of the patient is made the medium of connexion with a good conductor, as that of metallic cords, and the hand conductors applied to the part, and the balls (A and B, in Fig. 58) are in contact, no sensible effect will be produced, because the inner and outer metallic coating of the jar are thus in a state of equilibrium. The current being continuous, the outer coat of the jar being connected with the earth, the silent current is established; but if the balls, A and B, are separated, the intervening space being atmospheric air—a non-conductor—the electricity accumulates in the interior of the jar, and in quantity to the extent of the metallic surface; so that, if a large jar, the quantity will be in proportion, and at a certain degree of tension will readily pass from one ball to the other in the form of a condensed spark, making a report and sudden concussion as it passes through the intervening part of the body that connects the inner and outer coating of the jar. The severity of the concussion or shock, as it is termed, being in ratio to the extent of the metallic surface in the jar, and the distance the balls may have been separated. Hence, the separation of the balls determines the severity of the shock of a given-sized Leyden jar, or a battery of many jars—even to the fusing of the metal cord through which it is conducted.

The shock being used as a curative means can be definitely modified by having a lesser or greater surface of glass coated on both sides with tin-foil, so insulated as not to be connected by two or more inches from the metallic surface on both sides, by which the intensity of the accumulation can be obtained that will give the shock required. A glass tube, as seen at J, in Fig. 59, serves an admirable purpose for nearly all curative purposes where intensity is required with lessened severity. If care is not taken to discharge the Leyden jar before attempting an operation with it, there is a great liability of receiving a severe shock. The discharger, L, should always be applied to the ball and outer coating of the jar, and the jar should be removed a proper distance from the prime conductor of the machine.

These several modifications of static electricity, we consider as excitants, or sedative, and subject to diffusion or localization.

The sedative influence of static electricity is obtained by insulating the patient upon glass supports in the form of a table set upon glass legs as seen in Fig. 59. Upon this table a chair is placed for the convenience of the patient, and the electrical influence can be made subject to a variety of modifications.

Electrical Induction consists in concentrating in a person insulated electricity from an excited electrical machine. When the person thus insulated becomes surcharged with electricity no apparent physical effect is noted, so long as he is undisturbed; but if he be approximated by a metallic ball or the knuckle of a second person connected with the earth, sparks of electricity will be transmitted, and the sensation felt by both parties, *i. e.*, if the knuckle be the medium, will be equal; the severity of the shock will be in proportion to the amount of electricity accumulated by the machine.

Electrical Sedative Influence is obtained by withdrawing from a patient insulated, electricity, by means of a metallic point or points secured by a good conducting material. This depresses nervous energy to a very decided degree. Delicate persons, or persons in a feeble condition of health are liable to syncope if several points are used for withdrawal of the electricity. The first effect produced upon the person is an increase in the size of the veins of the part approached by the metallic points, and, after a time, a general depression, quite apparent in the person, will be produced; or, if suffering from local inflammation and pain, the congested condition of the part will be reduced, and the pain cease for a time.

For this purpose, the hand conductors, H or K, serve the purpose. The conductor, I, serves for applying or drawing off electricity from internal parts, the metallic, pointed wire being insulated within a glass tube and prepared with an insulated metallic cord to be attached to other electrical apparatus. Wooden points varied from sharp to that of an ovate form, as seen on the table of the electrical machine, Fig. 59, are used for modifying the intensity of the electricity when drawn from sensitive parts — an inflamed eye for instance.

Electrical Rubefacient. — This excitant effect is produced by a a disc of dry wood having one face sprinkled with metal filings, secured by varnish, and covered with cloth. The disc is so constructed as to connect the filings with the metallic conductor carrying off the electricity. The patient being insulated and the electrical machine excited, the operator passes this disc over the diseased part, exciting a warm, pungent sensation that can be increased to painfulness.

These varied conditions of the application of static electricity are only preparatory to various modifying effects to be obtained in accordance with the desire and judgment of the practitioner; differing from that of dynamic electricity in its availability to the practitioner as a means of controlling the various morbid conditions of nervous energy — as that of deficiency or excess, — without permanently impairing the normal conditions of tissues or organic functions. It is, apparently, congenial to organized matter and in consonance with the atmosphere, from which it is obtained, and which, when concentrated, it decomposes the air; eliminating ozone, a desirable disinfectant readily detected by the sense of smell. Static electricity is a powerful therapeutic agent, requiring not only a profound knowledge of the natural laws regarding it, but an equal knowledge of anatomy and physiology. If applied locally to the muscles, their origin and insertion must be known to the operator, and if made to act through the nerves, then their connection with the nervous centres, their course, anastomoses and termination, also, of the laws which regulate the electrical currents in the muscles and nerves, and the relation of these laws to the vital and physical forces. This applies, equally, to all the devices for generating electricity, and its application as a remedial therapeutic agent.

The recent discovery of laws that govern electricity as a therapeutic agent of determined influence upon the muscles and nerve tissues, and the relation of these laws to the vital and physical forces establishes it within the circle of scientific acceptation. Mettucci, Becquerel, and Du Bois Reymond have discovered and promulgated the methods of their discovery.

A single electrical shock may last only the one-thousandth part of a second, but the muscular shortening reaches its maximum and returns to its former state in about one-fourth of a second. If two shocks are given, the one immediately after the muscle has shortened to a state of rest, then there are two contractions. If the second stroke is given during the muscular movement caused by the first, and there is either contraction or relaxation, it causes increased shortening. But, if the second stroke follows very rapidly on the first (that is, within the one-thousandth part of a second), the shortening is not greater than with one stroke. If several shocks are given before a muscle has time to be relaxed, it becomes hard and permanently contracted, constituting tetanus. The less fatigued the muscle, the more rapid is the shortening. Interrupted currents, or shocks, of electricity, therefore, cause permanent or tetanic spasms in muscle, whether applied directly to itself, or indirectly, through a nerve, and the intensity of this will depend on (1) the intensity and rapidity of the current; (2) the amount of contractile power in the muscle; and (3) the mechanical resistance the muscle may have to overcome, as from the distance or weight of parts to which it may be attached. On the other hand, a continuous current of electricity only excites muscular contraction when the electrical current is closed or broken. In the interval it seems to flow through the tissues without causing any sensible effect.

When a continuous current of electricity is caused to pass through a portion of a nerve, it is thrown into a peculiar condition, which Du Bois Reymond calls an "electro-tonic" state. If this current is sent through a portion of a nerve in the same direction as its own proper current, then the latter is increased, as may be shown by the galvanometer, but if in the opposite direction, it is diminished. Again, where the nerve comes in contact with the positive electrode, or conductor, the electro-tonic state is diminished and the effect termed anelectrotonus; where it comes in contact with the negative electrode it is increased and termed "catelectrotonus." Between the poles, or conductors, at the point where the two opposite conditions of electricity meet (the point of indifference) the normal state of the nerve is preserved. Both the increased and diminished excitability of the nerves so caused bear a relation to the force and rapidity of the current. Further, the power of conduction in the nerve is diminished in the state of anelectrotonus, but, on breaking the current, the conducting power returns there, while it is diminished where formerly it was in the state of catelectrotonus. Hence, we can influence the contraction of muscles by the continuous current through the nerve according to its force and direction. It can also be easily shown that the farther from a muscle a motor nerve is irritated, the greater is its excitability, so that a feeble current applied to a nerve at a distance from a muscle will excite more contraction than a stronger one applied close to it. The stimulation of sensitive nerves by electricity excites their special function, on the forming and breaking a current, in proportion to its amount and rapidity. Hence, we can excite pain through the ordinary sensitive nerves, flashes of light through the optic, noise through the auditory, and taste through the gustatory nerves. As with muscles, also, during the interval, no sensible effect is occasioned. These laws of electrical influence on the animal system have been promulgated as the result of careful investigation made by some of the most reliable and noted electricians, and applied to electrical diagnosis tending to the most invaluable results in the practice of medicine.

ELECTRICAL DIAGNOSIS.

With regard to the proper method of applying electricity in disease, great difference of opinion prevails. Duchenne strongly supports the use of an interrupted current applied locally to the muscles, while Remak maintains the importance of a strong continuous current applied to the nerves and nerve centres. Both modes of procedure require to be more generally tested by experience. In one class of cases originating in the nervous centres, as in hemiplegic paralysis, Remak's plan may be most useful; whereas, in another class dependent on a primary morbid action affecting the muscles, as in lead paralysis, that of Duchenne may prove best. It has, also, to be ascertained what is owing to direct and what to reflex action during the topical application of electricity. A thorough knowledge of diagnosis should be possessed by him who undertakes the difficult task of employing so powerful, although manageable an agent for the relief and cure of diseases. (Bennett.)

Attempts have been made to employ the sensations produced by electricity in traversing diseased parts, as a means of diagnosis. Although nothing reliable has been accomplished in this direction, the fact is well known that the tissues and nerves which are inflamed or otherwise diseased are often more sensitive to the passage of electricity than those in a healthy state. Some discrimination may also, probably, be made as to the character of sensation in different conditions of disease. This has suggested the idea of determining the position and nature of internal lesions by the electrical current. Diseases of the spine and of the lungs are examples of cases in which this experiment has been made. The idea of thus probing deep-seated organs, is one of interest, and calculated to awaken inquiry. In some cases, the rapid administration of electromagnetism would be available for this purpose, but for nice distinctions, the continuous current of the battery would obviously be preferable.

Vitalizing power.— One of the most common objects of electrical application is to co-operate with vitality. This, indeed, may be considered the central principle in the medical use of electricity. In the nervous system it is illustrated in cases of exhaustion, prostration, enervation and paralysis; in the tissues, in gangrene, erysipelas, in indolent ulcers and deficient nutrition. It will be found, indeed, to enter, more or less directly, into every case of electrical application. The idea will be frequently suggested to the practitioner, by his own observation, that the agent electricity works in the direction of health, even in the most opposite affections. This is admitted simply by the fact, that its operation is to quicken the vital powers and natural functions of the part to which it is applied. Diseased action, when local, is perhaps especially controlled by the supply of nervous power, previously different, which is brought into action, connecting and harmonizing functions.

Reactive power. — Electricity constitutes, in the hands of the medical practitioner, a reactive power. According to Matteucci, the nervous system responds to electricity after all other stimulants have ceased to act. Thus it has a very important application in suspended animation, narcotism and stupor. Another equally important application, which has only begun to receive attention, is that to collapse, and the sinking stages of disease. This being admitted from positive experiment, no practitioner in medicine should fail to be supplied with electrical apparatus of the most reliable construction, and an intimate knowledge of its application to patients conditioned as above stated.

ELECTRICAL INFLUENCE.

ALTERATIVE ACTION.

An influence frequently exerted by electricity is that of changing the action of an organ, or the general tone of the system, thereby arresting a diseased condition. The application of electricity by the sponge-handle in cutaneous diseases, is, perhaps, founded on this principle, causing the skin to take on a normal action. So, also, in some nervous affections, and, perhaps, in application to the brain. As a simple alterative, electro-magnetism is most generally applied.

SEDATIVE ACTION.

The sedative action of electro-magnetism is of a secondary influence. The continuous current of the battery exerts often a tranquilizing influence, moderating and equalizing irregular action of the nervous system at the same time that it adds to its power. Thus convulsions are quieted by the steady current, though increased by the intermittent or electro-magnetic shock.

PROMOTION OF NUTRITION.

In deficient nutrition, electricity may co-operate in the vital transformation and organization of the nutrient matter by means of the nervous system or by direct action on the tissues of a part. To produce increased action in the latter case, static or galvanic electricity may be used. This will be done wherever a part needs to be nourished, or the waste of any organ to be replaced. As a general rule, the battery current should be feeble, and the application long continued and frequent.

PROMOTION OF SECRETION.

Wilson Philip says, " I cannot help regarding it as almost ascertained that in those diseases in which the derangements are in the nervous power alone, where the sensorial functions are active and the vessels healthy, and merely the power of secretion, which serves immediately to depend on the nervous system, is at fault, galvanism will often prove a valuable means of relief." * The immediate influence of galvanism on the tissue of secreting organs will also appear hereafter. The battery current is most efficient for this application, but nervous stimulation may be effected also by electro-magnetism. The currents should be sent along the nerves supplying the organ in the direction from the nerve centre, toward the nerve extremities, when it is desired to produce an immediate stimulation of the function, and, in the opposite direction, when it is desired to produce gradual and permanent stimulation. It may, also, be sent through and through the organ in different directions. The strength and stimulating character of the application will vary with the condition of the organ. In deficient secretion from the mucous membrane of the lungs a gentle and diffused current will be indicated.

PROMOTION OF ABSORPTION.

In effusions of serum or lymph in some forms of hypertrophy, in bony deposits, rheumatic enlargements, and every undue organic development, with the exception, perhaps, of some malignant

---

* "Vital Functions." 2d edition, page 331.

growths, the power of the absorbents needs to be quickened, and this may often be effected by electrical action. In this case, the application is usually made directly to the organ, though the rule still prevails, in acting through the nervous system, that the vital stimulus artificially supplied directs itself to, or principally perceived in, that function whose efficiency is suspended. In other words, the tendency of the nervous influence seems to be to harmonize the various vital functions, disproportionate action appearing thus to proceed from causes acting originally on the life of tissues. In serous effusions accompanied with inflammation, the battery cannot be used, but a cautious and gentle application of static electricity will be effectual. In other cases of effusion, the battery will be quite efficient.

CAPILLARY CIRCULATION.

The increase of the capillary circulation under the influence of the current has been referred to. This takes place in a very marked manner when the sponge handles are used in connection with the battery. The skin, in a short time after the application, becomes warm and red, especially under the negative handle where the current passes out in the direction of the nervous organic current. The excitement of the functions of the skin may be spoken of in this connection. It will be seen hereafter that perspiration breaks out and warmth is established in a part subjected to the influence of the battery. This takes place under the influence of static electricity or electro-magnetism. The arterial action is increased by all modes of application and is especially quickened in the smaller vessels, as may be presumed by the emotions of the shock.

DIGESTION AND MENSTRUATION.

The functions of digestion and menstruation are peculiarly under the influence of electricity in its various forms. They will be treated of in connection with other diseases. The application, in the case of diseases of women, is so simple that it can generally be made without any very great experience in the use of electricity.

USE IN INFLAMMATION.

The tendency of the battery current is to produce increased organic action; the tendency of a feeble electro-magnetic current is hardly more than an alterative stimulus, still, the application of dynamic electricity is contra-indicated. In that of static electricity with the patient insulated and the electricity drawn off by sharp metallic points, a sedative condition is rendered most effective in reducing active and progressive inflammation. Dr. Wilson Philip states as a result of microscopic observation, that the conditions of the capillaries in inflammation is one of distension and debility, while the arterial action is increased. Stimulation of the capillaries, therefore, in such cases, may relieve inflammation, especially when of a passive character, by the feeble electro-magnetic current, being an indirect influence, and the power of stimulating the capillaries of the surface of the body is peculiarly within the province of electricity in this form. In an advanced stage of inflammation, when the capillaries are so burdened as to stop the circulation, an increase of vital power, determined by electricity, may be of important service. So, also, electricity may be summoned to terminate a condition consequent upon inflammation which is only continued from want of reactive power.

USE IN CONGESTION.

The application of dynamic electricity to an organ in the early stages of congestion would be inadmissible; not so of static electricity, as it is of valuable service in relieving local congestion, as in cases of inflammation. Dynamic electricity in the latter stages of congestion may prove beneficial; aiding the vital powers in the resumption of the natural functions of the organ, if carefully used.

COUNTER-IRRITATION, REVULSION.

An organ that is inflamed or congested may sometimes be relieved by stimulating another organ connected with it by position or nervous association. It frequently happens in the inflammation of one organ that a neighboring one will be torpid. A double relief can, therefore, be obtained where it is possible to stimulate the latter without acting on the former. Electricity, in most of its surface applications, is easily capable of being converted into a rubefacient or irritant with the advantage of stimulating into activity all the functions of the skin. It can also be directed so as to excite specific internal organs by static electricity in the form of shocks. For irritating the skin, the sponge or metallic handle may be used with the battery, being kept near each other and moved over the surface, or, the metallic plates of Wilson Philip, consisting of two thin plates of metal, two or three inches in diameter, upon the surface of the body where it is desired to make electrical communication, may be used. The wires of the battery are brought in contact with these plates and constantly moved over them—especially the negative wire—to avoid injury to the cuticle beneath. The plates are previously dipped in water, or a solution of common salt—as should generally be done—in using the battery to aid the diffusion of the electricity. A more effectual means of producing this result is from static electricity by drawing sparks from the skin while the patient is on the insulated stool. It has been stated that suppressed eruptions have been brought to the surface by this means of revulsion.

ORGANIC CONTRACTION.

One influence of electricity, capable of important applications, is the restoration of organic contractility or tension in relaxed tissues. This applies, not only to muscular tissue, but especially to the ligamentous system, including fibrous and capsular ligaments. Examples of this will be given hereafter.

MUSCULAR EXERCISE.

Much advantage, at times, is derived from the effect of muscular contraction induced by electricity in moving organs one upon the other; and, probably, in an old inflammation, such as sprains, in breaking up adhesions. In the application of an interrupted current to the abdomen, the parietes are contracted in a variety of directions, as well as the muscular fibres of all the included organs which, in some forms of dyspepsia, is one of the most favorable of influences. In paralysis, the exercises given to the muscles is of great importance in preserving the contractility of the tissues and preventing the loss of organization and want of substance, as in the early condition of the paralyzed patient.

It is of great importance to the practitioner to understand the direction of the current of electricity. In static electricity, the current is from the prime conductor, and can be directed from the origin of a nerve to its termination by the application of the conductor attached to the prime conductor (the patient being insulated), by any connection with the patient and the earth from the extremities of the body. And, if it is desired to reverse the current, the conductor attached to the prime conductor should be applied to the extremity of a nerve, or the limbs, and the conductor connecting with the earth to some part of the spinal column.

In applying electricity from the galvanic battery, the pole connecting with the copper plate is the positive pole or the entering pole of the current; as that of the prime conductor in static electricity, as regards the body of the patient and the handle connected with the zinc plate, is the negative pole. With the electro-magnetic apparatus, the positive pole is that which produces least sensation when applied to an equally sensitive part of the body with the negative. The handles or electrodes may be conveniently distinguished by holding one in each hand, when the most pain and contraction will be felt in the hand grasping the negative handle, or that connected with the negative pole.

RESULTS OF STATIC ELECTRICITY.

The results of treatment by the several varieties of means for the accumulation of electricity determine, to a limited extent, their efficiency in the cure of certain pathological conditions of the body. Dr. Golding Bird, in Guy's Hospital, London, has applied the static electricity the most extensively of any of the practitioners availing themselves of its use in the practice of medicine; and upon whom reliance can be placed, because of his high attainments in his profession. The following is a synopsis of a few of the cases given in his valuable reports of treatment of patients with static electricity in Guy's hospital.

Static electricity has long been in the hands of adventurers in the treatment of paralysis, and in many instances with extraordinary success, inviting the attention of the learned and discriminating in the science of medicine. Dr. Golding Bird, from his large experience, classifies the following forms of paralysis that are relieved by the judicious application of static electricity.*

1. Partial paralysis from organic congestion or effusion, which has been removed. 2. Paralysis of the portia dura, of the seventh pair, from exposure to cold. 3. Paralysis of a limb from the same cause. 4. Paralysis of one side of the body, or a single limb, from exhaustion — as from lactation, and flooding. 5. Paraplegia from rheumatism, paraplegia from enervation.

He found the use of electricity most successful in recent cases, and contends against its use in many of the established organic

---

* London *Lancet*, June, 1846.

lesions; stating that he has known fatal apoplexy to follow its application in cases of ramolissement of the brain, or where indurated arteries existed. And, moreover, he remarks that he has never known electricity to do any good in rigid flexion of the thumb or fingers.

Under various modes of electrical application we find, however, that paralytic contraction has been successfully controlled in cases of diminished tone in certain sets of muscles impaired by long-continued extension, that impairment of the balancing muscular force having even tended to contortion.

Dr. Bird remarks: "From the want of exercise, the muscles of the affected limb become atrophied. The power of electricity in this respect is very remarkable; frequently restoring power to the paralyzed muscles in a very short time."

Dr. Todd, in a paper in the Medico-Chirurgical Transactions of 1847, from a large number of observations, arrives at the following conclusions: 1. That irritability of paralysed muscles is in direct relation to their state of nutrition. 2. It varies with the condition of nerves more than with that of the muscles themselves. 3. In a majority of cases of cerebral palsy, the contractility of the paralyzed muscles is less than that of the muscles of the sound side on account of diminished nutrition. 4. No diagnostic mark to distinguish cerebral from spinal paralysis can be based on any difference in the irritability of the muscles. 5. The irritability of paralyzed muscles under the influence of galvanism is an index to the state of their nerves. This applies equally to static electricity. These are indications to be carefully considered in the use of electricity as a therapeutic agent.

In regard to the efficiency of static electricity, Dr. Golding Bird expresses his favorable opinion of its use in the treatment of nervous diseases, of which we will give a synopsis. His mode of applying electricity is by insulating the patient. In regard to the drawing off of electricity, silently, in connection with the earth, he remarks: "During the discharge, heat is evolved, the circulation becomes quickened, the secretions generally become more active, and perspiration breaks out. A person thus situated is said to be in an *electrical bath;* and it is by no means improbable that this might be frequently employed with advantage in certain affections in which the functions of the skin and nervous membranes are deficient."

Dr. Bird's most common mode of applying the static electricity was by insulating the patient and drawing off sparks, which method is subject to much modification by drawing heavy or light sparks; and from various parts of the body and limbs, but more especially from the spinal column.

Paralysis of the extensors of the hands from lead poison, known as *dropped hands*, are mentioned by Dr. Bird in his reports in which he refers to eleven of these cases treated by static electricity. Five, he says, were cured, three relieved, and one improved; two received no benefit whatever. Sparks were generally drawn, in these cases, from the upper part of the spine while the patient was seated on the insulating stool in order to influence the axillary plexus. Four of these cases are most worthy of note:

1. A compositor, aged nineteen. Paralysis of the extensors of both hands, with amaurosis, preceded by an attack of lead colic. After four months of interrupted treatment the paralysis was cured, but the amaurosis remained, though the pupils, previously nearly insensible, contracted and dilated readily.

2. A cooper, aged twenty-nine, with recent and complete paralysis of extensors. Weak shocks from the spine down the arm. Within a month, able to resume his work.

3. A painter, aged twenty-seven. Complete paralysis of extensors. In fifteen days discharged — well.

4. A plumber, aged thirty-six, with total paralysis of the extensors of a year's standing. Shocks down the arms on alternate days. No improvement after twenty days. Sparks ordered to be drawn from the spine. In sixteen days, great improvement, and soon able to resume work.

We have treated many cases of these ailments, arising from lead poisoning, successfully, with static electricity alone. In obstinate cases we have given the patients, in divided doses, twenty grains, daily, of antim. sulph. aur., or the hydro-sulphuret of antimony, which we believe, from the relief afforded patients thus affected, to be of great benefit. In fact, patients were perfectly cured by this medicine, and without the assistance of electricity, but not in so brief a period as when electricity was made a part of the treatment.

As we have before stated, dropped hands are often the result of long continued position of the hands in a state of tension, as occurs in some occupations when closely applied to labor; shoemakers, clerks, and persons engaged on fine needle-work, or any similar employment requiring the fingers to be flexed and long retained in a state of tension. Of this class of patients, nearly every one is perfectly curable with static electricity, and often after dynamic electricity has failed.

In the treatment of this variety of partial paralysis of the forearm and hand, we but seldom insulate the patient, and simply pass light shocks from the elbow to the extremities of the fingers, having an elastic spring support applied, to keep the hand extended by elastic support. From this treatment of patients, under fifty years of age, laboring under this ailment, we know of no failures to improve the condition of the arm and hand, although a great number present for treatment, every year, at the Hospital for the Relief of the Ruptured and Crippled.

Dr. Bird speaks of the most remarkable influence from static electricity in rheumatic paralysis before the wasting of the muscle. Out of ten of these cases, only two failed to be relieved and cured. This, by far, exceeds any other treatment in its curative tendency. The treatment of four cases was as follows:

1. A boy of fifteen years, with paralysis of motion of right arm, preceded by pain and swelling of the neck. Medical treatment for nine months without benefit. Twelve shocks, daily, from the Leyden jar, to be passed from the cervical vertebræ to the fingers. In two months discharged cured.

2. A sailor, aged thirty-two. Paralysis of right arm and both hands from cold affusion in fever. Sparks to be drawn from spine and paralyzed muscles three times a week. In seven weeks discharged cured.

3. A man, aged thirty-eight, with entire paralysis of motion of right leg, following rheumatism. Sparks, alternate days, from the lumbar vertebræ and limb. In six weeks discharged cured.

4. A man, aged thirty, with paralysis of motion of both hands from effects of cold water. Sparks from spine and hands. After a few applications returned to his work.

Twelve cases of paralysis from various causes treated by Dr. Bird in Guy's Hospital resulted as follows:

1. A man, aged thirty, with hemiplegia of right side, of nine months' standing, induced by a fall in which he struck his head. Paralysis partial. December. Shocks to be passed twice a week from the spine down leg and arm. Cured in April.

2. A woman, aged twenty-six, with paraplegia, following a recent apoplectic attack. After several months' medical treatment, con-

dition unimproved. Sparks ordered from the lumbar vertebræ and legs. Rapid recovery ensued.

3. A woman, aged fifty-two, with partial paralysis of motion and of feeling of right arm. Under common treatment for three months, sensation had improved. December 20. Electro-magnetic current ordered from neck to fingers thrice a week. January 20. Motion much improved. February 20. Cured.

4. A waiter, aged forty-six, with paralysis of motion of right half of the body, with some loss of sensation, of three months' standing. September 22: Twelve shocks, on alternate days, down the back and limbs. October 31: Numbness of only one finger remaining. Returned to his work.

5. A smith, aged twenty-two, with recent complete paralysis of motion of right arm. August 4: Sparks to be drawn from the upper part of the spine and arm. September 1: Cured.

6. A boy, eleven years old, with complete paralysis of motion on the right side, of seven weeks' standing. Sparks were drawn from spine and limbs. After first application, walked back into the ward with the aid of a stick. In a few days completely cured.

7. A coal porter, with paralysis of right arm and face. November 29: Sparks from spine, face and arm. December 24: Arm cured. Electro-magnetic shocks to face in direction of branches of fifth pair. January 10: Much improved.

We have treated many similar cases with static electricity, differing somewhat in the application, but with quite equal success. Dr. Bird's experience fully confirms our own in nearly thirty years' practice in the use of electricity in its various modifications, and we have been most favorably impressed with static electricity as a curative means, though requiring great care in its application. We have seen patients irrecoverably injured by injudicious treatment with both static and dynamic electricity. Matteucci has pointed out the entire exhaustion of nervous power, similar to paralysis, resulting from an excessive use of electricity. If the improvement ceases under the use of electricity Mr. James recommends the discontinuance of the agent for a week or two. The progress in treatment by electricity is by this interim of rest rendered again susceptible to improvement, and more especially after a strengthening regimen, with tonics and friction, carefully guarding the patient from exposure to cold.

M. Bermond, of Bordeaux, relates a case of hemiplegia, following apoplexy, in a lady, aged twenty-six, in which the Leyden jar was successfully employed. After three months' medical treatment the hemiplegia remained nearly complete. The memory was slightly impaired, and there was unusual nervous irritability. At the first sitting, shocks from the jar were passed from the hand to the foot of the affected side. After fifty moderate shocks considerable improvement manifested itself. After the fourth sitting, four days later, the patient took some steps. At the tenth sitting, seven weeks from the commencement, the patient walked to the office of M. Bermond. After the eleventh application, a week later, the cure might be considered as almost complete. The shocks were increased in number toward the close, and directed, at times, to a single limb or to the tongue. This was a case of unusual discouragement.

M. Bermond relates another case, even more discouraging as to affording relief of a lady, aged fifty-six, who, when in full health, was attacked with apoplexy, resulting in hemiplegia, and which remained after relief from the apopletic seizure. Speech was difficult, the saliva constantly flowing from the corners of the mouth. Taste and hearing were both affected, deglutition difficult, the bladder distended, constipation at first obstinate, and cramps in the paralyzed limbs frequent for the first fifteen days. Œdema at length appeared throughout the left side. After a month, a slight improvement had taken place in other respects, when electricity was applied by M. Bermond. After the first application the patient was able to stand, and even to stoop slightly and recover the erect position. On the following day, the features had become more regular, the hearing improved, the œdema diminished, and an abundant perspiration had ensued upon the limbs of the left side. The application was then repeated. The circulation increased in force, and on the third application, which took place two days later, the pulse was greatly increased and plethoric symptoms induced, which yielded readily to treatment. After twenty applications, the patient had essentially recovered.*

We have related this case as a representative condition of a number of cases that we have treated, where the patients were of full habit and laboring under paralytic seizures. The electric treatment invariably increased the circulation of the blood to a plethoric condition, that in many instances 'hydragogue cathartics afforded no

---

* Bul. Med. de Bordeaux.

relief; but required for their relief, venesection. This plethoric disposition is a most favorable indication of recovery, indicative of an obscure internal congestion relieved by the diffusive effect of electricity and developed in the superficial circulation. Such cases advance to recovery upon the reduction of their plethoric condition. In a majority of these cases, hydragogue cathartics are all sufficient as derivatives that afford relief. Our treatment is the application of light shocks passed from the upper cervical vertebræ to the extremities of the limbs — the patient being insulated,—and, at the same time, drawing off the electricity from the extremities with metallic points; carefully observing the influence upon the patient, as we have observed in some patients a decided indication of prostration whilst under the direct influence of the electricity. The veins become greatly dilated in the feet and hands, apparently lessening the circulation about the vital organs — an effect not producible by dynamic electricity to the same extent. This equalization of the circulation of the blood by diversion from congested tissues in delicate patients, supersedes depletion, and thus avoids an expenditure of vital force, such as made in blood-letting, evacuants, and starving the patient.

For the relief of *sciatica*, static electricity has been a most potent remedy. Dr. Marchant, of Hemsworth, relates his own case, in which the pain extended from the sacrum to the hip. Leeches, blisters, opium and other remedies were employed with some amelioration of the symptoms, but not beyond the point at which a crutch could be dispensed with. Sparks were then drawn, night and morning, for fifteen minutes from the affected parts, which treatment was followed, in a week, by material improvement, and, in three weeks, an entire cure. It is stated in the reports of Guy's hospital that relief was often obtained in sciatica by drawing sparks from the seat of pain.

We have applied static electricity for the cure of sciatica by insulation, attracting sparks, and drawing off the electricity with fine metallic points, and, by light shocks; affording relief, in many cases, without auxiliary aid. But we have usually afforded the most immediate relief from the arsenite of potassa as an auxiliary to the efficiency of electricity. The patient being insulated, and a silent current drawn by a single metallic point and continued for fifteen minutes at a sitting, being repeated every day, usually affords relief. If not we prescribe from five to fifteen drops of Fowler's solution of arsenic, to be taken at three intervals during the twenty-four hours from day to day. If then not relieved, we diminish the interval and give, every four hours, until decided nausea is induced. The following prescription often affords relief independent of other remedies:

℞. Soda bicarb.   ʒ i.
    Pulv. opii.
    " ipecac. āā. grs. i.
    M.
Ft. cht. no. x.
One to be taken every fourth hour.

This relieves the kidneys, as they are most commonly in an abnormal functional condition in seizures of sciatica. Oxalate of lime may be detected by the microscope in nearly every case as in cases of dyspepsia, and so stated by Dr. Golding Bird, and described as oxaluria, associated with dyspepsia. For the relief of this he prescribes nitro-muriatic acid — quite a successful treatment. The soda, opium, and ipecac have served the purpose equally as well, and, we believe, have given a more decided and permanent relief.

Chorea is one of the ailments relieved by static electricity. During the past thirty years we have cured many cases of patients, aged from twelve to twenty years, and nearly all girls. In Guy's Hospital, Dr. Hughes reports a digest of one hundred cases, fourteen of which were treated by electricity. Of these, some were of long duration and severe form. Nine were cured. The Doctor remarks:

"The effects of electricity in chorea are sometimes very remarkable. On some occasions I have known it to effect a cure after a great variety of other remedies had been tried, for weeks and months in vain. The change has not been more beneficial than rapid. In the course of a week or ten days, the entire aspect of the patient has been changed. When electricity acts beneficially in chorea it produces its effects more rapidly than any other remedy with which I am acquainted."

A girl, aged sixteen, had chorea, with deranged catamenia. Aspect, fatuitas, occasionally wild, and almost maniacal. Disease of twelve months' standing. Sparks from the spine, and shocks through the pelvis for a month, terminating in a perfect cure.

Dr. Golding Bird furnishes a table of thirty cases treated by static electricity, in one of which was organic diseases of the spinal cord and failed in being relieved; one left from alarm; five were much relieved, and twenty-three cured. A few of these, he states, were under contemporaneous treatment of medicine, mostly mild purgatives. In many of them, every variety of treatment had been exhausted before having resource to electricity. The treatment was confined to machine electricity (*i. e.*, static electricity) drawn by sparks from the spinal column. The application was made every other day, for five minutes, or, until a papular eruption appeared.

In all cases of chorea that we have treated, we have given very decided purgatives to relieve the alimentary canal of any irritating fœcal accumulation that might exist, and bismuth and quinine as a tonic; believing the system to be in an asthenic condition. Conjoining these medicines we make use of the efficiency of static electricity in the form of drawing sparks from insulated patients, as described in the previous pages.

Amenorrhœa. By means of static electricity, this condition of the patient has been more readily relieved than by any other remedy. This has been our experience, and agrees with Dr. Bird's assertions in his lectures: "In electricity," he says, "we possess the only really direct emmenagogue which the experience of our profession has furnished us with. I do not think I have ever known it to fail in exciting menstruation where the uterus was capable of performing this function. The rule for insuring success in the great mass of cases of amenorrhœa is sufficiently simple. Improve the general health by exercise and tonics, remove the accumulations in the bowels by appropriate purgatives, and then a few electrical shocks, often a single one will be sufficient to produce menstruation, and at once restore the previous deficient function." A dozen shocks from the Leyden jar were usually passed through the pelvis from the sacrum to the pubis, and if the catamenia were not established in four or five weeks, he discontinued electrical treatment, and renewed it after searching for and removing the cause of the general derangement. A table of twenty-four cases was given by Dr. Bird, in which there was no relief in four, being well-marked cases of chlorosis, but a cure was effected in the remaining twenty.

We would here remark that dynamic electricity is very efficient in the relief of amenorrhœa, as well as in many of the ailments that are readily relieved by static electricity.

Tumors are ailments susceptible to the influence of both static and dynamic electricity. Vascular and erectile tumors are proper subjects for galvano puncture. In these cases coagulation has been effected, and absorption stimulated by this means, also, the absorption of scrofulous tumors. The general remark may be made, that wherever tumors or enlargements of any kind constitute the original disease, or are subject of direct treatment, galvanism may be resorted to with the hope of increasing the vital action of the part, and especially of the absorbent system. The current should be passed through the tumor from the surface, or, by electro-puncture needles.

Aneurisms have been treated successfully by the electric puncture. M. Petrequin, surgeon-in-chief of the Hôtel Dieu, of Lyons, to whom the priority of this application is generally accorded, performed his first experiment upon human blood immediately after its extraction. Blood, therefore, seems to be susceptible of coagulation while its vitality lasts. This gentleman communicated to the French Academy the first case of aneurisim treated by this method. This was a traumatic aneurism of the temporal artery of the size of an almond, of a soft consistence, and slightly sensible to pressure. Two fine needles of steel (gold or platinum should always be employed) were plunged about four-fifths of an inch into the tumor so as to cross it at right angles. A battery gradually increased to fifteen pairs was connected with these for ten or twelve minutes. Considerable pain was experienced, and the pulsation gradually ceased to the close of the operation. In ten days the tumor was in process of absorption. Other cases were, soon after, similarly treated, one, an aneurism of the ophthalmic artery and another of the right brachial artery, with decided success. It is necessary that the needles should terminate in the fluid blood, and not in the coats of the sac, to ensure a firm coagulation.

Treatment of tumors by the Voltaic pile is one of the most simple methods of applying electricity. Dr. Maurice H. Collis, surgeon to the Meath Hospital, and County Infirmary,* states his experience and mode of applying it, with pertinent remarks:

"It occurred to me, however, that something could be done toward procuring the absorption of tumors, or, perhaps, toward checking their growth, by using galvanism simply as a stimulant. A slight, occasional current, by its stimulus, will develop a wasted muscle; while the continuous use of the same current will, by over-

---

* Brit. Med. Journal, December 7, 1867.

stimulation, cause it again to waste. I, therefore, sought to apply this principle to tumors of various kinds, and to cause them to waste by keeping up a constant flow of electricity through them.

"For reasons, which I need not enter upon, I tried various forms of batteries, and, finally, returned to the simple voltaic pile composed of a dozen or more couples of zinc and copper, an inch and a half square, or of small cylinders, or, plates of wood covered with felt and wrapped around with zinc and copper wire. These simple batteries were excited by salt water or by sulphuric acid in the proportion of one part of acid to twenty of water.

"The mode of application was as follows: The tumor was covered with a plate of zinc, perforated zinc, silver foil, copper, or copper plated with silver. The positive pole was connected with this plate, the negative pole with a plate of copper which was brought into contact with the skin of the back, or other convenient part. The battery, tied up in a gutta-percha paper or oiled silk, lay on a table, or was tied around the waist of the patient.

"The results were as follows:

"1. Complete removal of secondary cancerous deposits from a gland in the neck.

"2. Rapid absorption of inflammatory deposit over and around an immense mass of strumous gland.

"3. Slow removal of said gland.

"4. Immediate check to the growth of a tumor composed of an aggregation of strumous glands.

"5. Considerable cutaneous and sub-cutaneous inflammation in the same case with softening and breaking up of the diseased glands.

"6. Decided absorption of inflammatory effusion around primary scirrhus of mammary region.

"7. Diminution and softening of a very firm fibrous growth attached to the periosteum.

"All these results, except No. 3, were obtained in from three to six days ; and all the cases had been previously submitted to a variety of treatment without benefit. Special results of a peculiar nature were observed, as follows :

"8. The zinc plate on the skin showed the usual tendency to decompose the skin, and produce ulceration in a few hours.

"9. Perforated zinc, employed to obviate this, produced phlyctenæ and pustules in the inter spaces not covered by it.

"10. Silver, whether as foil or plated on copper, produced redness of the skin, but with less rapid tendency to ulceration.

"11. Copper plates were comparatively slow to act on the skin in connection with the negative pole; but the copper connected with the positive pole, on one occasion, blistered the skin severely when a very strong battery was used.

"For the rest, the batteries of wire coiled on wood are much lighter and more convenient in proportion to their strength. They preserve their activity sufficiently, and do not wear out so soon as the voltaic pile of zinc and copper plates. They have the further advantage that as many or as few as desired can be used without delay in arrangement. I believe that a very strong current, whether as to quantity or tension, is not required. A small quantity is sufficient to excite the nerves of the blood-vessels; and there should be just sufficient tension to ensure that the current pass through the part to be acted on, and not merely round by the skin. I believe that, without claiming any miraculous power for electricity thus applied, we have in it an agent of considerable energy, and capable of yielding results of sufficient value to warrant us in resorting to it more frequently than hitherto."

MODES OF APPLYING ELECTRICITY.

The efficiency of electricity as a therapeutic agent depends much upon its application; in quantity, intensity and local application — thus modifying all the various means by which it is obtained. Electricity, when applied, should be localized as much as possible. Dr. Duchenne was the first to invite special attention to this important principle in its application, and termed his method "Faradization," in honor of Faraday, who discovered the important phenomena of induction. There are two modes of applying electricity as an excitant and deobstruent: that of direct and indirect. By concentrating the excitation directly on the muscle, we have the direct application, and if through the nerve, or branch, supplying the muscle, the indirect application. This requires an exact knowledge of the anatomy of the nerves and muscles. In the arm, the electric power can be limited to the median nerve on the inner and inferior third of the humerus, to the ulnar nerve, on the interval between the olecranon and the internal condyle. The radial nerve is accessible at the junction of the two upper thirds of the humerus with its lower third; the musculo-cutaneous in the axilla. On the thigh, indirect is easier. The crural nerve is to be found in the groin outside of the femoral artery; the two popliteal nerves, in the popliteal space. The sciatic nerve is only accessible to electricity on its origin in the pelvis, through the posterior wall of the rectum. On the face, the trunk of the seventh pair, covered by the parotid gland, is inaccessible to electricity, whatever may be the intensity of the current; but, it can be reached where it passes out of the stylomastoid foramen; here a hand director is to be placed in the external opening of the ear; in this point, the nervous trunk is separated from the excitor by only a small space. Its branches may be excited at the point where they emerge from the parotid gland. Contraction of the muscles, being under control of these branches, is the certain signal of their electric excitation. In the supra-clavicular region, the director placed immediately over the clavicle acts on the brachial plexus. On the summit of the supra-clavicular trough, they are in connection with the external branch of the eleventh pair. This is the respiratory nerve of Bell, and the most excitable of all the nerves of the human body. The lower half of the sterno-cleido-mastoid and trapezius is excitable to only a small extent. But, when a very feeble current is directed on the upper half of the sterno-cleido-mastoid, or, on the external border of the upper half of the trapezius, — a current, indeed, which would not be expected to produce any movement in the muscles at all — the head is strangely inclined to the side acted upon, and the shoulder drawn up by a violent and sudden movement.

The muscles, like the nerves, do not all possess the same degree of excitability. Some muscles are so extremely sensitive that the excitation can scarcely be induced. Muscular sensitiveness is most active in the muscles of the face, due to the ramifications of the fifth pair which excite these muscles. It is important to always avoid the points corresponding to the infra-orbitæ and sub-mental nerves, the excitation of which gives a very painful sensation. The most excitable muscles of the face are the frontal muscles and the orbicularis palpebrarum; the least so, the buccinator and the masseter; of the neck, the platysma myoides is just as excitable as the upper half of the sterno-cleido-mastoid and the external border of the upper half of the trapezius. The most excitable muscles of the trunk are the pectorales major and the muscles of the fossa infraspinatus — chiefly the rhomboidal; then follows the deltoid, and the muscles of the arm. The anterior are much more excitable than the posterior muscles of the extremities. The most excitable muscle of the leg is the tensor vaginæ femoris.

Galvanism has been applied in the form of plates known as *Mansford's* plates. In asthma, a small blister, the size of a dollar, being placed on the neck over the course of the phrenic and pneumogastric nerves, and another, on the side in the region of the diaphragm, and one metal placed over the vesicated surface on the neck and another over that on the side,—connection being made by means of wire—the normal nervous influence is, in this way, induced and is often signally beneficial. By a somewhat similar procedure with galvanic plates, ulcers in a torpid condition have been greatly improved, tending to a healing condition. There are various modes of obtaining galvanic influence; as galvanic chains, such as that of the Pulvermacher's hydro-electric chain battery — a modification of the voltaic pile, capable of being employed topically. The galvanic influence upon muscular tissues is greater than that of the faradic, often producing muscular contractions in cases where the faradic fails, and exerts a more decided chemical action, even to that of an actual cautery — known as the galvano-cautery. It consists of a continuous current, is developed by chemical decomposition and is characterized as "low intensity" so far as regards its action upon nerves and muscles, but of most extensive quantity and increase of temperature, termed "thermic results," not equaled by other means of accumulating electricity.

The continuous current is accomplished by introducing a part or the whole of the body of a patient into and making it a part of the circle of the battery and then passing the current through it. This will relieve spasms of a certain kind, and certain kinds of pain. In some cases this is accomplished in a few moments, in others a repetition at intervals will be required. It will also arrest tremor and spasms. There are various modes of applying the continuous current. To connect the positive end of the battery with a person's left hand, and the negative end with the right; the current, passing from the positive to the negative pole, passes up the left arm to the trunk and down the right arm to the battery again. This is termed the "inverse current application" when passing up the arm, and when passing down, the "direct." This tends to two opposite effects. The current passing up the left arm excites the *irritability* of the muscles and nerves, and that passing down the right arm allays irritability in that arm. This serves well for theory, but is not sustained as of any practical utility. Pain is as readily relieved by the current in one direction as the other, and the same is the case in the treatment of spasm, whether clonic or tonic, or merely tremor.

The continuous current is in degree painful in ratio to its strength even to that of being unbearable, or so mild as not to be unpleasant.

The interrupted current is made in various ways. By taking two wet sponges attached by wires to the two ends of the battery, and placing one on the upper part of the limb, and interrupting the current by occasionally lifting it, and the other sponge at a distance from or under the limb, makes a break in the current and constitutes an interrupted current. Or, any more complete means as that of apparatus so constructed as to break the current instanced in one of the ordinary interrupters. In the interrupted battery current, the direct application produces a more obvious effect upon the muscles in the way of contraction than the inverse or indirect, and is worthy of note when applied for the purposes of diagnosis or treatment. The various terms applied, such as "Faradization," "induced electricity," "magneto-electric," "voltao-magnetic," "voltao-dynamic," are all of which Faraday was the great exponent. It is electricity of very high tension and differs from the magetic current in the particular of not eliminating the sensation of heat, but, under ordinary circumstances, produces marked contraction of the muscles and a more painful action on the nerves of motion and sensation. It exists only at the moment of making or breaking the galvanic current. There are two terms commonly used in regard to different batteries—"primary" or "secondary." This is for chemical purposes. The difference is said to be that the primary will have a more distinct action upon one set of nerves, and the secondary upon another. But the most marked physiological difference between them is that the secondary is of greater intensity than the primary, and will sometimes proceed more deeply into the parts intended to be affected. The word primary is not intended to signify a primary current in the sense of being a battery current; it is, essentially, an induced current. The other is an induction from an induction, and is secondarily induced, the clinical difference between them being, mainly, the degree of intensity.

For confirmation, see London *Lancet*, March 5, 1870, p. 331; a paper by Dr. J. R. Reynolds, F. R. S.

In our experience, as well as in that of others, it has been discovered that electricity is only of utility in the treatment of a certain class of ailments, and requires modifications under varied conditions of the cases susceptible of relief from its skilful application. The first apparent difficulty that presents, is the varying sensibility peculiar to each individual in a normal condition, and which is apt to confuse the practitioner in his diagnosis. This can only be determined by close observation and the avoidance of hasty conclusions. If simply a failure to relieve, in some cases, should ensue, no great injury could result, but there is great liability to inflict injury in others, productive of the most serious consequences to the patient and the reputation of the practitioner, as the mistake made in the application of electricity is usually very apparent.

Sensibility is mainly dependent upon the nervous condition of the patient, being more or less exaggerated by mental impression, but, in some cases, one limb is more sensitive than its fellow, and most apparent in sensibility of its muscles to electric influence. This is valuable as a diagnosis of the lesion in that limb.

Partial insensibility is most readily detected by loss of tactile sense of the power of appreciating heat or cold, and presents in cases without paralysis, and may exist as a symptom of cerebral disease.

Increased cuticular sensibility, hyperæsthesia, presents in cases of paraplegia from lesion in the spinal column, even before the spinal projection has taken place, and is a diagnostic symptom. A change in the central nervous system is often indicated by painful muscular contraction, general or local, and, in some cases, is attended with increased cutaneous sensibility, often by neuralgia and false impressions. In these conditions, the sensibility is augmented by electric influence to a greater degree than in the normal condition, and can, in slight cases, be determined by comparison when limited to local parts.

The sensibility of muscles is usually diminished when their contractility is reduced, and in direct proportion to that reduction — as in lead paralysis.

Electricity can be made available to the arresting of morbid sensibility in nerve or muscle, or, to excite to activity a dormant nerve or muscle. It is a force to relieve paralysis, restore loss of sensation, or of contractility in a muscle, or, to diminish over action or spasm — whether tonic or clonic — by restoring nerve or muscle to a normal condition. It will relieve a peculiar under-tone of the muscles in a limb in which the movements are made by a limited force — the muscles being lax in texture, but equal in size to its fellow. To this we have an opposite condition — hardness of muscles, not to actual rigidity, and, in some cases, actual rigidity, in which it is difficult to flex or extend the arm or leg; in others, a tremulousness of a muscle; and, lastly, clonic spasm apparent in interrupted movements of the limb — all subject to relief from electricity.

The circulatory system is quite as subject to the favorable influence of electricity. The influence upon the muscular fibre in the walls of the blood vessels is the same as upon other muscular tissues; the capillary vessels being dilated, as they are in paralyzed limbs, and as indicated by the purplish color, and coldness of the limbs. To the relief of this condition, electricity can be made available. It tends to induce contraction of the vessels by simply acting on the cutaneous tissue, and can be so limited, and quite as subject to an apparent relief in the contracted condition of the blood vessels as in that of its contractile fibres.

The accomplishment of other desirable intentions requires knowledge and experience in the use of electricity as a therapeutic agent, and some general ideas on the subject may here be given.

An over-excited nerve, muscle or blood-vessel may be subdued by the continuous galvanic current, and, as before stated, from the shoulder to the hand, or from the body outward, not so strong as to give pain, and applied in the course of a principal nerve. Static electricity affords equal relief by insulating the patient and charging them largely with electricity, then drawing it off with a metallic point attached to a conductor passing to the earth. Faradization may be employed for this purpose in such cases as that of spasmodic torticollis. Relief may be afforded by passing through the sterno-cleido-mastoid muscle a weak, rapidly interrupted current, or by faradizing the antagonist muscle. This, also, applies to the leg or arm when in like condition.

In paralysis from cerebral lesion, electricity is inadmissible in its early stage or at any time when there is pain or congestive symptoms in the head. In progressive paralysis from whatever cause, with no pain or feeling in the head, electricity may be used freely and often, to great relief to the patient. There are certain conditions of paralyzed limbs in which electricity affords little or no relief. If, on applying the current, a ready response of the limb is found and nearly as strong as in the sound limb, very little relief will be afforded other than improving the circulation in the limb, no matter how long it may be applied or how complete the paralysis may be. All the benefit that can be obtained from the use of dynamic electricity is in exciting contractility in the muscular fibre, and this will often cure the patient; but it must be borne in mind that the contractile energy of muscular fibre is not the moving power of the limb.

As to the inverse or direct interrupted current in the application of electricity for the relief of paralysis, there is little or no difference. It is only when the continuous current is applied that the physiological effect on the nerve differs; the current downward or outward from the body diminishes irritability, and the current upward increases it; the current downward acts more strongly than it does upward. This is of some importance in relation to the influence of the electric current.

Much precaution must be used in the application of electricity as a therapeutic agent. A stronger or milder current must be used according to the direction in which the electricity is to be sent, and, in all applications, care must be taken never to use such force as to give pain. It is only necessary to produce contraction, and this is essential to the accomplishment of any beneficial influence. To be enabled to so limit the current to the condition of the patient, it is necessary to become familiar with their sensibility. As before stated they vary greatly in this particular, or in their peculiarity of constitution. To give pain from the application of electricity is to injure the patient, and should, by all means, be avoided, if not in the first of the application, it should be so modified as to terminate pleasantly, if possible. The application to a paralyzed limb to the degree of inducing a state of cramp is a most pernicious proceeding.

The duration of the application is another important consideration. It is possible to exhaust the vitality of the muscle, and there is a very great danger of doing so. The safer plan is to apply the electricity only for a few moments, at first, in order to test the sensibility of the patient, and repeat the operation only every other day —even less frequent when not much improvement is made. When much inconvenience or discomfort is produced it is better to discontinue its use for a time, and then return to it from time to time, until the patient can endure it without unpleasant feelings. When pain in the head, disposition to faint, or discomfort about the epigastrium are noticed, these indications are carefully to be regarded as dangerous symptoms when under the exciting influence of electricity. When insulated, and charged with static electricity, and a continuous current drawn by a metallic point from the patient, no ill effects will follow syncope, but, in many cases, much benefit in inflammatory conditions.

It is a well determined fact that abnormal conditions of the cuticle, nerve, or muscle, the result of cerebral lesion, are not relieved to any appreciable extent by electrical influence, and its application is often attended with serious consequences. Hence the necessity of a careful diagnosis; as we have motor paralysis and anæsthesia of a limb or some part of the body from impairment of the brain. Spinal lesion tends to similar conditions, and for this electricity is the most reliable therapeutic agent in our possession. We have a certain condition of patients, commonly considered to be given to hysterical affections — marked anæsthesia, in different parts of the body and limbs — and, as there exists a certain relative influence between the control of motive power and sensation, there is an opposite inability to control the movement of the limbs, amounting, in many cases, to an inability to use them. In these cases, the defective sensation in the cuticle is most at fault in the incipient stage of the ailment. The muscles become impaired in tone because the patient, being indisposed to exercise them on account of their awkward appearance, fails to do so; and this, in some instances, causes him to become *bed-ridden*, as it is commonly designated, from his refusal to leave his bed or chair.

Static electricity is the most reliable remedy for relieving the patient from his dependent condition. The patient being insulated, the electric rubefacient should be applied to the nonsensitive skin, with slow movement for about three minutes, and this treatment repeated every other day. We have seen patients, who have been confined to their rooms for a year or more, enabled, after a second or third sitting, to make a successful effort at walking. In cases where the muscles have become deficient in excitability from long-continued inaction, the metallic coated glass tube affords a sufficient concussion to arouse the dormant energy of the muscles; ten or twenty shocks at a sitting, in the direction from the origin to the insertion of the muscles, the patient being encouraged at the same time to make an effort to move his limbs.

In cases of "essential paralysis," or what we understand to be spinal paralysis of children, they are to be determined by the degree of irritability of the muscles as to their being susceptible to improvement from electrical influence when in the incipient stage ; that is, in a week or two after the seizure, as, in some cases, the little patient recovers from a complete prostration of all the limbs to that of an arm, or arm and leg, that may continue paralyzed. In very rare cases, the irritability is so much impaired as to refuse to respond to the electric current or spark, which should always be cautiously applied as a means of diagnosis. Electricity, we consider, in such cases, injures in degree to its continued use. These cases do recover, in time, to a degree of sensibility to the electric influence, when, with great deliberation in its use, it becomes an efficient auxiliary agent conjoined with means of warmth, manipulation, voluntary efforts at exercise, and the retention of the limb in normal position.

Cases of several years' standing may be made susceptible of much relief from the careful application of electricity, either static or dynamic ; one proving effectual after the other has failed. But in no instance is electricity admissible in treatment when persistent contraction has taken place. The limbs must be restored to normal form before electricity can be made available to the restoration of muscular power; as the effect is to confirm the contractile tendency of the already contracted muscle or muscles — that is, when normal position cannot be restored by gradual effort of the hands. In many cases, only one set of muscles are paralyzed, and the equilibrium of force impaired, and in many other cases there is much aberration from normal form that can be readily overcome by the hands. Suitable apparatus should then be applied to sustain the temporarily restored condition of the limb, and the electric influence localized as much as possible upon the paralyzed muscles. That its influence can be thus limited is not in accordance with our experience. We have treated many cases of persistent contraction, after faradization in the hands of experienced practitioners had been employed for months, with no other result than an increase of the deformity.

Restoration to normal form being first accomplished, then electricity serves an admirable purpose in restoring power to the paralyzed limbs; the influence on the sound muscles being resisted by the supporting apparatus. A muscle cannot maintain or obtain contractile force when fixedly or even partially extended. Faradization, skilfully applied, is an efficient remedy in these cases, but not more so than static electricity in the form of sparks drawn from over the paralyzed muscles, or light shocks passed from the origin to the insertion of the muscle or muscles, and is, in reality, the most localized form of applying electricity. For the restoration of nutrition in atrophied muscles — that is, in cases of several years, standing — static electricity is the most efficient in restoring activity in the nutrient vessels. Even where fatty degeneration has supervened, the muscle will develop and increase in strength. The sarcolemma maintains its integrity in the producing of muscular fibre quite equal with the periosteum in producing bone; requiring only an excitant to the normal functional condition that is possible until vitality is entirely exhausted.

There is a form of paralysis affecting the vocal organs and the voice, arising most commonly from an anæmic condition of the patient and termed by some writers hysterical aphonia. Static electricity in the form of sparks, or, light shocks drawn from the glass tube, will afford immediate relief and but seldom fails, when persevered in. The shock should not be passed through the larynx, as it is very painful; below or back of the larynx is quite as effectual.

Sudden and severe strains of the arms, more especially, are attended with severe pain and tingling sensation in the hand and tips of the fingers, and, when the pain ceases, great weakness and wasting of the muscles follows, particularly in the flexors, abductors and adductors of the thumb — they becoming exceedingly attenuated. A similar condition results from excessive exposure of the hand to cold. Static electricity in the form of light shocks and sparks usually affords relief and is the only means in these ailments to be relied upon.

There is a condition of progressive atrophy induced from slight injuries to the limbs. A case in point, to illustrate, is that of a young lady of eighteen years of age, having a sick mother and being desirous of lifting her, when in bed to change her position, kneeled on the bed-rail with one knee, when a sharp pain was immediately felt in the limb. In about a week thereafter, a severe attack of seeming sciatica followed, that was relieved by Fowler's solution. In about six weeks from the time of the injury to the knee, she experienced weakness in the injured limb, and an inability to get down the stairs, when we were again consulted. On examining the limb the muscles were found to be very flaccid in texture, and upon comparing the measurement of each thigh there was a difference of an inch and-a-half in favor of the sound limb. Faradization was commenced with wet sponges, but failed to improve the limb, which diminished in two weeks a half inch — to the great alarm of the lady and her friends. A change of treatment was desired. Static electricity, in the form of slight shocks and sparks, was applied, and measurement made of the limbs every two or three days, for two weeks, when it was apparent that the diminution in size had not progressed. It was then thought advisable to continue its use twice a week. At the expiration of three months the limb had increased in size one inch and was, apparently, as strong as its fellow. It continued to improve in nutrition, and in eighteen months after, it was half an inch less than the other thigh, but apparently quite as strong.

We have had analogous cases from slight injuries below and above the knee, and also of the arm, with like results of progressive atrophy of the muscles, and that has been relieved by static electricity, applied most commonly in the form of light shocks twice or thrice a week, requiring in some cases many months for an improvement in development, an increase of strength being more readily attained, which is remarkable, because of the deficient muscular tissue.

GENERATORS OF GALVANIC ELECTRICITY.

Believing it to be of special interest to the medical practitioner, we present for his consideration engravings of approved varieties of apparatus for generating dynamic electricity, the following have been obtained, with directions as to their use, from the Galvano-Faradic Manufacturing Company of New York, to whom we express our grateful consideration for the favor.

This machine is most generally selected by physicians. It has the advantage of a cell in reserve, should the other become exhausted. Also, the strength of both cells can be united in cases of suspended animation, or where great power is required.

The new and elegant hard rubber cells, though at first appar-

Portable Electro-Magnetic Machine.
Fig. 60.

ently, costly, are ultimately the most economical, as there is no breakage. They are also preferable in other respects, especially owing to their lightness.

By means of the hydrostat they are rendered portable, so that they can be carried around by city or country physicians, charged and ready for use, without danger of spilling the battery fluid.

Directions for use.—To prepare the battery fluid:

Take five pints of cold water, add thereto eight ounces of bi-chromate of potassa and immediately afterward seven fluid ounces of sulphuric acid. Mix and dissolve. Fill the battery cell with this liquid two-thirds full, which will be about eleven ounces. Connect the battery or cell with the machine by means of the hooks. Each binding screw must then be well tightened.

Fasten the poll-cords into the arms of the current charger, except on the No. 1, where they are to be connected with the binding posts marked P and N, which letters indicate the *positive* and *negative* poles. To the former the red cord is attached, the green to the latter, the easier to designate the poles while in use. When it is not especially intended to apply the metallic discs, sponges should be fastened to the extreme points of the handles of the electrodes. Previous to each application these should be thoroughly moistened in warm water. All being thus prepared, the machine can be put into action, by elevating the graduated hinge-jointed rod and then lowering it down in the battery cell. A knock on the box will cause the spring to vibrate and set it going.

The primary current (mild) is obtained by inserting the T-shaped current charger into the opening marked *primary*. Its strength is regulated by the sliding coil. Drawing it outwards increases, and pushing it inwards decreases the intensity.

The secondary current (strong) is obtained by moving the current charger into the orifice marked *secondary*. *The sliding coil should be fully drawn out to the right before this change is made.* Pushing it inwards increases, and drawing it outwards diminishes the intensity of the current.

In addition the power of either current can be increased or diminished by lowering the rod more or less down into the battery cell, being greatest when entirely down. It can be affixed in any position by the stay screw.

Even when in daily use the machine will continue in working order, without any additional charge, for a considerable period, depending on the frequency of its use. Subsequently when the battery fluid becomes weakened, undo the screws on battery and hydrostat, empty the cell and refill and replace all as before. When the zinc plate is worn out, a new one must be substituted.

N. B.—*Do not forget to draw the rod up and rub it dry, and lay it over horizontally immediately after use.*

PORTABLE GALVANIC BATTERIES.

To obtain all the results electricity is capable of producing, it is requisite to be provided with both an electro-magnetic machine and a galvanic battery. The effects derived from an *interrupted* and a *continuous* current being in many respects different. The electro-magnetic machine is the instrument most frequently required for general practice, while the galvanic current is especially used in certain classes of nervous disorders, and is alone applicable for electrolysis or the resolution of tumors. This is owing to its possessing chemical powers, of which the faradic or induced current is almost devoid.

These batteries, an improvement on Stöhrer's, are also exceedingly simple and very easily managed. They are powerful, reliable, efficacious, continuous, always ready for immediate use. There are three sizes:

No. 1. Eight cells, suitable for eye and ear cases, and where a strong current is not requisite.

No. 2. Sixteen cells, for general use.

No. 3. Thirty-two cells, double strength, where increased power is necessary, and for electrolysis.

GALVANIC BATTERY.
Fig. 61.

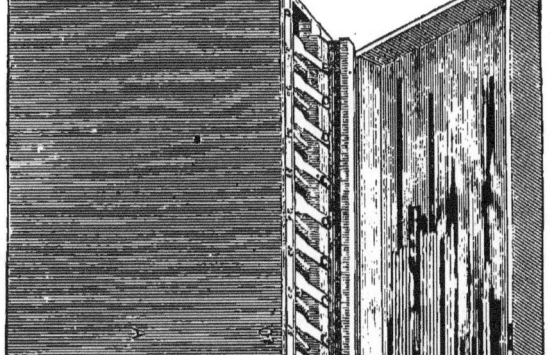

These instruments can be had with or without either the hydrostat or commutator.

Directions for using the Portable Galvanic Battery — To prepare the battery fluid: To 5 pints of cold water add eight ounces of bichromate of potassa, and immediately afterward 7 ounces of sulphuric acid; mix and dissolve. Charge each cell 1, 1, 1, with the liquid as follows: Lift the inside box B, by means of the keys 2, 2, and retain it in position by turning the keys half round. Remove the slats *d d*. Place a glass funnel successively into each cell, and pour in the fluid. If the battery is provided with a hydrostat, the cells may be charged two-thirds full, otherwise only two ounces to each. Turn the keys, lower the inner box. Insert the electrode pole cords into the binding posts P and N, on the slide or commutator, which is movable along the central beam *C*. Replace the slats. It is then ready to be brought into action.

When required for use, elevate the inner box as before and secure it in position. The elements are thus immersed in the exciting fluid and the battery put in working order. When the slide or commutator is brought opposite Fig. 2 on the front slat, the current from two cells is obtained; when opposite Fig. 4, that from four cells, and so on to Fig. 32. By thus moving the slide, to the left or right, the intensity of the current is increased or diminished. It can be further modified by raising the keys to the fullest extent or only partially, and turning them at the desired elevation.

The improved commutator or polarity charger for the galvanic battery is used with the utmost facility. When the plate marked N, rests over the letter Z, no current passes. Wherever the plate N, be turned, whether to the right or left, that side becomes the *negative*, and of course the other *positive*.

N. B.—Always lower the cells immediately after use.

When the battery fluid becomes weakened renew it as before. The exhausted liquid can be withdrawn by an India-rubber tube, using it as a syphon, by a battery syringe, or by carefully lifting the beam with its carbon and zinc plates attached, raising the inner box, emptying and replacing one cell after another, filling as before, and refixing beam and plates.

Sponges should be fastened to the extreme points of the handles of the electrodes and thoroughly moistened in warm water previous to each application.

CABINET REGULATOR AND BATTERIES.
Fig. 62.

The above engraving represents the cabinet galvanic regulator, with drawers for the batteries. This apparatus is complete, with current selectors, commutator No. 1, rheotome, galvanoscope, rheostat (water), and faradaic coil. The galvanic cells (Siemen's and Halske) are placed in the drawers of cabinet, and are connected with the regulator.

To the physician of large office practice, it presents an instrument possessing the valuable accessories of those costly apparatuses which have been imported, and for convenient hospital and dispensary use, it is almost indispensable, as it can be carried from ward to ward, and, being on castors, it can be easily moved to the bedsides of patients.

The cabinet is of black walnut, thirty-eight inches high, seventeen inches wide, thirteen inches long.

BARTLETT'S REGULATOR.

Fig. 63.

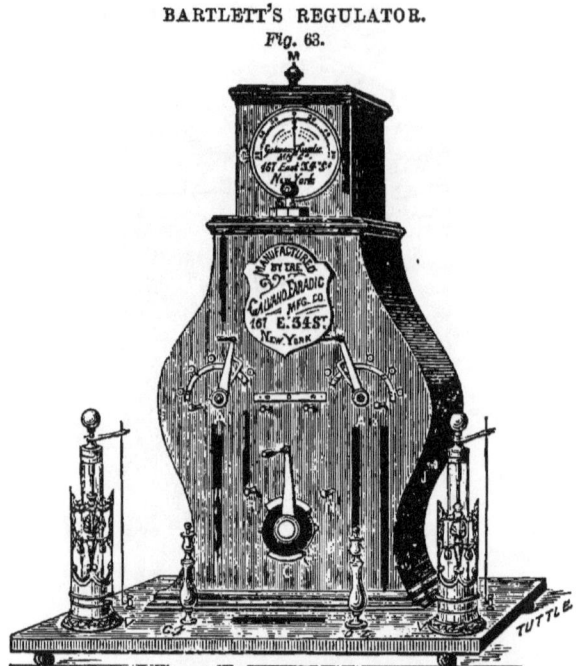

The above engraving represents Bartlett's Regulator. The current selectors A 1, A 2, enable the operator to unite for effect any desired number of elements, from two to sixty. To obtain the current from 28 cells, turn the winch A 2 to 20, A 1 to 8. While if we desire only 4 cells, the winch A 2 is moved to the letter O, the winch A 1 to 4. When both winches are at O, no current passes.

The rheotrope or commutator C changes the *polarity* of the current, so that it can be instantly broken or reversed. When it points vertically the current ceases to pass. When the winch of the rheotrope is resting on N, the *positive* current passes through the binding post c, and the *negative* through the binding post z. If the winch rests on R, the *positive* passes through z, and the negative through c. When the battery is not required for action, let the rheotrope stand vertically, and both the current selectors rest on O, and by this means the elements will be preserved and maintained.

The rheotome, or current interrupter J, causes and controls fluctuations of the current. When the knob J, rests on the right stop, no break is experienced. If it be turned gradually to the left, slow intermissions are felt; so that the greatest rapidity is reached when it is fully turned to the left. Thus we have the means of increasing or diminishing the interval between each interruption — an object of the greatest therapeutical importance.

The current modifiers V V, weaken or intensify the strength of the currents. The glass tubes contain water, through which the rod or piston passes to the bottom of the tube, and there remains in contact with a metal conductor. If the rod be elevated from this contact, the water intervenes, and being a much poorer conductor, the current is proportionally weakened. The rod should be raised by a kind of rotary motion, at the same time bearing down with the other hand on top of the tube.

The galvanoscope G. If the winch S, stands perpendicularly, no current passes through the galvanoscope. If it be turned toward either side the needle will be accordingly deflected; the strength of the current will be denoted by the extent of the deviations of the needle from the point O. One quarter of the circle is the full extent to which it can be deflected. Having obtained the desired information, the winch S, should immediately be replaced, as otherwise the sensibility of the magnet would be impaired. If the circuit be open and the needle deviates on either side from the point O, it can be replaced by turning the knob M, to the right or left.

THE GALVANO-CAUSTIC BATTERY.

The following engraving represents a full-sized transportable galvano-caustic battery, an improved modification of Stöhrer's and Grennet's. There are four capacious glass cells, each containing four large zinc and three carbon plates.

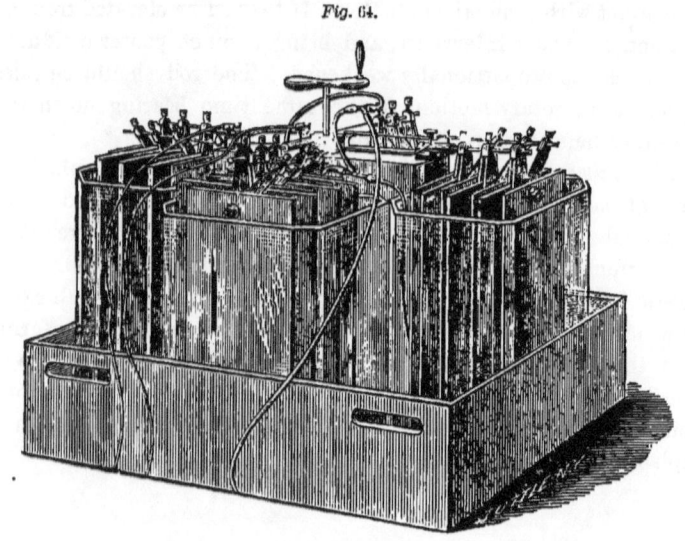

Fig. 64.

TO PREPARE THE BATTERY FLUID.

Pour eight fluid ounces of sulphuric acid into one quart of water. After this has cooled, add to it three ounces of bichromate of potassa roughly powdered, or more if required to form a saturated solution. If after stirring and dissolving, a few crystals remain undissolved, it will be advantageous, as they will be, subsequently, utilized. Ten quarts of this fluid are required to charge the four cells of the galvano-caustic battery. The front of the tray bears our name-plate. Cell No. 1 is on the left; No. 2, back; No. 3, right; No. 4, front.

TO ARRANGE THE BATTERY AND EACH PART IN POSITION.

Place the glass on the tray on each side of the lifting partition; insert the arm marked No. 1, which supports the elements (zinc and carbon) into the opening on the left end of the lifting partition, marked No. 1, which secures it in position. Then place arms marked

2, 3, 4, in their position, seeing that the numbers on arms correspond with numbers on partition. The small, crooked brass rods are for the purpose of connecting the cells together, and all but one are notched on each end to correspond with the numbers on elements.

The blank one is inserted through the holes in these protruding connections on first cell, which connect with zincs, and connected to the small permanent binding post on the lifting partition. Rod notched I and II, is connected from the zinc connectors on second cell, and through the hole in arm of cell No. I. Rod notched II and III, in a corresponding way with Nos. II and III. Rod notched III and IV, from III to IV cell.

See that all the binding screws are screwed firmly down. Elevate the elements by means of the central lifting screw to within one inch of the top of cells. Fill each cell about three-quarters full of battery fluid. Attach one pole cord to the nipple on permanent binding post on lifting partition. The other to the nipple on the arm of the fourth cell. Thus bringing the whole number of elements into action. When only three cells are needed for use, disconnect the rod from the fourth cell and attach cord to second cell. Same with one cell. When the zincs or carbons become crystallized, place them in boiling hot water and allow them to remain until the water cools. Then dry them thoroughly.

Fig. 65.

The above is a representation of a case containing all the galvano-caustic electrodes *in situ.*

1. Handle for looped electrodes.

2. Stem for loops.— Prepare this for action by passing the wire through its dome; tubes at each side of stem; holes at extremity of handle No. I, and those in revolving barrels; to which secure the ends of the wire. The flat extremity of the ivory stem and tubes should then be simultaneously pressed into their respective openings at the end of the handle, and securely fastened by the small binding screws; place the ratchet in position; when the circuit is closed turn the barrel gradually by its handle so as to draw tight the loop and constrict the tumor.

3. Long tubes for laryngeal application, nasal polypi, etc.

4. Short tubes for loop. The wires should be arranged in 3 and 4, as in No. 2.

5. Universal handle.— The upper end of each of the other cauteries can be inserted in the socket of this handle. Be particular that the brass pieces of the electrode, to be used, are so placed in the handle that they rest on the brass sides of same, in order that the opening in the handle, and the ivory in centre of electrode should be on a line; the insulated ring should then be forced down to firmly fix them in position.

6. Platina point cauterizer for concentrated cauterization, counter irritation, etc.

7-8-9. Dome cauteries for cauterization of sinuses, bleeding surfaces, bases of ulcers, places where cicatrical contraction is desired for the destruction of small growths, etc.

10. Long curved stem cauterizer for cavities, tracks, etc.

11. Straight stem cauterizer.

12. Galvanic knife cautery. To be used where it is desired to slice off, or remove.

13. Right angle loop. May be used as a knife.

14. Galvanic curette. For removal of small fungoid masses or small tumors.

15. Beak cautery for counter-irritation. To be used with low heat.

16. Short curved stem for cavities, tracks, etc.

17. Pocket for platina wire and porcelain domes.

On the handles 1 and 5 are slides for making or breaking the connection.

GALVANO-CAUTERY.

Extracts from an Article by Drs. Allan Hamilton and Chas. A. Warner, New York:

"The advancement of surgical science, particularly in the direction of Electro-therapeutics, has placed a most valuable remedial agent in the hands of the profession.

"There has long been need of an instrument that would have the advantage of being used in inaccessible parts of the body, where the use of the knife is attended by inconvenience, or forbidden altogether. This surgical force is evidently so valuable, so efficient, so universal in its applicability, that it will eventually to a great degree take the place of the knife and the actual cautery.   *   *   *

"It is found that if a poor conductor of the galvanic current is placed between two exceedingly good ones, it will undergo a marked elevation in temperature.

"From a knowledge of this fact we find that if such a substance as platinum is made the terminal arch of a powerful circuit, where there is *quantity* with *intensity*, it will be instantly raised to red heat, and if the current is sufficiently strong, it will become incandescent. This power has therefore been recognized and the galvano cautery devised.   *   *   *   *   *   *

"These instruments have been modified, and we now have a number of varieties of form, adapted to the several operations we wish to perform. The most important variety is the galvano-cautery loop or écraseur, which may be used for the removal of pediculated tumors, such as polypi, excrescences, folds of tissue, and bands of adhesions. It may be used for the removal of polypi of the larynx, uterus, or rectum; internal hemorrhoids, cauliflower excrescences, or for amputation of the cervix uteri. In fact, whenever we can use the écraseur for the removal of the soft parts, we can employ the Galvano-cautery loop.

"The second variety includes the galvanic moxa, or cone of porcelain, which is heated to incandescence, by a coil of platinum wire, wound about it. This form of apparatus is used when the desire is to produce cicatrical closure, or contraction, as in the vaginal wall, where prolapse occurs, the basis of old ulcers, the removal of small growths, the cauterization of bleeding surfaces, for counter-irritation, and for bites of rabid animals or reptiles.

"The third form is the *galvanic knife*; this is employed where the desire is to slice, or remove masses of tissue, degenerated or indurated, or for the extirpation of growths where the loop is contra indicated.

"There are several modifications of this form, one of which is the *galvanic curette*; this may be used with good effect when the aim of the operator is 'to hoe,' agriculturally speaking.   *   *   *

"In the performance of all the operations, the following steps are to be observed:

"1. To connect the Electrodes with the battery.

"2. To observe that the connections are all clean.

"3. To immerse the battery plates in the exciting fluid to the depth desired. This must be determined by the color of the heated wire, or platinum extremity. A dull red heat is to be exceeded, as the cautery is very apt to drag or tear off portions of tissue and cause hemorrhage. A white heat accompanied by scintillations is to be avoided, as the wire is very apt to break or melt, and the marginal inflammation and consequent sloughing will be serious.

"4. The wire should always be tested upon a piece of raw meat, and the degree of "severance power" gauged.

"5. The patient should be anæsthetised if the operation is of considerable magnitude.

"6. If the loop be used, it should be passed about the part to be removed cold, and when *in situ* the connection made. Then gentle traction should be exercised by the rachet wheel in the handle.

"7. After the operation, the parts should be treated with simple cold water dressing, or if hemorrhage follows, it may be stopped by the per-sulphate of iron. * * *

"The advantages of galvano cautery are: that there is seldom any hemorrhage; that there is never any shock; that the cicatrix is always smooth; that the operation is easily performed, and the separative process is quick; that the convalescence is short; and, finally, that the procedure is simple, and almost always successful."

In relation to the several electrical apparatus now in use, it would be unjust were we not to mention specially those of Dr. JEROME KIDDER, which have deservedly gained the Prize Medals at the annual fairs of the American Institute, and at the Vienna Exposition, and are continually receiving the highest encomiums for excellence, efficiency and simplicity of construction from the medical fraternity in this country, as well as in Europe. His instruments commend themselves to all who use them, and mark a new era in Electrology.

CHAPTER VI.

CONTRACTION OF THE HANDS, FINGERS AND TOES.

Treatment of this class of affections hitherto most unsatisfactory. — Section of tendons a signal failure. — Ordinary causes of contraction. — Muscular paralysis alone accessible to influence of electricity and tenotomy. — Froriep, of Berlin, the first thorough demonstrator of the affection in all its details. — Baron Dupuytren's lectures on the subject. — Method pursued in dividing the prolongations of the palmar fascia. — Contortion and ulceration of the toes. — Bunyons. — Remedial apparatus. — Contortion of the toe-nails. — Ulceration of the foot. — M. Brachet's practice. — M. Sommé's treatment of morbid growths. — Sir Astley Cooper's method of procedure. — Our own treatment. — Onychia maligna. — Spastic contortions. — Treatment.

The treatment of this class of ailments has been the least satisfactory of all the ordinary aberrations from normal form. Fixed, or elastic extension has afforded little or no relief to the contracted wrist or hand. The toes and metatarsal joints have been much more readily relieved by appliances.

The section of tendons in the hand has signally failed us to produce good results. The indications for the operation are obscure, indefinable; and, when attempted, have, in many instances, only impaired what little remaining usefulness there was in the fingers.

The most common causes are rheumatic diathesis, bruises, injury to the bones, fractures, wounds, abscesses, cutaneous eruptions and paralysis of antagonizing muscles. The latter variety, alone, present a favorable condition to be relieved by electricity and tenotomy. The tendons are resisting and prominent beneath the skin, and, with a knowledge of the anatomy of the part, may be severed without risk of injury to the patient.

The impairment is rarely limited to the muscles, and commonly presents a combination of the various movements of the hand — flexion being accompanied with extension of the fingers, or lateral inclination. The muscles of the fore-arm frequently participate in the affection, and are more or less flexed or pronated. The palmaraponeurosis, when restricted, may aid in the flexion of the phalanges by means of fibres which it supplies to each side of the fingers.

Attention was first attracted to this fact by Froriep of Berlin. This may be relieved by severing the tense bands of fascia, and in some cases, the partially anchylosed joints may be relieved by forcible extension. We have practised both with success; but in the latter, have encountered most serious results, and consequently, have in a measure abandoned the practice.

When the ailment is confined to the lateral deviation of the hand, the severing of the palmaris longus and brevis, and, in some cases, the flexor carpi ulnaris is necessary. This affords almost immediate relief, if it is not due to paralysis of antagonizing muscles, when a greater duration of time will ensue before a cure is obtained. The subsequent treatment consists in making an effort to extend the hand by means of suitably contrived elastic extension.

Experience has deterred Orthopædists from severing tendons at the base of the first phalanx, and limited the practice to the forearm for the relief of contracted fingers. Mr. Franklin, who has had the most extensive practice in these ailments, says: "In cases that arise from injuries to the fascia itself, and in which the tendon becomes imbedded in the cicatrix, the division must be effected at that point. You cannot, however, in this instance do more than straighten the fingers, as in every case that I have operated upon, which has arisen from this cause, the power of flexing the finger has been altogether lost, and in some cases, when the division has been performed in the finger itself, in which the integrity of the sheath has not been lost, a loss of power has been the consequence." This accords with our own experience in dividing tendons at the articulations of the phalanges.

For the most practical information ever given on this subject we are indebted to Baron Dupuytren in his lectures published by the Sydenham Society in 1854. He says, "Most of those who are thus affected have been in the habit of using force with the palm of the hand, and of handling hard bodies, such as a hammer, anchor or plough. The affection usually commences in the ring finger, when it extends to the others, especially to the little finger, and increases by almost insensible degrees. The patient, at first, feels a little stiffness in the palm of the hand and a difficulty in extending the fingers which soon become a quarter, a third, or a half bent. The flexure is sometimes carried much farther, and the extremities of the fingers are bent into the palm of the hand. From the very beginning, a cord is felt on the palmar surface of the fingers and hand, which is drawn tighter when an effort is made to straighten the fingers, and disappears entirely when they are quite bent. It is of a roundish form, and its most salient part is at the articulation of the finger with the corresponding metacarpal bone, where it forms a sort of bridge. Its extremities are lost, insensibly, at the second phalanx of the finger and about the middle of the palm of the hand; or, perhaps, short of this point.

"The skin covering the finger is thrown into arched folds, the concavity of which is below, and the convexity above. This condition is for a time limited to the finger primarily affected, but at a later period, the other fingers are, though in a less degree, involved in the deformity.

"Notwithstanding all these appearances of deep-seated mischief, the joints of the affected finger show no trace of anchylosis, and, without excepting even the first phalanx, they cannot be straightened with any amount of effort, and I have seen from one hundred to one hundred and fifty pounds weight suspended by the hook, which the finger forms, without changing its position a line.

"The ring-finger is incapable of extension, and the others will not admit of it completely; the patient, consequently, cannot lay hold of large objects; if he presses anything tightly, he experiences a sharp pain; the act of gripping is obstructed and causes a painful sensation. When at rest, there is no pain; and none is felt till he endeavors to straighten his fingers too suddenly.

"The following short description of the palmar fascia will show how it acts in producing these effects: The superficial portion of the fascia is derived partly from the insertion of the palmaris longus tendon, and is partly a prolongation of the anterior annular ligament of the carpus. At its commencement it is very strong, and then thins off gradually toward its lower edge, dividing into four parts, which are severally directed toward the heads of the last four metacarpal bones. Here, each of these processes bifurcates to admit of the passage of the flexor tendons, and either subdivision keeps along the sides of the phalanges and not in front, as many anatomists have described. It is these prolongations which are more stretched than the fascia, and which require to be divided. When the skin and the fascia are dissected, there is some difficulty in separating them, owing to the cellular tissue and some fibrous prolongations which arise from the fascia. These adhesions account for the puckering and movement of the skin. In cutting through these fibrous prolongations it might be apprehended that the nerves and vessels would be involved, but, when the fascia is in a state of tension it forms a sort of protecting arch across them, so that it may be cut without danger. The use of the palmar fascia is to support the tendons of the flexor muscles, to preserve the palmar arch, and generally to protect the various parts of the hands. In roosting birds it is very largely developed, and is remarkably elastic. Such are the functions usually attributed to the palmar fascia; but it has others, whereby it tends to keep the fingers semi-flexed, which is their natural position in a state of repose; and this function, when exaggerated by disease, produces contraction of the fingers."

The operation of dividing these prolongations of the palmar fascia consists in supinating the hand and extending the fingers as much as possible. A transverse incision is made with a bistoury, dividing the skin opposite the metacarpo-phalangeal articulation of the contracted finger; then the tense prolongations of the palmar fascia, which will respond with a crackling noise and the straightening of the finger. As a matter of precaution against wounding the nerves and blood vessels, the tense fascia should be divided slowly, and its division determined by the yielding of the restrained finger — each finger to be operated on separately. Tense bands may be observed in the palm of the hand, in which case they should be divided in a similar manner; as the thick integument covering the palm of the hand is more or less implicated in the abnormal condition. Means of extension, elastic if possible, should be applied to the back of the hand, and simple water dressings to the palmar surface. The wounded parts heal slowly, and, at times, with considerable pain and swelling, but finally without any very serious difficulty. Flexion of the fingers may be attempted in about fifteen days, and the treatment completed in about twenty days.

CONTORTION AND ULCERATION OF THE TOES.

Contortion of the toes may arise from various causes, but more especially from wearing tight shoes or boots. A most painful ailment arises from inflammation of the joint of the metatarsal bone with the first phalanx, in which case more or less thickening is found in the joint, and, of course, impediment in walking. The individual under such circumstances instinctively retains the toe in the flexed position, which terminates in a permanent contortion of the toe as connected with the foot. Lateral flexion is most common, and terminates in an abnormal development of the metatarso—phalangeal joint of the great toe, which becomes exceedingly sensitive, and is commonly known as a *bunyon*. Upon extending the toe, the long flexor will be found rigidly tense, and may be severed in the sole of the foot, introducing the knife outside of the tendon, passing it beneath, and dividing from below upwards; in this way the puncturing of the internal plantar artery can be avoided, and then an apparatus should be applied as seen.

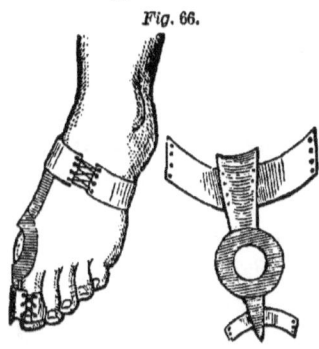

Fig. 66.

This consists of a delicate spring plate, shaped to conform to the parts to which it is to be applied, and a point of steel extending the length of the toe, to which is attached a piece of soft leather to pass around the contorted toe and restore it by the lacing to a normal condition. This apparatus will cure the ailment when the severing of the tendon is not so directly indicated, and an ordinary shoe can be worn over it without causing much of an unseemly appearance.

For the relief of the other toes, when contorted, an apparatus is employed, comprising:

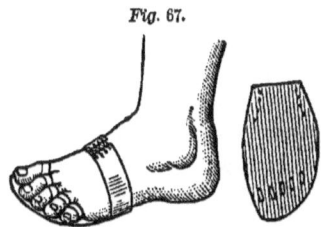

Fig. 67.

A steel plate shaped to the anterior portion of the sole of the foot, having openings or slits made to correspond to the projections of the toes; soft leather slips, four inches in length, with the portion to cover the toe about one inch in width, tapering off so as to be tied on the under side of the plate. By this the toes can be brought down and retained in *situ* without pain to the patient, and be worn with the shoe or boot. In severe cases the tense tendons should be severed to facilitate the case; as no injury will follow the operation — water dressing for a week, and the plate then applied completes the cure.

The great toe-nail and its connection, or matrix, is subject to disease of a most painful character and of frequent occurrence. This disease is never found in the hands, and is usually confined to the great toe. Contortion of the nail inward, the result of the direction of pressure, as from that of the surrounding shoe, may be considered as a common cause, and in a degree equal to its close approximation and cramping of the toes. This pressure forces the angles of the nail into the fold of the skin on the inner side of the toe, extending laterally by growth, and from irritation, the fold becomes swollen as well as aggravated by the adjoining toe pressing upward and over the nail. This irritation is increased by walking, and a very painful inflammation ensues. The flesh being injured, a sero-purulent oozing takes place, and, from walking, the whole foot eventually becomes implicated, while the oozing becomes more abundant and fetid, as it mingles with the perspiration of the foot. If not relieved, the ulcerated condition degenerates, under an unfavorable condition of health, into a cancerous sore, or fungoid excrescences; and, in some of these conditions, the inflammation extends to the periosteum, giving rise to caries or necrosis of one or more of the phalanges.

Various modes of treatment have been practised, one of which is, that of attempting a cure by elevating the nail, pressing lint under the edge, applying caustic to the morbid growth and cutting the nail frequently. These proceedings must be considered as only palliative rather than curative. The more severe means has been to cut away all of the overgrown flesh, thus permitting the nail to extend down the side of the toe. This last was the practice of M. Brachet. M. Sommé, a surgeon of Antwerp, cut off the portion of the nail growing into the flesh and then filled the cavity with powdered alum, which he said dried up the ulcer and prevented the growth of the nail. Sir Astley Cooper recommended the thinning of the nail with a bistoury, and the introduction of a small roll of lint beneath the point disengaged from the flesh; when active inflammation, however, prevented the introduction of the lint, he tore off the nail with forceps.

Our practice has been, in the incipient stage of the ailment, to cut the free or distal portion of the nail straight across, thus forming two free angles, under which lint, to a considerable amount, can be forced without inducing much pain. The nail, being thus elevated, overlaps the flesh, and, if carefully persevered in, arrests the tendency to this painful disease of the great toe. This practice serves well for children, whose toes should be frequently examined, and if the least tendency toward impingement of the nail upon the flesh of the inner side of the great toe be perceptible, the nail should be pared as described above. Children of five or six years of age, wearing shoes, are very liable to this painful condition of the toe. In exaggerated cases of considerable duration, and attended with much swelling, the patient is placed at rest, and the toe poulticed until the pain and swelling have somewhat subsided; then, with a pair of strong sharp-pointed scissors, divide the nail by pressing one of the points to the root of the nail and about a line from the overlapping flesh. Then seize the segment that is penetrating the flesh with a strong pair of forceps and tear it away from its attachments. Plain water dressing is all that is required for the subsequent treatment in ordinary cases, and, in a week or ten days, the patient will be able to wear a shoe without pain — nor is the operation so painful as might be imagined, and when the nail is removed there is no return of the disease in ordinary cases, if treated as in the incipient stage of the ailment.

ONYCHIA MALIGNA.

In 1814, Mr. Wardrop wrote an article upon the treatment of this ailment,* which appears to be an ulceration of the matrix of the great toe-nail, quite different from in-growing nail, requiring special treatment. It is a peculiar morbid condition of the integument at the root of the nail, tending to the alteration of the structure of the nail and results from violence done the toe — inflammation following sooner or later. Baron Dupuytren writes as follows upon this disease:

"These cases may frequently be due to the action of syphilitic virus, and occurring at the matrix of the nail, they have been named *onglade*.

"1st. It affects the nails of the feet and hands, indiscriminately.

"2d. It attacks several at once.

"3d. It shows itself in small ulcers at the intervals between the fingers, whence it spreads to the circumference of the nails.

"4th. The nail is detached spontaneously from the integument.

"5th. It resists all anti-syphilitic remedies; and, setting aside all the peculiarities I have named, the most experienced surgeon would

---

*Med. Chir. Trans. Vol. V, page 129.

have difficulty in distinguishing it from common inflammation of the matrix of the nail."*

The ailment may be considered as an indication of some constitutional derangement, curable by a strict regimen, rather than medicine, as, in some cases, the nail being cast off, the cure is effected, while in others the cure is only accomplished by the removal of the nail, and even the matrix of the nail. In this ailment fungous growths will be found at the base of the nail, and not at the side as in that of the ingrowing nail. The diseased member is of a violet red color and very painful; the nail is shortened and reduced in breadth; sometimes, indeed, it has wholly disappeared, and, instead of a nail, there are only some traces of a horny substance; the nail, too, is frequently concealed beneath the fungous flesh. Its color is gray or black, and, in some cases, it is not adherent, as in the normal state, and the morbid growths discharge a serous pus of a most fetid odor, bleeding freely when the patient attempts to walk. When in this pitiable condition, which medicine will not relieve, their presence cannot be endured in the room with others, because of the stench, and nothing but the knife and a generous diet will afford relief. The matrix of the nail and all of the diseased growths must be carefully removed. In performing the operation the patient reclines on a bed and receives an anæsthetic, as the pain is necessarily severe; the foot is then drawn over the side of the bed and is held and supported on the knee of an assistant. A deep semi-circular incision is made with a straight bistoury, about three lines from, and parallel to, the elevated fold at the base of the nail. The diseased parts are then taken hold of by the forceps, and, with the nail, are all carefully dissected out so as to remove all the diseased flesh in that region. Simple dressing is all that is required; the surface of the denuded parts are soon covered with healthy granulations which must be controlled with nitrate of silver. In fifteen or eighteen days, usually, the patient can wear a shoe, and, in time, a smooth thick skin assumes the place of the nail.

SPASTIC CONTORTIONS AND THEIR TREAMENT.

Children laboring under contortion, from spastic contraction of muscles, usually attributed to convulsions before and after birth,

---

* Published by Sydenham Society, 1854, page 254.

are subjects for relief in many instances. It is not safe, however, (although usually not so considered) to give a favorable prognosis in these cases, as some patients may be greatly benefited by a judicious perseverance in treatment, while others fail to be improved. The spastic influence is manifested in nearly all the muscles of the body and limbs. The arms are folded firmly to the body, and the hands flexed in a pronated position which they are inclined to constantly maintain. The contortions of the muscles of the face give the patient an idiotic appearance, though the mind is in a normal condition. The legs are partially flexed with irregular tension; the limbs adducted, and the heels elevated; presenting a seeming case of talipes equinus, which greatly inconveniences the patient upon attempting to walk, and, in some cases, subjecting him, after arriving at the age of eight or ten years, to the constant use of crutches; while in extreme cases, to the dependent condition of being unable to convey food to the mouth or to move about in any other way than by hitching upon the floor.

The treatment of this class of ailments is by no means as satisfactory as that we have been treating of. In this we have contortions and irregularities of motion, the result of involuntary and uncontrollable movement of all the muscles; not shortened muscles, from a want of normal antagonistic force, but muscles impaired by the loss of mental control, resulting in a persistent condition of the sound muscles. The equilibrium of action is restored in some cases, by severing the tendons of the shortened muscle, thus permitting the extended muscle to contract to a normal condition, such tendency existing in the extended muscles. The division of tendons, however, in this peculiar condition of the patient often, in time, restores to them the control of their feet. From this result it would seem rational to suppose that a dynamic influence is thereby imparted, tending to a more normal condition of muscular movement by an impression made upon the motor nerves. We do not hesitate, in this ailment, to divide the tendons that appear to be the cause of contortion of the limbs, and in some cases to a great improvement of the patient's previous condition, applying, of course, supports as in other ailments requiring them. This treatment applies mainly to the relief of the lower extremities. The arms and hands are much improved by systematic exercises; always limiting the amount to the accomplishment of one feat before another is attempted. Patients unable to feed themselves have thus been enabled to perform for themselves more of their necessary personal requirements than would be supposed possible for them to do in their decrepit condition, and the result obtained will, in a majority of instances, amply repay the effort of teaching them. They may, also, be much improved by passive exercise; as that of manipulating the tense muscles, shampooing and inducing them to perform some agreeable labor that will require a control of the muscles; when, it will be observed that by perseverance and time, seeming impossibilities will be overcome. Many whom we have treated, years ago, are now enjoying all the immunities of social life as though they had never been so afflicted; others have been much relieved from their dependent condition, and only a very limited number remained that were not improved. The encouragement we have had from great perseverance in treatment, induces us to advise that, in all cases where the mental faculties are not greatly impaired, this afflicted class should be afforded the benefit of a decided effort for their relief; and more especially as their mental faculties are usually in a normal condition. The impediment of speech caused by the spasmodic action of the muscles of the tongue, and interrupting the correct expression of thought, tends to apparent imbecility of the patient, and should, of course, be carefully considered; and bear in mind that there is not co-ordinate action of mind and muscle in this abnormal condition, but that tuition can be made available to much improvement, by directing the patient to concentrate an effort of the will upon one movement of one limb at a time; and when the movement is perfected, another may be commenced, but not before, as confusion will be the result, tending to failure in the treatment. There is no other ailment that we have to treat, susceptible of any degree of improvement, that requires the same amount of perseverance and length of time to accomplish. Five, and even ten years of constant effort has, in some cases, enabled helpless individuals to a tolerable condition of independence and self-assistance.

CHAPTER VII.

LATERAL CURVATURE OF THE SPINE — TORTICOLLIS.

Congenital lateral curvature of the spine. — Co-existence of this contortion in fœtal monstrosities; how caused. — Symptoms usually observed. — Prognosis of cure. — Non-congenital curvature of the spine. — Various conditions of aberration. — Oblique abdominal muscles largely contribute to the contortion. — Direct or reflex irritation, under certain circumstances, tends to increase curvature. — Pathological condition tending to lateral curvature. — Mr. Barwell on internal organic lesions, their causes and treatment. — The female more subject to this affection than the male, though men, women and children are liable to it. — Pneumonia and pleurisy as predisposing causes. — Diagnosis of lateral curvature of the spine. — Two cases rarely found alike in diagnosis — First stage of curvature. — Second stage. — Third stage. — Treatment of lateral curvature of the spine. — The general opinion that curvature of the spine is less amenable to treatment than other contortions, erroneous. — Accurate diagnosis of abnormal developments and general health of patient the primary requisite. — Parallel bars, their utility. — The supporting apparatus. — Description of extension chair. — Barwell's description of the advantage resulting from the use of chair. — The patient's health a primary consideration. — The skilled medical practitioner the most competent to give primary relief.— Beneficial effect of calisthenic exercises. — The use of electricity not beneficial, and objectionable as a therapeutic agent in spinal curvature. — Tenotomy as a means of cure. — Guerin, of Paris, the first to attempt dorsal myotomy. — Want of perseverance in treatment the most serious difficulty.— TORTICOLLIS, what it is, and how it should be treated. — Electricity applicable. — Division of the Sterno-cleido-mastoideus muscles resorted to two centuries since. — The sub-cutaneous method now generally adopted. — Bonnet's method of procedure.

CONGENITAL LATERAL CURVATURE OF THE SPINE.

The co-existence of this contortion in fœtal monstrosities has tended to the conclusion that the muscular contraction resulting in lateral flexion of the spine of the fœtus is caused by tonic spasm from lesion in the nervous centres, and not malposition, as asserted by Mr. Tamplin and others. And, to sustain this hypothetical view, supposed evidence of the fact is presented, that the contortion of the spine in the fœtus is accompanied with strabismus, club-foot, torticollis, and convulsive appearances of the face and other spasmodic affections, as that of the limbs, and even epilepsy.

Whatever the cause may be, latent influence certainly exists that tends to this abnormal condition in, not only the fœtus, but children of from one to ten years of age; as we have had such cases to treat in the Hospital for the Relief of the Ruptured and Crippled, and the most careful investigation has failed to give satisfactory information even as to the probable cause. The parents state that they have not observed from the birth of these children any symptoms of convulsions or spasms — only a gradual tendency to the contortion, which they have made most strenuous efforts to correct by opposing position, bandaging, and stiffened corsets; to the apparent infliction of pain to the child that counteracted all their efforts.

It will be observed that in most cases the liver, spleen, and mesenteric glands are hypertrophied, and a morbid sensitiveness to pressure in the concave side of the iliac region, indicating some internal lesion. The general appearance of these patients indicates a feeble organization, impaired by some occult disease that has gradually developed as a sequence — lateral curvature of the spine.

The prognosis of cure in these cases is always of doubtful consideration. The contorted condition of the body is of the least importance in the treatment — other than to arrest the increasing tendency to deformity. This is, in some cases, very embarrassing to the practitioner; as it is attended with pain and actual injury to the patient, whose morbidly sensitive condition countervails the efforts that may be made to sustain the body by steel supports, which, if persevered in, in some cases, tends to serious impairment of the general health of the child. All firm supporting braces under such circumstances are, more or less, inadmissible, and can only be useful when a more favorable condition of health is obtained.

NON-CONGENITAL LATERAL CURVATURE OF THE SPINE.

Lateral deviation of the spine becomes most frequently observable in girls at the age of from thirteen to eighteen, and, in rare instances, in boys. This is one of the pathological conditions of the afflicted, that has been, through the indifference of medical practitioners, generally submitted to adventurous treatment, as cases in which they had but little confidence in the possibility of even relief being afforded. This apathy, it may be said, has existed from the days of Hippocrates to the time of the introduction of successful practical tenotomy. This inestimable discovery in practical surgery of restoring to normal form contortions of the limbs, incited the efforts of many of the most eminent surgeons of the day to make the novel means available to the relief of nearly every description of contortion to which the human form is subject.

There are three conditions of the aberrations to be considered in the progress of this ailment: tortion, lateral curvature, and recurvation of the spinal column; the latter being a compensating tendency to the maintenance of the vertical status of the trunk and head; presenting, in some cases, several distinct curves in the spinal column from near the base of the lumbar to the upper portion of the cervical vertebræ.

The antero-posterior normal curves of the spinal column, invite very special consideration in relation to their tendency to the abnormal condition of lateral curvature of the spine. We consider the primary tendency of lateral inclination in the spinal column, to be the result of *tortion* produced by a pathological condition of certain muscles, having their direction of force obliquely from the hypochondriac to the iliac region; which, subsequently, involves all the muscles that sustain the spinal column when in a normal condition; and their relation of sustaining force impaired, is in direct ratio to the contortion resulting in the shortening of the body.

The *inferior serratus posticus* muscles having their origin from the spinous processes of the three last dorsal and two first lumbar vertebræ and dividing into four fleshy divisions,— passing obliquely upward and attached to the ninth, tenth, eleventh and twelfth false ribs, near their cartilages; have their direction of contraction most favorable to rotating the trunk, and of exceeding great power when contracting. Their action is to depress and tend to approximate the ribs toward the crista ilii; and when these muscles contract without the co-ordinate counteracting restriction of their co-operators, rotate the spinal column upon its vertical axis, which is indicated by the advancing of one shoulder, as the whole spinal column is subjected to this influence, thus determining a positive diagnosis in the incipiency of the ailment.

The external and internal oblique muscles of the abdomen of one side, contribute largely — when acting without counteracting force — to the contortion, by their oblique direction of force; tending to rotate the spinal column, and to the right because of the more active tendency and inclination of that side of the body, as in writing and similar occupations requiring an advanced position to the right side.

Direct or reflex irritation tending to abnormal uterine function in the growing girl, contorted position, or whatever may tend to the impairment of normal antagonism of the muscles described, and, subsequently, by shortening of the muscles that sustain the normal status of the spinal column, in ratio of their retraction, increase the curves, to the diminishing of the stature. Thus it will be observed, that the compensatory curves or lateral direction are only the result of the depressed normal curves rotated in a lateral direction.

PATHOLOGICAL CONDITION TENDING TO LATERAL CURVATURE.

There is in many individuals a seeming predisposition to lateral deviation of the spine, and limited to certain degrees of exaggeration — in young women especially — the tendency to the aberration being most apparent during their menstrual periods. The prominent hip and advanced shoulder will then be perceptible to even the casual observer, and in some individuals only during that period — the patient suffering from pain in the side and back. This may be considered as reflex irritation affecting the inferior serratus posticus, obliquus externus, and internus abdominus muscles, which is greatly increased in all cases where the patient labors under painful menstruation; the pain being more severe in one hypogastric region. This incipient stage of spinal tortion will be observed more especially when the individual is walking; and young women thus conditioned can be seen, daily, in our thronged streets, though from stratagem in dress (as these persons are usually cognizant of their contorted form) the prominent hip and depressed shoulder may not be noticeable, yet, the obliquity of their personal appearance is very perceptible and cannot be readily controlled when walking — the depressed, advanced shoulder being quite apparent.

Of persons thus afflicted, many are never more unfavorably conditioned, whilst others, from various inimical causes too numerous to mention, in this limited work, and upon which many respectable volumes have been written, yield progressively to the most exaggerated contortion of the body, accompanied with pain and great inconvenience. The incipiency of lateral curvature of the spine most frequently presents in the young girl, when approaching the catamenial flow, a seeming predisposing cause, and thus it may be assumed that the female is rendered more subject to this ailment than the other sex. Lateral curvature of the spine is, by unfavorable influences, as that of local irritation, or long-continued position of resting on one leg, induced in persons even to the age of forty years; but, after the age of twenty, there will be found as many males as females thus afflicted; it being then, in more than a majority of cases, the sequence of organic lesion of the liver, spleen, heart, lungs, or kidney; affecting only the one side; inducing a lateral position that may become permanent; and, in some instances, increasing in tendency, but rarely to the same extent as those having the earlier impression.

In regard to these internal organic lesions Mr. Barwell, in a treatise on the causes and treatment of lateral curvature of the spine, published in London, in 1868, page 74, says: " Causes of internal origin assume a variety of shapes from the gradual long curve occupying all or nearly all of the dorsal region to the well marked, short and sudden aberration of two or three vertebræ occurring anywhere between the first and tenth. These latter, particularly if they be high, are generally connected with tubercle of the lung. Pneumonia induces, as a rule, the low, long curve which, even when the functions of the lung have been restored, continues for a long time,— perhaps permanently. Pleurisy produces several forms of curves, but chiefly a high curve, longer and less sharp than the consumptive curve." This eminent and experienced practitioner also says in regard to these causes: " They do not arise from contraction on the diseased side, but from the fact that when any portion of the lung becomes unfit to perform its office, or, when disease renders such performance painful, the rib, or ribs, over that part of the organ cease to move, whilst those on the other side continue, under the action of the serratus, to act unopposed upon the vertebræ and twist them round." With this view of the etiology of the ailment our experience is in accordance with the statement of Mr. Barwell, that these causes affect both sexes, children and adults equally ; and also that the primary influence of the catamenial period, in enfeebled constitutions, gives the preponderance of cases to girls from eleven to eighteen years of age.

DIAGNOSIS OF LATERAL CURVATURE OF THE SPINE.

The aberration from normal curving of the spine to that of lateral curvature is involved in three conditions :
1st. Tortion.
2d. A single lateral curve.
3d. Compensatory curves that sustain an equilibrium or vertical bearing of the body.

In the incipient stage of this pathological change in the spinal column we but rarely find two cases presenting similar appearances, yet certain curves are more apparent than others. The patient being stripped of clothing and standing at ease in as erect a position as possible, the surgeon standing on the left, will readily observe the advanced right shoulder; a convexity to the right above and to the left below. The principal curve commonly occupies the dorsal or dorso-lumbar region, which is explained by the fact that the centre of motion of the vertebral column, and of the lateral flexion in particular, is situated at the point of junction of the dorsal and lumbar regions, and is due to the anatomical disposition of the articulation uniting the eleventh and twelfth dorsal vertebræ. The natural curve in the lateral movements of the spine decreases from the loins upward, and so perceptible is the change from a normal condition, in the incipient stage, that the indications are obscure and render the diagnosis difficult.

Tortion, being the first impairment, advances the right shoulder and is distinguishable from the hurried gait of the individual from the obliquity of the advancing body. When this condition is determined other indications are always present, as tortion changes the normal curve of the spinal column which gives prominence to the ribs of one side — the deviation, however, being obscured by the muscles, though in a very attenuated person a single curve may be detected. Other more apparent indications present, as may be seen in engraving 68.

LATERAL CURVATURE OF THE SPINE. 179

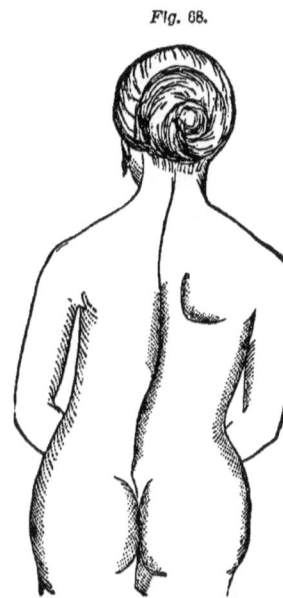

Fig. 68.

The hands interlocked in front of the abdomen will present a closer approximation of the right arm to the body than the left, and an increased prominence of the apex of the scapula of the same side. Also, an increased elevation of the longissimus dorsi in the dorsolumbar region of the left side, and a resultant cincture of the right, that leaves a comparatively limited angular space between the arm and body. This presents the apparently elevated hip; one of the most marked indications of incipient lateral curvature of the spine.

In the second stage of spinal flexion, exaggeration of form increases to a very perceptible deviation from a vertical line, and presents a more widened space between the body and the arm of the left side, as seen in the following engraving.

Fig. 69 represents the second stage of lateral curvature of the spine, in which there is as yet but a very limited change in the form of the hips.

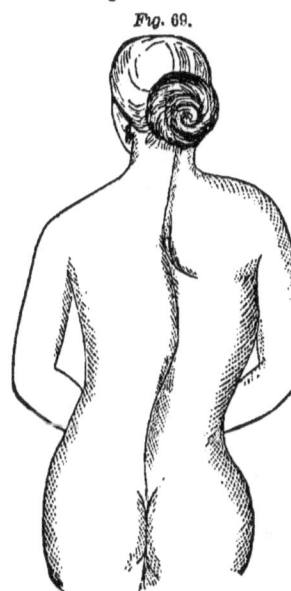

Fig. 69.

These two stages frequently exist without attracting the attention of the associates of the individual thus affected, and the ailment advances to the third stage, being then, probably, first discovered by the dressmaker, who observes the unequal formation of the hips and chest in the attempt to fit the dress. This unfortunate oversight often precludes the possibility of effecting a perfect cure; the two first stages, other circumstances being favorable, are susceptible to successful treatment.

In the third stage we have recurvation or the compensating curves very marked. The spinal column is then forced into opposite directions to the sustaining of the vertical bearing.

By this adaptation the erect position of the person is maintained. As soon as a part of the vertebral colum deviates from a perpendicular, another portion institutes a curve in an opposite direction; shortening the body, and causing additional curves. For this reason, in reality, a single curve never exists; more commonly two are found unaccompanied by a third. Three are very common, and four occasionally met with as seen in the accompanying engraving.

Here we have four curves. The first the dorso-lumbar, maintained by the sacro-lumbalis, and longissimus dorsi; central curve, the same muscles with the spinalis and semi-spinales dorsi; and the cervical curve, the complexus cervicalis ascendens, and transversalis colli. In the dorso-cervical region, two curves, in extraordinary cases, are definable; and the result an exaggerated condition of the contortion, from the effort to maintain the head in an erect position. See Nos. 1, 2, 3 and 4, in Fig. 70.

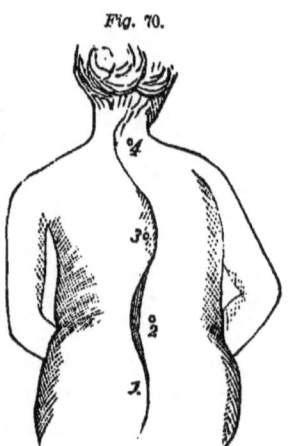
Fig. 70.

TREATMENT OF LATERAL CURVATURE OF THE SPINE.

It is the general opinion of medical practitioners that lateral curvature of the spine is less amenable to treatment than contortions of the limbs. This impression we attribute to the failure to discover the ailment in its incipiency; and that, when advice is sought, it has advanced to the third stage; requiring much practical skill to relieve the patient, even in a moderate degree, from the pain and inconvenience naturally resulting from so lamentable a condition.

The first effort should be to obtain a satisfactory diagnosis of the indications that present in the abnormal developments, and condition of health of the patient; and to ascertain whether any inimical influence is tending to the increase of the ailment; such as occupation or habits of unfavorable position of the body. The treatment should consist of a proper regime associatied with special exercise of the muscles of the body. The latter may be obtained most readily in the use of the parallel bars.

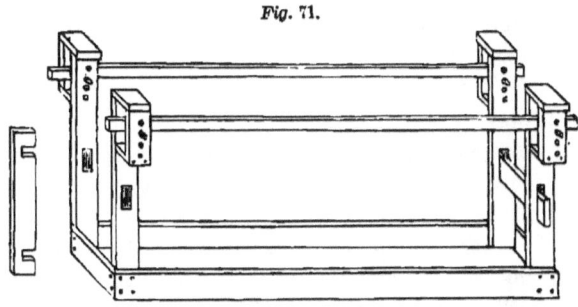

Fig. 71.

The patient being relieved of all clothing that would in the least restrict muscular exercise or freedom, and having entered the frame the longitudinal bars should be elevated so as to only, at first, permit the person to sustain the upright position upon the extreme ends of the toes; the body and lower limbs being, as nearly as possible, suspended by the vertical support of the arms upon the bars. This position having been attained, the patient should then be directed to flex the legs to a right angle with the body, and to keep up an alternate extension and flexion of the body and legs until fatigue gives notice of the necessity of rest, when a suitable elastic support of equal elevating force should be applied to the body while in the suspended position. The supporting apparatus we have devised for the treatment of the first stage of the ailment is represented in

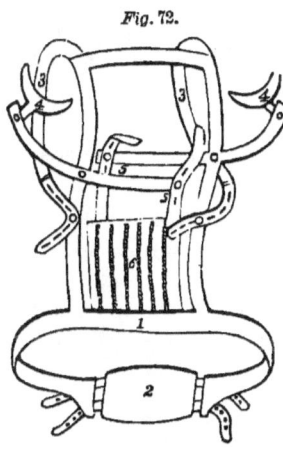

Fig. 72. Represents a back view 1, a steel belt to rest upon the pelvis; 2, a front plate of thin metal covered with soft material and having straps attached which enclose the steel belt above the hips and support the pendant portion of the abdomen; 3, shoulder straps to be passed over the shoulder in front, brought back, and secured to studs riveted into upright steel bars that support the back; 4, axilla supports, having free movement in their attachment to curved, horizontal lever bars, with studs, 5 5, on each, to which are but-

toned straps fastened to wire spring cloth, and by which elastic extension is made; 6, the wire-spring cloth firmly secured to the steel belt, 1. This apparatus affords simply elastic lateral extension to the dorsal muscles, admitting also of increasing elastic force that can be adjusted to the elevation of the depressed shoulder. By this means the enfeebled muscles are assisted, and an equilibrium of action established, without, in the least, arresting normal expansion of the chest, being a most admirable therapeutic agent in the treatment of incipient cases of lateral curvature of the spine.

From the exercises on the parallel bars, followed by the application of the elastic support, the patients having improved in general health, in ordinary cases, will in a few months be enabled to maintain, for a limited time, the normal position of their back without support. When this improvement has been attained, cold sponging, or the douche, may be applied to the back once a day, a most salutary means of perfecting the cure. Persons thus predisposed must avoid all sedentary pursuits, and avail themselves of active, out-of-door exercises, so as to invigorate the muscular system, which, when improved in tone, will maintain their normal integrity.

Cases of lateral curvature, or the first stage of contorted spine, may be favorably influenced by maintaining an appropriate position, which can be obtained by having a chair so constructed that the child's elbow, of the contracted side, shall rest upon an arm sustained upon coiled wire springs, of sufficient height and appropriate form, tending to elevate the shoulder, and thus extend the contracted muscles. In addition to the elevating arm, the seat of the chair, consisting of two portions, one over the other, that will admit of being separated by means of a wedge or screws, may be arranged so as to elevate one side or the other, as may be desired. See Fig. 73.

This is a valuable device, and one most favorably considered by Mr. Richard Barwell, in a treatise published in London, in 1868, who states therein, page 110, that, "When the patient sits on this stool, with the feet stretched out in front so that they do not influence the trunk, and when the end on the convex side (of the patient) is slowly lifted, one observes the following changes: Firstly, and previous to any perceptible change in the lateral bend, the lumbar vetebræ begin to relinquish their tortion — to untwist themselves; the parts on the convex side become less

LATERAL CURVATURE OF THE SPINE. 183

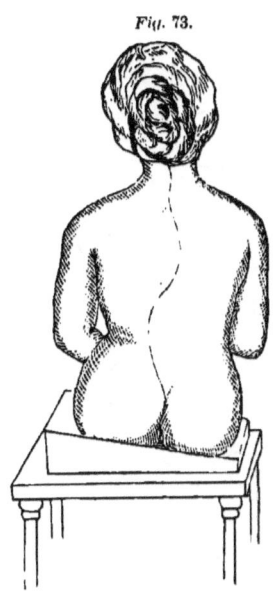

Fig. 73.

hard, those on the concave side more so and the transverse processes sink deeper — are not so evident; the lateral inflexion then also becomes affected, and in all but severe cases it will disappear." This is only auxiliary, however, to the treatment in this class of ailments. Constitutional, remedial means must be very carefully prescribed even to give promise of relief; and spinal supports, when they can be endured, must be worn.

The first and most important consideration is the condition of the patient's health, not the contortion, which is of minor importance to this, and cannot be successfully treated whilst the patient is laboring under the pathological condition that induced the aberration of form. The skilled practitioner in medicine is the most competent person to give the primary relief. Judicious regimen and auxiliary therapeutical means may be made available to advantage; as that of calisthenic exercises and sustaining support to the spine. Careful avoidance of all pursuits requiring an unfavorable position of the body. The practice of extending the body, reclining upon the back upon a bench or any firm plane surfaced support, and remaining in that position for several hours at a time, the head not raised above the line of the body, contributes to the means of cure; and, as before stated, the body support only to be removed or loosened when taking exercise on the parallel bars.

Electricity in the advanced stage of the ailment has, in our experience, been of no benefit, and objectionable as a therapeutic agent, though it has been highly recommended by some writers because of its tendency to maintain, if not increase, the tensity of relaxed muscles on the prominence of the curves on either side of the back. However carefully it may be applied, there is great difficulty in confining it to the elongated muscles; a serious objection to its use in all cases of impaired muscular equilibrium. The spinal support to be worn in the advanced stage of the ailment, is required to encircle the body and afford decided lateral support to the prominent ribs.

It will be observed that lateral pressure is obtained by the lacing in front, and the advanced shoulder is controlled by the shoulder straps. This lateral pressure on the protuberant ribs and advancing shoulder, maintains the extension of the body obtained from exercise on the parallel bars, by the lacings of the textile fabric of the fronts, which only cover about one-third of the circumference of the chest. The steel bars and

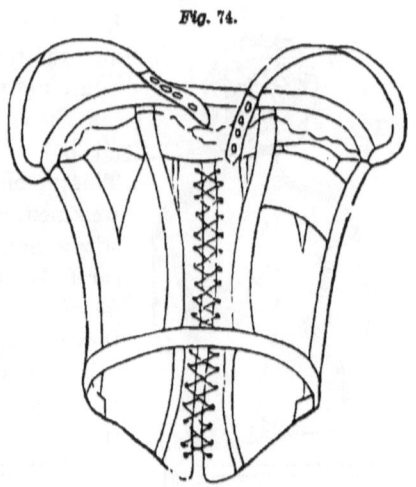

Fig. 74.

plate, restrict and tend to restore to normal form the salient parts, thus tending to the unfolding of the heavy encasement, and giving freedom to the lungs. This gives relief from pain in the side and chest, and impeded respiration. The patients will, in nearly every case, express great satisfaction in the relief that has been afforded them, and this relief alone is commensurate with the wearing of this support in the confirmed cases of the ailment.

No objection can be offered to the support, as to the wearing of corsets, as it cannot compress the chest because of the irregular contour that limits the pressure.

TENOTOMY AS A MEANS OF CURE.

This adventurous procedure of dividing the spinal muscles was introduced by the eminent orthopædist, Jules Guerin, of Paris, and is one, we believe to be, of no advantage in the treatment of confirmed cases. And, as to the first and second stages of the ailment, they are as readily cured without tenotomy as with, being so determined by the indubitable authority of a committee of eminent surgeons. Carefully conducted exercise tending to extension, and maintained by suitably devised apparatus, is required if the tense muscular fascicula have been divided, and, as it is impossible to divide all the muscles in the compensating curves, there is but a very little compensation afforded in the division of a few tense bands.

The patient is placed in such a position as will render the retracted muscular fasciculus (as it is the hardest of the fascicula that present) most tense — usually laid upon a table or bed, back upward, and directed to raise the head, which action brings the dorsal muscles into play, and the retracted fibres are made tense. A fold of skin is then pinched up at the outer edge of the extended fasciculus, and a puncture being made, a strong tenotomy knife is introduced flatwise at its base, at a point which will recede to the distance of an inch from the external border of the muscle. The knife being turned upon the mass, the fibres are divided by a slow sawing movement.

The contraction, we are informed, in some cases, after six or eight months from the time of the division (having been subject, in the meantime, to mechanical treatment), becomes again so decided as to require severing once more to complete the treatment.

Guerin, of Paris, was the first to attempt dorsal myotomy in the treatment of lateral curvature of the spine, and to greatly extol the practice; thus inviting the attention of members of the Academie de Médécin, in Paris, to the investigation of its merits.

On the 12th of November, 1844, after a most excited discussion in the Academy, the matter was referred to a committee. Velpeau and Roux, of this committee, in due time, read a most elaborate report upon the subject, having carefully investigated the matter in all its bearings. This extraordinary scientific investigation resulted in an exposé of the actual anatomical deviations, tending to the true pathological indications to be met by treatment, which has since been carefully considered and more fully determined, and has resulted most favorably to this afflicted class, by introducing a more scientific treatment.

Probably one of the most serious difficulties to be encountered in the treatment of lateral curvature of the spine, in the advanced stage of the ailment, is indiscretion on the part of the patient, in failing to perseveringly pursue the treatment. The patient having been afflicted for several years, and approaching the age for introduction into society, the temptation to indulge in social enjoyment is so strong that temerity often gets the ascendancy of prudence, and all restraint is placed in abeyance. This indulgence is so frequent as to seriously interfere with the treatment, and is eventually abandoned because of the lack of apparent improvement.

The regimen to be enforced is not essentially different from that of the first stage. The frequent use of the parallel bars, and constant wearing of the spinal supporter, day and night, contribute most largely to the cure.

TORTICOLLIS.

There are various causes that tend to this distortion, mainly the result of the retraction of the sterno-cleido mastoideus muscle. Bouvier speaks of a condition of wry-neck caused by inflammation of the fibrous tissue of the cervical vertebræ, and terms it *articular torticollis*. It is either acute or chronic. Contortion results from long-continued effort of the patient to relieve the tense and painful ligaments by displacing them in the direction which the head ultimately retains; abscesses and cicatrices in the cervical region; tumors and glandular engorgements; or, paralysis of the muscles of one side — the head yielding to the same for want of antagonizing force.

A remarkable feature in this ailment (as in that of lateral curvature of the spine) is, that the deviation is most frequent to the right side.

The appearance of the patient is that of having the head rotated, and depressed upon the shoulder of the affected side, the chin, apparently, elevated on the extended side, and the whole features of the face changed. In the region of the sterno-cleido mastoid muscle a firm cord is felt, which is made more prominent if force is applied in a direction opposite to its action. A very common cause of this deformity is paralysis, which is readily diagnosed by the lifting of the head with the hands. The next most common is inflammation of the cervical fascia from exposure to cold drafts of air; the fierce pain retaining the head in a contorted position. Extensive burns tend largely to the production of wry-neck, and are actually irresistible in the cicatrization of the denuded surface.

TREATMENT OF TORTICOLLIS.

In torticollis, except in the incipient stage, the whole vertebral column becomes more or less implicated, and requires a somewhat similar treatment to that of lateral curvation of the spine, namely, extension support, as in that of the second stage of lateral curvation. To the spinal supporter (Fig. 74) is attached a steel bar extending upwards and over the top of the head, to which is attached a cross-piece of three inches, curved to conform to the head, and made to rotate upon a swivel. To this cross-piece supporting straps are attached to elevate the chin; and, with the assistance of a supporting strap to the occiput, this tends to elevate the head and extend the neck. This form of apparatus is also useful in the treatment of caries of the cervical vertebræ, and is represented in the chapter in which that ailment is treated.

Manipulation of the contracted muscles is of much value in the treatment, and, if from a paralytic seizure of the muscles, the application of strychnine ointment, twelve grains to the ounce. Electricity, also, carefully applied to the relaxed muscles will restore their contractile ability. The main reliance, however, is in constant extension of the neck. The division of the contracted muscle is only a very limited auxiliary to restoration of the normal condition; but by patient perseverance the ailment is curable by means of properly constructed mechanical support.

DIVISION OF THE STERNO-CLEIDO-MASTOIDEUS MUSCLE.

It is nearly two centuries since the sterno-cleido-mastoid was first severed for the relief of wry neck. The operation consisted in first dissecting out the muscle, then dividing it on a grooved director. This mode of operating is at the present day practiced by some surgeons in preference to the subcutaneous, apprehending danger of wounding some of the large blood-vessels of the neck.

The subcutaneous method is now the most commonly adopted, being the least painful, and, with ordinary care, not in the least dangerous; each surgeon differing slightly from others both in the management of his patient, and in the division of the muscle. A common procedure is to have the patient seated, the head thrown back and to the side opposite the contracted muscles, and so held by an assistant, while a second assistant depresses the shoulder of the affected side. This arrangement renders the muscle prominent. The operator then lifts the muscle with his thumb and finger, and passes beneath it a small curved tenotome at a short distance above the sternal origin. When the point is felt beneath the skin on the opposite side, the knife is made to slowly and carefully divide the muscle by a slow movement. As soon as the knife is withdrawn, the blood should be pressed out, and the puncture closed with adhesive plaster. A compress and roller complete the dressing; and special attention must then be given to diet and repose. The wound commonly heals by the third day, when the supporter and extension bar should be applied, and the patient permitted to sit up, and, in two weeks' time, to walk about.

The section of the sternal origin of the muscle suffices in some cases, but it is quite as well to divide both fascicula. It is usual to sever the most prominent of the two if intending to divide but one, to test the tenseness of the other, and to wait a few days before dividing the other. This procedure is suggested by Bonnet.

CHAPTER VIII.

RACHITIS.

Osseous growth and development of cartilage cells the first indication of its presence.— Dr. Rindfleisch's description of symptoms.— Kölliker the first to observe the transition process of cartilage to bone.— Peculiar formation of the joints in rachitic subjects.— " Double-jointed " persons.— Anterio-posterior curvature of the spine a common indication of "rickets."— Usually confined to children under one year old.— Treatment of anterio-posterior curvature.— More decided treatment, with mechanical support, necessary for children of advanced age.— Cod-liver oil as a curative medium.— Bismuth an effective remedy.— Projecting sternum.— Dupuytren on its origin.— Description of the patient's condition.— The effect of the chest, impairment on the lungs.— Treatment of projecting sternum.— Truss for projecting breast.— Genu extorsium, or bow-legs.— Often cured without instrumental aid.— General treatment of genu extorsium.— Various apparatus.— Genu valgum, or knock-knees.— Diagnosis in children.— Mr. Tamplin's theory of the deformity.— Treatment of genu valgum.— Mechanical support the best remedy.— Apparatus.— Efforts at walking of great utility in treatment.

Rachitis, the abnormal condition of the formative process in the growth of bone, tends to great aberration of the normal form of the skeleton, and as a pathological condition invites special investigation from the practitioner in medicine. The first impairment presents in the osseous growth and development of the cartilage cells by an increased number of divisions. Dr. Rindfleisch states, in his valuable text-book of Pathological Histology, that from ten to twenty and more layers of cartilage cells unite simultaneously upon the proliferating process, and these again divide, forming cell groups of thirty to forty elements, which are placed in long drawn columns somewhat bent and compressed, here and there, by the material flattening vertically toward the surface of the bone, while upon the normal bone we scarcely observe the proliferating zone of the cartilage with the naked eye, except as but an exceedingly narrow, reddish grey stripe ; upon the rachitic bones, it expands as a broad, translucent grey, and very safe cushion between the cartilage on the one side, and the completed osseous structure upon the other.

That which takes place in the cartilage is repeated in the periosteum. The young, vascular, germinal tissue which is produced by the periosteum at its surface lying toward the bone, under normal circumstances, only presents a thin stratum that is scarcely perceptible to the naked eye. The rapidity with which it, though just produced, is transformed into bone does not permit a large accumulation to take place. It is otherwise in rachitis; under the influence of the morbid process this transition substance often accumulates in very great layers of a line in height. The great abundance of the thin-walled and wide capillaries occasions the coloring of the young connective tissue. The deposit, as a rule, is drawn in the form of a broad, flat, bed-like elevation over the surface of the bone. Upon the flat bones of the cranium they are sharply circumscribed, which is not the case, to such an extent, at the extremities. Kölliker was the first to observe the transition process of cartilage to bone in rachitic subjects. The cartilaginous bone consists of largely separate medullary spaces which are considered by Rindfleisch as not to be regarded as calcified cartilage, but as osseous tissue; and he further remarks: "It is true we cannot doubt that, excepting the impregnation with lime salts, only a very moderate change of position of the single cells has led to the transformation of the proliferated cartilage into the texture in question."

The amount of calcification may equal the normal quantity in the formation of bone, but in rachitis it is distributed over a very much larger space than in normal circumstances, leaving the osseous tubercle very thin and the transformation of the sub-periosteal exudation in an exceedingly loose, vascular osteophyte, covering the bone in an undue thickness and larger than in the normal condition — compact bony substance being retarded for an indefinite time.

Rachitic aberration of the normal form of the bones of the skeleton is supposed to arise from mechanical force, and mainly from the broad layers of the proliferated cartilaginous tissue which passes in between the epiphyses and the diaphyses of long bones, swelling them out laterally and forming *roundish rolls surrounding the bone.*

In this peculiar formation of the joints in rachitic subjects, they are said to be "double jointed"—*articuli duplicati*. And the swelling of the collective costal cartilages at their points of contact with the bone is termed "rachitic garland," and presents in the articulations of the ribs, tending to what is known as "chicken breast"—*pectus gallinaceum*, the sides of the chest being drawn inwards by respiration. Deformities of the pelvis are attributable to the great mobility of the cartilages of the synchondroses sacroiliacæ. The sacrum, under the weight of the head, trunk and arms, is pressed down upon the yielding synchondroses, forcing the promontory of the sacrum, forward into the pelvic strait; narrowing it from above, and tending to the kidney shaped configuration.

To the abnormal periosteal growth may, principally, be attributed the curvatures and fractures of the shafts of the long bones. The growth of the bone in thickness, the continued opposition of compact substance at the periphery is constantly and everywhere accompanied by resorption of the compact substance at the inner surface toward the medullary cavity,— a resorption which keeps pace with the opposition, and which, also, does not stand still during the rachitic process. The rachitic condition interrupts the peripheral opposition of the compact substance; the consequence is a decrease of thickness of the already thin shaft, and which is not sustained by the osteophytic layers. The result is that the bones bend, and often break, producing great deformity, and lessening the stature from curvature of the bones of the legs.

Anterio-posterior curvature of the spine is a very common indication of rickets,— a uniform curving of the spinal column posteriorly from the cervical to the last lumbar vertebræ; the entire spine being implicated in the aberration.

The ailment is most commonly confined to children under the age of twelve months, laboring under constitutional debility, and to failure of the dorsal muscles to support the spinal column. The weight of the head and upper extremities inclining forward makes an undue pressure upon the anterior portion of the inter-vertebral substance, tending to elongate the posterior ligaments of the spine. In adults, this becomes a confirmed condition and the spinal column so fixed as to limit the head to such a dependent condition as to entirely deprive the individual of the privilege of looking upward. These, however, are very rare cases, commonly the sequence of inflammation that has induced permanent contraction of the abnormal muscles. A strumous diathesis predisposes to this ailment in adults. In children we do not know of a case in which this form of curved spine has become a permanent deformity.

In this ailment, a seeming concomitant of struma, the patient presents a pallid and anxious countenance, attenuated limbs, greatly enlarged joints, a voracious appetite, tumid abdomen, symptomatic of mesenteric disease, the alvine evacuations clay-colored, the intervertebral is more or less absorbed from undue compression forward; the ribs are prominent and project posteriorly more than is natural, and the scapula becomes raised and appears, upon examination, to be separated from the thorax. The head and neck appear to be sunk between the shoulders, and, on examination of the spine, the processes will be found very prominent,—more especially in the lower dorsal region; presenting to the inexperienced practitioner an embarrassing aspect as to a correct diagnosis and one that has been mistaken for caries, commonly known as Pott's disease of the vertebræ.

TREATMENT OF ANTERIO-POSTERIOR CURVATURE.

Infants of about six months of age, and dependent on their mothers' breast for nourishment, should be given a teaspoonful of beef tea once or twice daily. As a medicine, we have found the sub-nitrate of bismuth to be the most remedial in tendency, if continued for months, of any other within our knowledge — three grains of bismuth rubbed up with gum acacia, and given morning, noon and night. The body should be encircled with a flannel belt from arm-pits to hips, and firmly secured; making atmospheric exposure available as much as prudence will allow, and close, hot rooms carefully avoided. The child's body and limbs should be frequently bathed with a wet sponge, and chafed dry with the warm hand, and, if before an exposed fire, so much the better. For bathing, spirits of any kind should be rejected as pernicious, the evaporization robbing the child of its wonted normal heat. The same may be said of its use upon delicate adults. Older children require a more decided treatment, as that of bitter tonics, an increase in the dose of bismuth, and a liberal diet carefully limited to stated periods. Mechanical support will also be required in these advanced cases, as that of the spine and head supporter used in the treatment of caries of the cervical vertebræ.

The patient, when in bed, must dispense with the use of a pillow, and make efforts to recline upon the back. This will, at first, be both painful and difficult, but after a time, with perseverance, the desired object will be attained without inconvenience, and greatly to the patient's benefit. This treatment applies to all rachitic ailments. Cod liver oil has been greatly extolled as a valuable medicine in the treatment of rickets; German and French physicians consider it as the most reliable in the treatment of the ailment. Our experience has not been so favorable, having, even in small doses given to patients, discovered the undigested oil in the alvine evacuations. Bismuth, in a measure, corrects this tendency, but serves an equally efficacious purpose when administered with a nutritious diet, as the promotion of the digestive process appears to be the most certain means of cure.

PROJECTING STERNUM.

This is a very common ailment, tending to serious impairment of the general health of the individual thus afflicted, from its limiting the capacity of the thorax by lateral pressure. Dupuytren advances the theory, that it is the sequence of deficient stamina; the offspring of lymphatic, scrofulous or rachitic people, dwelling in low, cold and damp places, and brought up on unsubstantial food. The deformity is, in some cases, congenital; in others, it occurs during childhood, or later. Dr. Copeland states that his experience leads him to believe that it generally comes on gradually after birth, owing to deficient inflation and development of the lungs, arising from weakness of the muscles of inspiration and the flexibility of the ribs at the time of birth. He, also, says that, in cases of this description, the vital energy of the lungs is insufficient for their healthy action, and the respiratory mechanism unable to accomplish their full expansion or sustain the continued pressure of the atmosphere, before which the soft and imperfectly formed thoracic parietes gradually yield. And that it has appeared to him very frequently, to be greatly increased, if not occasioned, subsequent to birth, by the very common practice among nurses of lifting a child by pressing the palms of the hands on the sides of the chest immediately under the armpits. That both of these most eminent gentlemen, in their profession, are correct in their views as to the predisposing tendency to the aberration, is most rational, and agree perfectly with our own experience. The condition of the patient is as follows: The sides of the chest are very much flattened, one side sometimes more than the other; the ribs occasionally appear as if straightened, and the sides form an acute angle, one with the other, the whole serving to give to the sternum, undue prominence, which deformity is known as "pigeon breast." See Fig. 75.

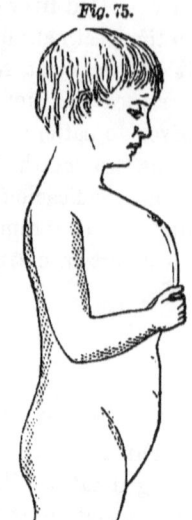

Fig. 75.

The unfavorable condition of a chest thus contracted must seriously impair the functions of the lungs and the impairment is in ratio to the diminished capacity of the chest. From the depression of the sides of the chest the heart's action is embarrassed, and is observable, strongly pulsating against the ribs; the circulation turbid, the breathing quick and often difficult, being performed through the mouth, a wheezing, and short, dry cough; the nostrils obstructed by enlarged tonsils, an almost constant accompaniment in this ailment, more clearly defining the primary cause of the deformity — constitutional deficiency, a true rachitic condition.

TREATMENT OF PROJECTING STERNUM.

Again we must refer the constitutional treatment to the skilful practitioners in medicine, as it is wholly unnecessary to even hint at a treatment that is already familiar to them. Not that we would intimate any incapacity on our part of presenting a judicious course of treatment, but that we consider it would be presumptuous, and digressing from the tenor of this work, which is mainly intended to introduce therapeutic agents not in common use as auxiliary to the relief of resultant deformity, such as lateral curvature of the spine and projecting sternum, the sequence of enfeebled organic functions.

It is, however, important to consider the influence of the deformity upon the impaired organs and the interference in their functions tending to the impairment of the general health of the patient. It is reasonable to suppose that a great compressing cover like this, oppressing the centre and source of the circulatory and respiratory system, should produce an infinity of functional derangements. Hence, to the relief of the patient, an effort must be made, in the incipient stage of the ailment, to arrest and prevent an increase of the deformity, and, in established cases, to overcome the abnormal condition by whatever means the practitioner may consider the most favorable. Mechanical appliances are objected to by many surgeons for various fancied reasons. Our experience impresses us favorably toward their use, as we have found them to be effectual in restoring the chest to normal form, in cases of long standing, and patients are invariably benefited by their application. It is to be regretted that improperly constructed apparatus should have tended to discourage their use, and this is wholly the fault of the surgeon because of his reliance upon the instrument-maker.

The construction of more skilfully devised surgico-mechanical appliances should be within his sphere of professional knowledge, but is too frequently referred to the mechanician.

Fig. 76 represents our truss for projecting breast. A semi-circular steel spring, having attached to either end cushioned plates; to the back a circular pad, and to the front an oblong pad which is slightly concave, to coaptate with the sternum; the back of this pad having two studs at suitable distance to receive the encircling straps that retain the truss and pads *in situ* when properly applied. By means of this apparatus we obtain elastic compression of continued force concentrated upon the projecting sternum from counter-pressure on the back. The elasticity of the encircling belt and spring admits of lateral expansion of the chest, the principal force being concentrated upon the two extremities of the steel springs, and their tending approximation being resisted by the anterio-posterior extension of the yielding body.

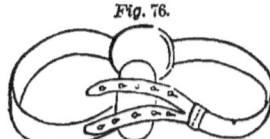
Fig. 76.

The eminent Dupuytren's practice was as follows:* "placing the child sideways and the hand, or knee, against his back, or, still better, his back against the wall, and placing the palm of the other hand upon the most projecting part of the sternum, and in pressing or pushing the anterior part of the chest toward the posterior part by alternate movements, which, after some days practice, accords so well with the movements of respiration, that the little patients and

---

* " Repertoire d'Anatomie, V. p. 198.

those who exercise the pressure, soon learn to exercise it during the time of expiration and to suspend it, so as to allow the breast to develop itself during the movement of inspiration. During these movements, a sound is heard similar to that made by the air in alternately entering and escaping from a bellows.

"I have often attentively observed the immediate effects of this exercise. Their effects are a flattening of the projection of the sternum, a greater or less tendency outward of the ribs, the momentary return of the chest to a more natural shape, respiration much more strong and perfect than in general, and, when the pressure is removed, the *immediate return* of the parts to their ordinary state accompanied with a strong inspiration.

"These pressures should be repeated ten times — a hundred times a day, if it were possible, and continued for several minutes each time; their efficacy will be in proportion to their frequency and duration."

This experienced surgeon also gives some additional valuable advice : " The practicing of these pressures must not be indifferently consigned to any one. A mother's affection, alone, is capable of the perseverance requisite for success; with this ally, there is scarcely any malformation of the kind we have described that cannot be remedied; and I have seen children who were dreadfully afflicted, become eventually strong and well constituted." Indifference to the advice of the surgeon is the most common cause of failure to cure chronic ailments of whatever character; but more especially deformity of limbs or body.

The constant vigilance and actual labor required in treatment, as advised by Dupuytren, would not be readily obtained from even the majority of mothers. The great difficulty, and almost impossibility of obtaining this extraordinary attention upon a patient to effect a cure induces a more favorable consideration of treatment that can be made readily available, and to the cure of the patient, with as much certainty as the above almost impossible requirement, which if not judiciously practiced must tend to the compromising of the patient's life from undue pressure upon the chest.

GENU EXTRORSUM.

Bow-legs are among the curable contortions occurring among children whose constitutions are defective — mainly rachitic, but invariably fleshy and clumsy in appearance; the body rocking from side to side each step that is taken, and the curved leg yielding like a spring, the vertical bearing defective. Hence, it is commonly said, that the child's legs were straight when born, but became curved from its weight when it commenced to walk—which was at an advanced age—the child, also, having been late in teething. See Fig. 77.

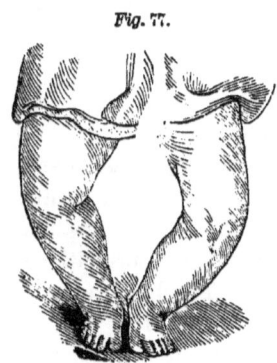

Fig. 77.

Represents the deformed legs. The tibia and fibula of the legs are curved in a lateral direction—in some cases to an extraordinary degree, commencing from the external malleolus, which in some children nearly touches the floor. The legs diverge from each other in one continuous curved line from the hips to the ankles. The femur, however, is but seldom implicated in the curve, other than adaptation to the lower curve. True diagnosis of the condition of the limbs is of much importance in the treatment of bow-legs in children.

In many cases they can be cured without instrumental aid; the lateral divergence of the legs resulting from a relaxed condition of the ligaments about the joints (the bones having maintained their integrity), and, as the child increases in strength from a prescribed regime, or, improves in strength without interference, the deformity disappears.

It is this favorable tendency that disparages treatment in the early stage of the ailment, often to the patient's great disadvantage. If, after having arrived at the age of six or eight years, effort is then made for their improvement, the patients will not only suffer severely, but their health will be endangered thereby, because of the force required to overcome the curved and hardened bones.

If, on examination, the tibiæ are found to curve laterally, and the knee-joints to maintain their integrity of adaptation, and, when efforts, in some form, are made to straighten the leg, the limbs maintain a persistent resistance to the force; the earlier instrumental force is applied the less inconvenience and pain will ensue to the patient, and this treatment they must have, or remain bow-legged through life.

It occasionally occurs that one limb will curve laterally, and the other posteriorly; presenting the appearance of one leg being straight and the other curved, when the patient is standing in front of the observer. The parents, of the child so impressed, are, in such cases, very apt to insist that only one leg requires treatment and this arises from the fact that a very decided curve posteriorly, in the leg below the knee, is not nearly so apparent as the lateral deviation. Both are equally amenable to relief when suitable appliances are devised to meet each indication.

TREATMENT OF GENU EXTORSIUM.

The child laboring under relaxed ligaments of the joints will be, most commonly, found to have a tumid abdomen and an enlargement of the ankle, knee and wrist joints,—indicating the rachitic diathesis of the patient, and requiring constitutional treatment for this ailment, which may be left to the family practitioner. By this means a cure in some instances can be effected in the course of time, but in much less time by the auxiliary aid of suitably devised light steel supports, although these are much objected to by eminent authority,— which objection we believe to be founded upon an erroneous impression, gained from the observation of defective apparatus devised by the instrument makers and not that of the skilful surgeon. Support tending to establish a vertical bearing upon the limbs would surely tend to relieve the relaxed ligaments and establish an equilibrium of muscular tension that cannot be otherwise than a therapeutic aid of considerable merit.

In cases of lateral deviation of the shafts of the tibia and fibula, while the knee joints maintain their integrity, there is muscular retraction on the inner side of the legs, to which force the bones have yielded; whether abnormally soft or not, we believe them to be in a pathological condition, and the indications can only be met by mechanical support and that most readily; as the long bones of children are very flexible and can be easily overcome by moderate, continued force to the improvement of form. That there exists a faulty functional condition of the child thus afflicted is quite apparent from its condition. Nervous irritability limited to certain muscles would tend to induce this condition of the legs, without impairing the function of nutrition; as children thus conditioned are usually, as before stated, very fleshy, and this irritability having subsided, mechanical treatment, alone, is indicated.

Fig. 78. Represents our apparatus applied. It consists of a straight bar of steel extending from the internal malleolus to the inner condyle of the os femur, cushioned at either end, and secured

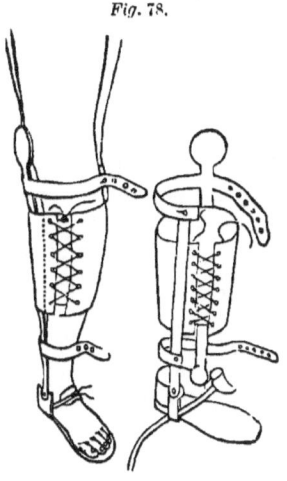

Fig. 78.

by semi-circular steel bands and straps to an outer bar, limited in length — as seen in the apparatus not applied — between which a broad belt of strong woven fabric is made to encircle and lace in front, passing over the inner bar and inside of the outer one. This has a direct broad bearing against the outer surface of the leg, and, when laced, tends, from the counter resistance of the inner bar, to straighten the leg. The outer bar and encircling bands sustain the appliance *in situ*. The apparatus shown as not applied represents the form of that applied to the leg in which the outer bar is not seen because of the position of the leg. As before stated, the broad belt is passed free *within the outer bar* and, as seen, *over the inner*. It will be observed that there is an instep pad and tapes that secure the foot to the sole-plate and heel-cup. The latter should be made the size of the foot, so as to be worn at night — an essential necessary to the obtaining of a speedy cure.

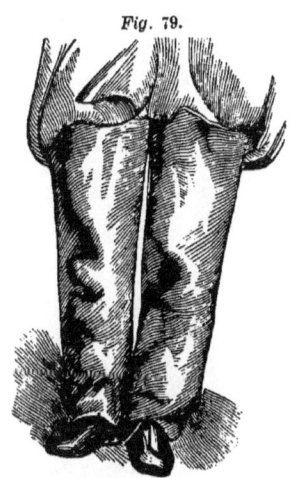

Fig. 79.

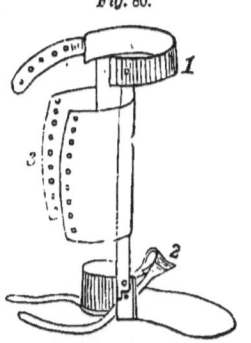

Fig. 80.

Fig. 79. Represents a child clothed and having the braces on.

Fig. 80. Represents the form of the brace when applied for the cure of the posterior curve of the tibia and fibula. 1, a steel band to be applied over the head of the tibia. 2, the instep pad that secures the foot in the heel cup. These two points of support maintain the fixed position of the leg for counter-pressure which is made by the lacings (3) over the gastrocnemius muscles tending to straighten the leg. With this form of apparatus we straighten the irregular posterior bending of the bones of the leg.

The legs should be examined twice a day, and tender points protected with ravelings from worn linen fabric. This is much superior to lint, which will become impacted and hard when firmly compressed. Bathing the limbs frequently, and manipulating them with as much pressure as the patient will bear is of decided advantage, besides affording actual comfort because of the support given to the yielding leg when walking.

GENU VALGUM.

In this difficulty, as in many others, we have indications of an enfeebled constitution. Knock-knee, the reverse of bow-legs, is a subject of study, for the condition that has tended to one or the other is very obscure and not definable with a comprehensiveness that would decide the pathological condition of either. In the bow-leg aberration there is an apparent deficiency in osseous pabulum, and, from the attenuated condition of the patient having knock-knees, a deficiency in muscular nutrition.

In the incipient stage of this ailment in children, there will be found great mobility of the joints; and, if the knees are slightly inclined inward when the child is standing they will become apparently straight when lifted off the floor, or, after the inclination has assumed a decided form, the legs can be extended quite straight with the hands without giving pain to the child.

Fig. 81. Represents a case of knock-knee in a child.

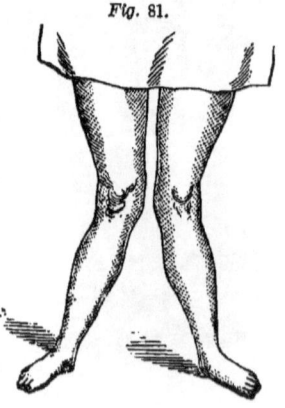

Fig. 81.

Children having knock-knees are, usually, greatly attenuated in muscular development, presenting a most delicate appearance compared with that of children having bow-legs. Their general health is greatly impaired; the secretions unhealthy; the appetite, in some cases, voracious, and in all precarious; the surface dry and contracted; the abdomen large, and the anterior spinous processes protuberant. This ailment is not limited to infants, but children of ten, twelve, and fourteen years of age are alike subject to the deformity. In some cases the patient will have knock-knee of one leg and a posterior inclination of the other; and, if not relieved by mechanical support, finally become unable to walk from its increasing tendency.

Mr. Tamplin states in his lectures that this is not a congenital deformity, — at least, he had never seen a congenital case nor could he imagine how it could occur provided we admit that congenital distortion arises from malposition in utero. We have treated two cases of congenital knock-knee, and, what was peculiar in both, the feet were turned inward and upward; being cases of exaggerated talipes varus conjoined with genu valgum. In ordinary cases, more especially of young children, valgus is the most common concomitant.

TREATMENT OF GENU VALGUM.

In the treatment of children we have never availed ourselves of myotomy, or found any necessity for it; the cases being, usually, readily redressed by mechanical supports. In cases of advanced age, the severing of the biceps flexor femoris is necessary. In this operation, care must be taken to avoid dividing the peroneal nerve — an unfortunate occurrence that has been the lot of several orthopædists. Paralysis of the flexors of the foot is the result of such an accident, and great alarm is given to patients and friends, though it is said that, in time, the use of the paralyzed muscle is recovered. To avoid the accident, press the knife close under the tendon, and not beyond it, and make no attempt to divide the other tense fascicula which will be observed after the tendon is divided. Immediately after dividing the tendon we apply the steel spring appliance for redressing the deformity; constitutional treatment being implied, as in all cases of like condition.

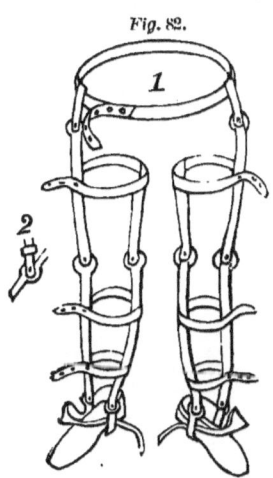
Fig. 82.

Figure 82 represents the construction of the apparatus, having steel soles, and cups in which the heel is to be secured; 1, a steel belt to encircle the body; 2, a joint that can be controlled by a slide — useful in severe cases.

Fig. 83 represents the apparatus applied. Here the strong, outer bar is shown, to which the leg is tending by means of the encircling bands; the inner bars cushioned at the joints bearing against the inner condyle of the femur. Gaiters should be worn over the foot support, to keep it clean and render it unobjectionable to be worn in bed — in fact, not only gaiters, but other clothing, as seen in that used in the treatment of bow-legs, for the braces must be worn constantly, day and night, and no attempt made to stand upon the feet without them. The patient may have them taken off several times during the day, and have the limbs manipulated and bathed, but must always at such times be in a recumbent position.

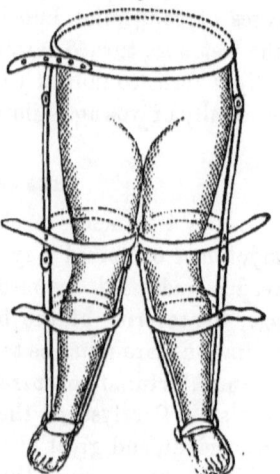

Fig. 83.

Efforts at walking are of great advantage in the treatment, although distressing at first, as the knees will be thrown forward to a half kneeling position, in severe cases, but this very uncomfortable and unavailable effort to straighten the limbs contributes largely to their restoration of normal form, as the patient endeavors to extend the contracted muscles. The slide rendering the joint immovable can only be made available during the night, when it retains the legs in a straightened position, tending largely to the restoration of tone to the weakened muscles and ligaments, the result of the continued extension. When the normal form is attained, and the necessity for extension appliances no longer exists, and the general health of the patient is improved, strength is readily restored.

CHAPTER IX.

HERNIA — PROCIDENTIA UTERI — ECTROPION VESICÆ — RELAXED ABDOMEN.

Derivation of the term.— Formerly defined as "protrusion of any of the abdominal viscera."—Dunglison's definition.— Treatment.— Abdominal Hernia, its Varieties.— Scarpa's Exposition.— More recent authorities on Hernia.— Causes of Hernia.— Varicose Veins and Aneurism.— Symptoms of Abdominal Hernia.—Various Denominations of Hernia.— Divisions of the various Abdominal Hernia.— Strangulated Hernia.— Diagnosis.— Glandular Tumors.— Symptoms, as distinguished from apparently similar affections.— Hydrocele.— Congenital Hernia.— Circocele.— Sir Astley Cooper's mode of treatment.— Femoral Hernia.— Glandular enlargements and Psoas abscess sometimes mistaken for Hernia.— Hernia complicated with a retained testicle.— Essentials to a correct diagnosis of hernial protrusion.— Prognosis.—Liability to stricture.— Possibility of gradual reduction.— Taxis, or reduction by the hand.— Should not be attempted in the case of an enlargement of the omentum or intestine.— Symptoms of Hernia.— Incarcerated Hernia.— Treatment.— Advantage of application of Taxis in femoral Hernia.— Sir Astley Cooper's advice as to treatment in strangulated Hernia.— Statement of Dr. Philip Crampton Smyly, of England, on treatment of hernia.— Testimony of Dr. Thomas Bryant, surgeon to Guy's hospital.— His novel means of reducing Strangulated Hernia.— The construction and application of trusses for the relief of Abdominal Hernia.— There is no *one* form of truss or uniformity of pad adaptable to all the varieties of hernia.— Patients suffering from hernia in an exceptionally perilous condition.— Palliative and preventive treatment, while in a reducible condition, the only proper course.— Usual origin of Hernia.— Predisposition to Hernia evinced by relaxed condition of abdominal muscles.— Forms of truss, single, double and children's.— Rationale of cure of hernia in adults.— The author's truss for the relief of reducible Inguinal and Femoral Hernia.— PROCIDENTIA UTERI.—Pathological condition similar to hernia.—Primary cause. — Symptoms.— Derangement of system consequent upon the ailment.— Escharotic applications only palliative and often injurious.— Treatment of Procidentia Uteri.— Anteversion of the Uterus.— Prolapsus Vesicæ.— Necessary appliances.— Ectropion Vesicæ, or Extroversion of the Bladder.— Diagnosis.— Mode of relief.— Relaxed abdomen.— Prolapsus Ani and Hemorrhoids. — Causes.— Diagnosis.— Treatment.

HERNIA.

In this chapter upon Hernia, we shall not confine the limits of this ailment, as former writers have done, to the three general divisions, the head, chest and abdomen; as analogous conditions exist, and are quite as common in the limbs as in the parts formerly designated. Hence, in addition to these boundaries, we include the limbs; they being subject to similar pathological conditions — as in that of varix and aneurism.

This seeming innovation upon the established limits of hernia is assumed, because of the derivation of the term *hernia*, a branch, that is, a limb or portion to protrude from; including in a pathological sense of the term, all protrusions similar to those usually considered as confined to the limited boundaries of the head, chest and abdomen. Hernia was formerly defined as protrusion of "any of the abdominal viscera from the cavity in which they are naturally contained into a preternatural bag, formed by the protrusion of the peritoneum, constituting a hernia or rupture, according to the most common acceptation of the term."* This definition of hernia is less tenable than that of the limited boundaries before stated; as in protrusion of the bladder, *hernia vesica*, there is no hernial sac, nor is there in congenital hernia within the tunica vaginalis. In Dunglison's Medical Dictionary we find designated: Hernia arterium, hernia venarium, varicose vein; hence, we include these pathological conditions of the limbs under the designation of hernias, requiring somewhat similar therapeutic agents for treatment; considering the term hernia to include all protrusions from within the body and limbs and sustained by superficial integument, being the result of a yielding of the supporting tissues and, when reducible, amenable to relief from the application of surgico-mechanical appliances.

Hernias, in regard to treatment, agree with other deformities; they being also aberrations of the body and limbs requiring for treatment a similar class of agents.

ABDOMINAL HERNIA — ITS VARIETIES.

Scarpa remarks in his *Traité des Hernies:* "No anatomist would believe that the intestine cœcum, naturally fixed in the right ilium, and the urinary bladder, situated at the bottom of the pelvis,

---

* Lawrence's Treatise on Ruptures, and additions by J. Parrish, M.D., p. 3.

could undergo, without being torn, so considerable a displacement as to protrude through the abdominal ring, and descend even into the scrotum; that the same intestine, the cœcum, could pass from the right iliac region to the umbilicus, so as to protrude at this opening and form an umbilical hernia; that the right colon could have been seen protruding from the abdomen at the left abdominal ring, and the left colon through the right one; that the liver, the spleen and ovary could sometimes be the parts contained in the umbilical, inguinal, and femoral hernia; that the cœcum could engage itself within the colon, and even protrude at the anus; that the stomach forced through the diaphragm could form a hernia within the chest; that the omentum, or intestine, or both these parts together, could sometimes make their escape from the belly through the foramen ovale or sacro-ischiatic notch of the pelvis; that a noose of small intestines after being engaged in the abdominal ring, or under the femoral arch, could suffer the most violent strangulation without the course of the intestinal matter being intercepted; lastly, that in certain circumstances the intestine and omentum could be in immediate contact with the testicle, within the tunica vaginalis without the least laceration of this latter membrane."

This very celebrated anatomist carefully obtained these facts, that have been since proved by numerous observations on individuals affected with hernia, and very elaborate accounts given of them by such eminent authority as that of Sir Astley Cooper and Mr. William Lawrence, who have given us the most reliable, modern works upon the subject of hernia, up to 1861, when the Jacksonian prize of the Royal College of Surgeons, London, was given to John Wood, F.R.C.S., for an essay on "Rupture, Inguinal, Crural and Umbilical."

CAUSES.

The causes of hernia have been resolved into two conditions; those that impair the resistance of the abdominal muscles — a relaxed and debilitated condition of tissues, as in varicose veins and aneurism, and those that increase the pressure of the viscera; such as violent exertion or accident.

The first condition may be considered as predisposing to hernia, and is admitted by the most reliable authority to be an elongation of the attachments of the viscera, tending to an undue descent upon the equally unfavorably conditioned boundaries of the abdomen. Whatever influence may tend to impair the tenacity of the system; as that of sickness, arduous and continuous labor, poor diet, and long continued mental depression, may be considered as a first cause, and these unfavorable influences, to a limited extent, reach all classes of people, hence the exceeding prevalence of this almost incurable ailment, abdominal hernia, only amenable to palliative treatment. Hernias are the most common of physical ills to which the human family is liable, and the above stated conditions of the system tend most to their production. The ailment is apparently hereditary in many families, their constitutionally lax fibre predisposing them to these protrusions in various parts of the body and limbs. It is a most common circumstance to find a person laboring under hernia, to be also subject to hemorrhoids and varicose veins.

Causes tending to produce an increased pressure upon the viscera are various but comprehensible. In the powerful action of the abdominal muscles and diaphragm on the viscera in feats of agility, such as tossing an immense weight, and jumping, the pressure upon the contents of the abdomen is sufficient to produce a hernia in the most robust persons, as we have witnessed in many instances, and violence to the person, as in case of concentrated force from a fall upon some projecting point of hard substance or pointed instrument thrust against the abdominal walls — more especially after a full meal — and yet not penetrating the skin, but separating the muscular fibre. If not apparent immediately, a hernia presents in time, and denominated ventral, or adventitious, from its not protruding from the ordinary localities of hernial protrusions.

SYMPTOMS.

The ordinary symptoms of abdominal hernia are, an indolent tumor at some point of the parietes of the abdomen — most frequently protruding out of the abdominal ring, or, from below Poupart's ligament, or of the navel cincture, and, occasionally, from various other situations; the swelling of the integuments, suddenly, and subject to a change of size — being smaller in some cases, and entirely disappearing at times, as when the patient assumes a recumbent position during the night, and increasing when standing up or holding the breath, which is the most positive indication of the existence of a hernia. The viscera, if reducible, will diminish when pressed, and after a time reappear. Fluids follow on the removal of pressure — tumors from indurated tissues are not commonly attended with nausea and sickness, are not reducible. The size and tension often increase after a meal or when the patient is flatulent. Patients afflicted with hernia are subject to colic, constipation, and, often vomiting, in consequence of the displacement of the bowels — or, if replaced after having been protruded in large quantity for some considerable duration of time. The protrusion, in many cases, produces but comparatively little inconvenience to the person, other than exposure to sight when very large.

In ordinary cases, a patient affected with even a small hernia will, when laughing or sneezing, find the tumor instantaneously increase in size, if not supported, or, involuntarily, as it were, pressed upon by the hand. Its tendency to increase in size, in many instances, very rapidly, renders it quite apparent that the mesenteric attachments are elongated and in direct ratio to the quantity of the viscus protruded, dragging, as it were, the most distant viscera (in the normal state firmly fixed by the folds of the mesentery) to the descending hernia and becoming a part of the protrusion.

The contents of an ordinary hernia, in most cases, present distinguishing indications by which the viscus filling the sac may be determined. If, by intestine, the protrusion is denominated *entrocele*, and if by a portion of the small intestines, the tumor is usually small and liable to become painful and tense from any distending cause, and will resist efforts at reduction. However, it is, in most cases, very readily redressed even when large portions protrude. Upon coughing, the intestine will feel as if blown into, and when pressed into the abdomen will be attended with a gurgling sound.

If the hernia consists of omentum it is denominated *epiplocele*, and the contents distinguished by the sensation of pressing a mass of dough, having greater weight when largely protruded, and when being reduced passes in without sound.

DIVISIONS OF THE VARIOUS ABDOMINAL HERNIA.

Hernia has also its local designation; as when the viscera pass directly from the abdominal cavity, taking the course of the spermatic cord, or round ligament, it is denominated *Oblique Inguinal Hernia;* and when it passes directly through the internal abdominal ring, so that its direction from the external protrusion is direct, and not oblique, it is known as *direct inguinal hernia*. When situated over the internal ring in the form of a small tumor, it is named *bubonocele*, and *congenital* when the protruding viscus is found within the tunica vaginalis; *ventral,* when protruding at irregular parts of the abdomen — usually the result of violent penetration of the muscle and not the skin; *exomphalos* or *umbilical* hernia when protruding at the navel. These are the most common varieties of hernia, though there are others — as *vaginal* and *perineal* hernia, and all, if reducible, are amenable to relief — and, in infants, nearly all curable by means of well devised surgico-mechanical appliances.

It is essential to have a practical knowledge of the various conditions of hernia, as they present in different persons, and under every peculiar circumstance, to give even a promise of anything approaching to successful treatment, as it is one of the most intricate pathological conditions to which man is subject. And yet, patients thus afflicted are often, without proper consideration, submitted by the medical practitioner to the uneducated in medical science for treatment. We allude to venders of trusses — mere adventurers, having some patent truss intended principally for sale. This is much to be regretted, as it is estimated that every tenth person is so afflicted, and every one thus conditioned, liable to a violent death if not skilfully treated by the application of a suitable truss meeting the indications that may present.

The eminent Scarpa says, in his treatise on Hernia: "There are indeed a certain number of surgical operations for the prompt and safe execution of which mere anatomical knowledge will suffice; but in many others the surgeon cannot promise himself success, even though he be well acquainted with anatomy, unless he has particularly studied the numerous changes of position and alterations of texture of which the parts upon which he is about to operate are susceptible. If he has not the requisite information upon all these points, false appearances may deceive his judgment and make him commit mistakes, sometimes of a very serious and irreparable kind. In order to have a convincing proof of this truth, it will be sufficient to take a view of the different species of hernia and their numerous complications."

These remarks, it is true, are made upon the surgical operation for the relief of strangulated hernia; but, nevertheless, are applicable to the treatment of reducible hernia; because of the different species and complications, and the necessity for security against the liability to the unskilful use of the knife. And it matters not how skilfully the operation is performed, it is attended with doubtful premonition as to the saving of the patient's life: there is greater safety in a truss properly applied, and varied in form so as to coaptate to the hernial parietes, and retain the viscera *in situ* — thus protecting the patient from the serious condition here alluded to.

DIAGNOSIS.

When an inguinal hernia is first formed, the two rings maintain their relative position. The internal ring is to be found at equal distance of the space, between the anterior superior spinus process of the ilium, and the angle of the pubis. From this point the inguinal canal descends obliquely downwards, between the aponeurosis of the external oblique and the fascia transversalis.

The viscera having entered the internal ring, may be detected in the canal, where it remains during the erect position of the patient — presenting a somewhat oblong tumor that disappears when the patient is in a recumbent position for several hours, or, can be reduced, when not painful, by taxis. This stage of hernia is liable to give much inconvenience to the patient — a dull pain in and about the epigastric region, tending even to nausea and a general distressed feeling. In many instances, because of its obscurity, the ailment remains unobserved by the patient. This condition of hernia is the same in the female as in the male, and the round ligament has the same relation to the swelling as the spermatic cord in the male. The viscera, if not reduced and retained by a truss, will eventually descend and pass through the external ring, when it will be observed as a circular tumor; or, the viscera may protrude directly through the external ring from within the abdomen, and present a similar appearance — the condition only being known when the hernia is reduced, the opening passing directly into the abdomen, and not obliquely as when through the inguinal canal.

Glandular tumors are not reducible, and are not found in the location of the internal or external rings, or in the direction of the the canal, hence, not readily confounded with hernia if the anatomy of the inguinal region is carefully considered.

When the viscera has descended into the scrotum it must be distinguished from other complaints. If a swelling of the scrotum is uniform on its surface and commenced in the most pendent portion, gradually ascending, and we cannot feel the testicle, but are able to discern the spermatic cord of its natural size, and in a healthy state above the tumor, and, particularly if we can distinguish a fluctuation, or discover a degree of transparency in it by holding a lighted candle in front of it, we are then assured of an effusion of fluid into the cavity of the tunica vaginalis testis. A protrusion of the viscera is first observed emanating from the abdominal rings and gradually descending, when the spermatic cord cannot be felt but the testicle may be distinguished, and when the symptoms described above as belonging to a hydrocele, and the indication of hernia also. They may both exist at one time. This condition is found in infants, and serves to perplex the professional attendant if not aware of the possibility of the co-existence of the two ailments.

The origin of enlargements within the scrotum, if below or above, is alone to be relied upon for a diagnosis. Hydrocele, in some cases, extends along the cord as high as the ring, the swelling at the same time being so tense that fluctuation is not perceptible. Then the origin of the swelling from below and its gradual ascent, its being constantly of the same size, and the impossibility of distinguishing the testicle is indicative of hydrocele. Again, in congenital hernia, there is difficulty in distinguishing the testis, as it is surrounded within the same tissue with the protruded viscera. The tumor in this case is irregular and extends from the ring, becomes more tense from coughing, can be separated from the testis and permanently reduced; that is, it can be returned with the finger when the patient is standing, but water will descend and fill the sac even under the pressure of the best-fitting truss. This is what deceives in a complicated case, when hernia and hydrocele both exist; but, upon careful examination, it will be found that the neck of the sac is compressible, and the sac filled from below, upward. Watery cysts in the spermatic cord may be distinguished from hernia by the uniform surface of the enlargement, the fluctuation, and the invariableness of the size. Circocele is often mistaken for hernia, by practitioners, and it is, in some cases, difficult to distinguish between an omental hernia and a varicose state of the spermatic veins, as they possess all the peculiarities of a hernia. Here, again, a circocele enlarges from below upward, and the convoluted veins impress a peculiar vermicular sensation, that a person having determined a few cases is not liable to make a false diagnosis.

Sir Astley Cooper advises that the patient be placed in a recumbent position and have the swelling reduced, and that the surgeon should then press on the ring with his finger and allow him to rise. This precludes the possibility of the viscera coming down, but not the passage of the blood through the spermatic artery.

Femoral hernia protrudes from under the crural arch, and can be readily defined by the impulse given from the femoral artery as it passes from under Poupart's ligament, through the crural arch formed by the inferior edge of the aponeurosis of the obliquus externus abdominus. Under the crural arch an oval depression is found on the front of the thigh on the surface of the pectinæus muscle. At the upper, outer and lower sides, this hollow is bounded by a sharply defined edge of the fascia, there not being such boundary internally. Where the attachment of the fascia to the crural arch terminates, it forms a distinct semi-lunar or crescent-shaped fold, covering the femoral artery and vein.

The descending viscera under the arch — in some cases the tumor — is circumscribed and soft, and the neck, seemingly, bound down by a well-defined encircling edge of fascia. There is another form of femoral hernia in which the viscera are contained within the sheath of the blood-vessels; the protrusion, in that case, is more obscure, and the defined edge of fascia not so apparent. This condition of femoral hernia is rare. The protrusion in this form of hernia is usually small and confined to women — men being but seldom afflicted with it.

Glandular enlargements are liable to embarrass the inexperienced practitioner, and even those who have had the opportunity of examining cases of crural hernia. Patients' lives have been compromised by the surgeon having supposed a strangulated hernia to be an inflamed gland. Psoas abscess has been mistaken for hernia, as it partially disappears upon pressure and will give an impulse on coughing. A varicose condition of the saphena vein tends to deceive as it can be reduced by pressure and enlarges when the patient is in the erect position. It must be observed, however, that femoral hernia, like that of other hernial protrusions (if not incarcerated) can be reduced and maintained *in situ* by means of a truss; and, if strangulated, however small, it is attended with a most painful disturbance of the bowels and induces nausea and vomiting. These indications should be carefully considered, and the patient relieved by taxis if possible; if not, there should be no delay in giving relief by the knife.

HERNIA COMPLICATED WITH A RETAINED TESTICLE.

It has been observed in dissections, that a small fold of peritoneum is continued from the upper part of the testis to a portion of the ilium or the cœcum in some subjects, and forms a preternatural connection between the parts. The descent of the testis under such a condition would inevitably bring the intestine with it, and, being so connected, preclude the possibility of wearing a truss, as the pressure could not be borne on the testicle. The descent of but one testicle into the scrotum determines in a measure this condition of the patient when he cannot endure the pressure of a truss. In some cases the whole mass can be so perfectly reduced within the abdominal cavity as to admit of wearing a truss; in others, it can only be reduced within the inguinal canal, and remains there for a time, forming a bubonocele. In other cases a bubonocele may be observed and the descent of only one testicle. If no inconvenience arises, a truss should not be applied, as this is usually observed in children, the testis may descend into the scrotum, or so plug the canal as to prevent a descent of the intestine. Mr. Pott describes these cases as of frequent occurrence and most difficult of treatment; and this agrees with our experience in the treatment of several hundred cases of hernia, annually for the past twenty years.

To obtain a tolerably satisfactory diagnosis of hernial protrusion, it is essential to obtain a history of the ailment, then the symptoms and peculiarities attending the case, and, if no urgent symptoms present, and a tumor exists that can be reduced and retained by means of a well-fitting truss, no further efforts are required for the safety of the patient.

PROGNOSIS.

The necessity for wearing a truss is imperative, in most instances, to the saving of the life of the individual laboring under a hernia — even of the smallest dimensions. Small protrusions are the more liable to become strictured, and this is the opinion of our most reliable writers on the subject of hernia: as Scarpa, Potts, Hey, Astley Cooper, Lawrence and Wood.

To obtain a tolerable prognosis of a case of hernia is to ascertain the age and condition of the patient, the date of the protrusion, if it is free or not free from stricture or inflammation, the symptoms which attend it, and the possibility of its reduction — including all, if any, peculiar circumstances attending it.

If the subject is an infant, the case is seldom attended with any serious consequences. The pliability of the tissues admit of the ready reduction of the viscera, yet there is danger in permitting them to remain without a truss. The tendency, even when very young, is for the aperture to become small if the protrusion is frequently redressed and partially retained by the tight fastening of the diaper; from this circumstance, there is danger of strangulation of the contents of the hernial sac, which may be produced by some violent forcing, as that of choking, coughing, or crying vigorously. Strangulations have taken place in children of six months, and have caused their death — not having been discovered by the mother, as the protrusion was very small, and only detected by the surgeon when too late to save life by an operation.

Adults in the vigor of life, laboring under hernia, are in a more critical condition, if neglectful or subjected to mal-treatment, than the very young or the aged. The danger arises from the susceptibility to inflammation in the protruded intestine resulting in stricture. In people in the decline of life there is not that degree of danger in permitting the protrusion to remain. Their languid circulation renders them less liable to inflammation, and their rupture is, mostly, of long standing, the aperture enlarged, and the edges indurated from frequent reduction of the contents.

In recent rupture, or in cases where it has been of long duration, and secured by wearing a truss, the protrusions are extremely liable to stricture, because of the diminished size of the aperture. Protrusions of large size and of long standing, it is not always well to reduce. It is only necessary to sustain, and prevent the increasing tendency to enlargement, as in cases of scrotal hernia, which, we are credibly informed have extended to the knee and were of extraordinary size. To attempt to retain a hernia of much less size even, would require such force from the springs of a truss as to be unendurable. And another difficulty arises from an attempt at reducing large protrusions of long standing: nausea and vomiting is produced to an unbearable extent. There is a possibility of reducing, by degrees, large hernias, unbilical more especially; and we have succeeded in such cases to the great relief of the patient from the chafing and excoriation of the projecting mass, by the gradual increase of a supporting bandage.

The contents of the hernial sac invite special attention in regard to a proper diagnosis, tending to the relief of patients who have had little or no inconvenience from large hernias of long standing, when they are suddenly seized with symptoms of a strangulation in the part. This condition of the patient arises from the existence of omental protrusion only during the time previous to the severe symptoms. This condition predisposes to an escape of a portion of intestines, which descends and becomes more or less constricted and painful, and which if not nearly reduced by ordinary methods no time is to be lost in having the parts relieved by the knife.

TAXIS.

The reduction of hernia by the hand, requires much practical experience; as the attempt at reduction may increase the unfavorable condition of the patient.

Patients seldom call upon a surgeon to reduce a hernia if not alarmed at their inability to reduce it, and the suffering caused thereby; hence, the parts are in an irritable condition, and an increase of that tendency must be avoided if possible.

The difficulty may exist under various phases of the ailment. The enlargement of a portion of the omentum or intestine in a contracted aperture, and attended with suffering from some cause of irritation is one, and under such circumstances, an attempt should be made to allay the excitement in the part, and not increase it by handling. The symptoms are a swelling in the groin, scrotum, or crural arch, sensitive to the touch; if an intestinal hernia it is usually most painful and the pain increased by coughing, sneezing, or standing upright. These are the first indications of an increasing tendency to a more serious condition; as nausea, and inability to evacuate the bowels per anum. When in this condition the hernia is said to be incarcerated, and if relief is not soon afforded the suffering patient, death will soon ensue. The prolapsed viscera are constricted and greatly congested, and the relief is dependent upon the return of the viscus, freed from compression. This can only be accomplished by returning the bowel into the abdomen by taxis, or dividing the parts which form the stricture.

For the reduction of hernial protrusions much advantage is obtained from a favorable position of the patient, which, by the way, is the preparatory step. The patient having been stripped of all clothing that could interfere in the handling, should be so placed upon his back as to admit of the hips being elevated above the shoulders, about twelve or fifteen inches. This position was first attempted by Winslow, and greatly approved of by Sir A. Cooper. The thighs are then flexed towards the body; this position relaxes the fascia of the thigh, also the aperture through which the viscera has passed. The tension being, in ever so small a degree, relieved tends to abate a part of the cause of irritation. Having obtained this position for the patient — the suffering being great — we have two towels doubled and wet in water, at a temperature of 60° Fahr., and applied over the lower portion of the abdomen. The following mixture should then be administered in half drachm doses repeated every ten or twenty minutes, to meet the urgency of the case:

℞
Antim. tart........ gr. i
Aq. bullien......... ℥ ii
Tr. opii............ ʒ i
Ft. sol.............. ℳ

One fourth of this to be given every fifteen minutes.

This treatment seldom fails to relieve the pain before extreme nausea ensues, and the pain having been relieved, we consider our patient in a condition to have slight pressure made upon the protruding mass. We then, first, make a gentle effort to grasp the entire protrusion with the thumb and fingers of the right hand, holding the mass tolerably firm, if not giving pain, for some minutes, and then an effort to draw it back from the strictured ring. If all this is accomplished without complaint from the patient, the fingers and thumb of the left hand are applied to the neck of the sac, or upper portion of the protrusion, and gentle pressure made upon only a small portion. If the protrusion is observed to diminish under pressure after a few minutes, the patient is safe; the whole mass will be reduced after a reasonable time, if this procedure is carefully observed. The attempt to press the whole mass up at once usually fails. If the patient complains under the efforts made at reduction, we desist for a longer time, keeping the towels wet and giving the solution until nausea tends to vomiting. The patient may remain in position for several hours, and if relief is not afforded by these means, we then have resort to an anæsthetic, and the taxis again attempted — carefully avoiding the use of much force, as nothing is gained thereby and injury is to be apprehended. When the protruded mass is grasped it must be retained firmly, and the pressure can, in most cases, be increased without giving pain if steadily maintained — in fact, the grasp should be continued as long as the strength in the fingers will admit, as we have often witnessed the reduction progressing favorably when we were quite exhausted, and observed it again increase in size upon the withdrawal of pressure. An intelligent assistant is of great service under such circumstances. The direction of pressure in the application of taxis to inguinal hernia, other than upward, we have never found to be of any advantage, although obliquely outward is advised by Sir A. Cooper, Lawrence, and others. In femoral hernia it is of some service to lift, as it were, the mass off of Poupart's ligament and bear it outward and below. If pressed upward, in the position it is usually found, it is to prevent its return by keeping it doubled over the ligament. Much advantage is obtained in the reduction of femoral hernia by adducting the flexed thigh. This relaxes the femoral fascia, and tendon of the external oblique muscle, and, in some cases, relieves the patient of pain. The same care to avoid severe pressure from the hand should be observed in reducing femoral hernia as in that of inguinal.

Mr. Hey remarks ("Practical Observations in Surgery") on the use of the warm bath for the reduction of strangulated hernia: "I have often seen it useful; but I have also often seen it fail of success. Whenever it is used in this disease the patient should be placed, if possible, in a horizontal position. Gentle efforts with the hand to reduce the prolapsed part are perhaps attended with less danger and with greater prospects of success while the patient lies in the bath than in any other position. The free use of opiates coincides with that of warm bathing, and under some circumstances, these means deserve to be tried."

One noted authority among the old writers on the subject of hernia (Wilmer), strongly recommended the application of cold to the protrusion, and this procedure is approved by most of the modern surgeons, often with that of tobacco injections, which ought to be used with great caution; as patients have died from its poisonous influence. Cold applications in the form of ice were particularly recommended by Benjamin Bell. The best way is to pound the ice, tie it up in a bladder, and place it on the rupture. When ice cannot be procured, Sir A. Cooper recommends a mixture of equal parts of nitre and sal ammonia — ten ounces of the mixture to be added to one pint of water in a bladder. This eminent authority gives this further reliable advice: "That if after four hours the symptoms become mitigated and the tumor lessens, this remedy may be persevered in for some time longer, but if they continue with unabated violence, and the tumor resists every attempt at reduction, no further trial should be made of the application." [On inguinal and congenital hernia].

This acknowledged skilful surgeon (skilful more especially in intricate surgical operations) advises much greater perseverance in the use of other means than is generally practiced. We have labored three and four hours to reduce a strangulated hernia without, apparently, making the slightest favorable impression, and finally succeeded in our efforts — in fact, but seldom fail, except in cases where adhesion to the surrounding tissue had taken place. As patients have been relieved of all of the painful symptoms of strangulated hernia by treatment — previously stated — and free evacuations followed the relief, the protrusion only rendered less tense and void of pain, and the size apparently not diminished, therefore it is obvious, that it is possible to relieve strangulated hernia when incarcerated, and that the effort should be made when perfectly convinced of the incarcerated condition of the protrusion.

As before stated, the hernia may consist of omentum that has become incarcerated, and at no time previously found to be troublesome, but from some unfavorable circumstance a portion of the intestine has been forced through the neck of the sac, and becomes inflamed and swollen, exciting the ring to contract. In this condition, induced by accident, from persevering in the treatment the inflammation has been subdued, the swelling and consequent irritation diminished, to there turn off the intestine; but for the accomplishment of all this several hours of anxious labor is required in most cases. The several means we have described for the reduction of strangulated hernia by taxis, we have witnessed and practiced, and speak of them as being more or less efficient, but there are others worthy of consideration and trial. In the British Medical Journal, December 23, 1871, p. 724, Dr. Philip Crampton Smyly, Surgeon to the Meath Hospital, England, makes a statement as follows: "The objects to be attained in the treatment of hernia in a state of strangulation, are the release of the protruded parts from stricture, and their replacement within the abdomen, provided they are in a suitable condition. These objects are usually sought to be accomplished either by taxis, or operation with the knife. Some years ago a nurse in one of the medical wards in the Meath Hospital had a reducible femoral hernia. She neglected to wear a truss, and one day it consequently became strangulated. My father, being the surgeon on duty, tried taxis, as did also other surgeons, without success. After consultation, an operation was decided on, but every argument failed to persuade the patient to submit. She would rather die than be cut. After the surgeons had left, the clinical clerk (since a very distinguished medical officer in the army) and I, thought it a good opportunity to study the relation of the ring to the sac. The result of our examination not a little surprised us. In withdrawing my finger from the ring into which I had inserted it, we heard a distinct gurgle. My fellow student pressed the tumor, and it passed into the abdomen. The patient lived for many years afterward, and performed her duties in the hospital. I have since frequently tried to repeat this happy manœuvre, and with most satisfactory result.

"For inguinal hernia in the male, the index finger is applied to the lower part of the scrotum. This is invaginated (as in Wützer's operation for radical cure), the finger being passed behind the testicle and carried up to the external ring. The hernial tumor is then pressed downward over the finger toward the back of the hand, so as to make the strictures in the ring tense, and consequently smaller. The invaginating finger is then forced firmly upward and outward in the direction of the internal ring. As soon as the finger is firmly grasped, the hand should be slightly turned, and the finger pushed toward the middle line. Considerable force may be safely applied in this way, as all the delicate structures are behind the finger, which acts mainly on the stricture. On withdrawing the finger, the hernia can usually be easily returned. The same principle is equally applicable to femoral hernia. This plan may have occurred to others; but if so, it is perhaps not generally known, and any suggestions by which a cutting operation may be safely avoided is acceptable to the practical surgeon. My colleague, Mr. Porter (surgeon to the queen in Ireland), was much pleased with the success of this plan in a case of inguinal hernia strangulated four days; and he has since tried it himself with satisfactory result.

"The advantages which I claim for this procedure are: 1. The strangulating portion of the ring is dilated before any pressure is applied to the bowel. 2. Much greater force may be applied to dilate than could safely be brought to bear when the intestine itself is employed for dilatation, as in ordinary taxis. 3. There is much greater probability of returning the bowel into the abdomen in a good condition, and, consequently, in a number of cases avoiding a dangerous surgical operation."

Another auxiliary tending to the reduction of strangulated hernia is presented by Thomas Bryant, Esq., surgeon to Guy's Hospital, in the *Medical Times* and *Gazette*, April 20, 1872. This gentleman was called upon by Mr. Kelson Wright, of Kennington, to see with him a case of strangulated hernia in an old man, aged 71. "He had been the subject of a right scrotal hernia for thirty years, and had worn a truss. He had had occasional difficulty in its reduction after its descent; but Mr. Wright had always succeeded in reducing it. On this present occasion the same effort had failed, and when I saw him vomiting had existed for two days, and a large hernia existed in the right side of the scrotum, one portion of it felt tenser than the other. Chloroform was given, and the taxis employed, but without success; consequently herniotomy was performed, it being necessary to expose the bowel. When this was done the cœcum escaped, dragging down with it some three inches of small intestine covered with peritoneum — the external ring pressing firmly upon it. With some difficulty the bowel was returned, the wound brought together, and the whole carefully bound together by means of a pad and spica bandage. A morphia suppository was given.

During the night, however, the old man would get out of bed, and in the attempt he tore off all the dressings. As a consequence, the bowel came down again; vomiting returned, with abdominal pain. Mr. Wright was sent for, but all his efforts to return the intestine were fruitless. I was consequently sent for. I found the old man lower than when I saw him before. The hernia was larger than ever. I gave him chloroform, and attempted reduction, but failed. I then increased the opening at the internal ring; but on doing this more large intestine came down, and no effort of mine could reduce it. I consequently punctured the intestine in four or five places with a grooved needle, and let off the wind; this measure enabled me to do what, under other circumstances, I could not do — reduce the hernia. The wound was then re-adjusted, and a good pad firmly secured on with strapping, opium being given; and I am pleased to add, no one bad symptom followed these rough measures, and a good recovery ensued. Mr. Wright tells me the wound united without a drop of pus appearing — the whole uniting by primary union.

"It was interesting to note that when the bowel was punctured nothing but wind escaped, except in one spot, where the smallest drop of blood oozed out, evidently from the congested intestinal wall. None of the contents of the intestines escaped even after the rough manipulation to which they were subjected."

This eminent surgeon further remarks: "I punctured four or five times (that is the bowel), until I had reduced the whole of it. No fæces escaped whatever, although during the whole of this time I was so manipulating the intestine that if it was possible anything could come out it would have done so. After the patient recovered I asked myself the question: If the large intestine, exposed to view as that was, and not supported in any way by the abdominal walls and contents, could be tapped without any escape taking place, surely, where the intestine is supported, the risk of extravasation would be greatly diminished by the natural support given to it in the abdominal cavity? I think we must draw this conclusion, then: that there are cases in which you may puncture the intestine freely, and with every prospect of affording great relief. In hernia, this case clearly proves that you may resort to it, and I believe you had better adopt it if there be much trouble in returning a very large hernia. Had I learned this lesson earlier, I should have tapped long before I did. But might we not employ this treatment much earlier? If, after operating, we can reduce a strangulated hernia by pricking, which we could not reduce without, is it not possible that we might reduce some hernia without any operation at all? Consider a small enterocele. See the intestine bulging beyond the neck of the sac and becoming congested. See how it becomes strangulated and congested if reduction is not effected. If you tap it you let out all the wind, the whole of the knuckle of intestine collapses, and you get only a little flaccid lump. Under these circumstances, what is there to prevent the bowel being replaced by natural agency? I confess my liking for the idea from the surgical point of view. I think it is scientifically correct, and I see nothing to prevent it having the desired effect. I cannot recommend it practically, because I have had no experience of it; but I intend to test the value of it when a suitable case comes before me. And what would constitute a suitable case? Not those slight cases where you can return the intestine without opening the sac, because we know, as a rule, they do well. But I certainly will try it in those cases of hernia that we not infrequently get in hospital, and sometimes in private practice — large scrotal hernia, and large umbilical hernia, the interference with which is nearly always followed by death. It is quite exceptional for a large inguinal or umbilical hernia to recover after herniotomy when the sac has been opened. By opening the sac I mean not only just at the neck, but complete exposure and manipulation of the contents."

Entertaining, as we do, a most favorable view of this novel means of reducing strangulated hernia, they both having succeeded under the most trying circumstances, present to us a consideration of sufficient inducement to give them a trial in apparently suitable cases, and without entertaining much apprehension of inflicting serious injury to the patient.

We would, however, suggest that, in the invaginating process, the propriety (if not an improvement) of using distending forceps for the overcoming of the constricting ring. An instrument made for the purpose, that would do less violence to the implicated tissues than that of thrusting the finger within the constricted aperture. For this purpose, we would suggest the construction of a pair of forceps, for the distension of the constricting parts in strangulated hernia, and of the following form: A pair of hinged forceps about six inches in length, terminated with thin rounded ends a third of an inch wide, slightly hollow or grooved within the blade, of equal thickness; and thin, but not sharp to endanger the cutting of the invaginated integument. And at about one-third of the length of the blades from the hinged end, to be of such thickness as to admit of the adjustment of a screw through one of the blades. By means of this screw a graduated distention of the ring could be carefully controlled. And in cases of small protrusion of viscera, or when relieved of distension from flatus by tapping, one blade could be entered above and the other below; and the instrument made to distend the aperture without impinging upon the protruding viscera. This would afford perfect safety to the patient, and a means of making ample distension for the reduction of the bowel to within the abdomen.

The effort by taxis in the ordinary way of forcing the protruded viscera, congested and inflated, through the irritated, and consequently constricted ring, must be considered as a violent procedure even under the most skilful treatment. The very pressure of the protruded viscera made against the ring increases the recently constricting tendency to the increase of the obstructed circulation in the protruded mass. This may be avoided by these new and apparently feasible methods, and in many instances save the patient from the hazardous operation of herniotomy that has proved fatal in many cases.

THE CONSTRUCTION AND APPLICATION OF TRUSSES FOR THE RELIEF OF ABDOMINAL HERNIA.

Therapeutic agents for the retention of replaced viscera have not been considered by surgeons in general, of as much importance as the pathological condition actually demands. The surgico-mechanical appliances commonly denominated trusses, are mainly devised and applied by the mechanics that construct them. These adventurers obtain favorable notice, most frequently from some novel device in the construction of a peculiar pad or the means of attaching it to the spring that encircles the body, and, by the favorable notice given them from eminent surgeons, amass fortunes from the sale of their device to the great number thus afflicted with hernia — about every tenth person. This immense traffic is, because of its equally large profits, truly a very great inducement to efforts at introducing useless complications, and the fabrication of absolute falsehoods concerning them, as a certain means of radical cure. By these means they obtain extravagant prices for appliances of no peculiar merit. This is greatly to be deplored when we consider the critical condition of the patient, and their reliance upon their medical adviser's statement, that the greatest possible security will be afforded them in the purchase of one of these supposed improved trusses.

There is no one form of truss, or uniformity of pad, that will meet all of the varied conditions of patients laboring under the several varieties of hernia. This must be readily comprehended by every experienced and reflecting medical practitioner. Yet, the patient is often inconsiderately recommended to the truss-maker that has attained the greatest notoriety for the peculiarity of his truss of fixed form — precluding the possibility of adaptation to an extraordinary case and, consequently, a failure in retaining viscera *in situ* when applied. Some of these trusses having small obtuse, conical pads, are applied to patients greatly attenuated and the hernial aperture thereby greatly distended from this concentrated force. This form of pad is also applied to inguinal or femoral hernia, or to an infant, or, most corpulent person that may require a truss. This is by no means, in a surgical sense, meeting the indications that present, as in all other ailments that obtain the skill of the surgeon. And why not? The condition of the patient having a hernia, is considered by the most eminent in the profession, as truly perilous, and much more so than many other ailments, to which a surgeon would give very especial attention, and that do not so imperil the patient's life as that of a reducible hernia; which, from unskilful treatment can become strangulated. When that alarming condition has ensued, *then* the utmost skill is devoted to the patient, when it might easily have been prevented by the use of properly devised surgical appliances.

In hernia, as in all other pathological conditions of the system, certain indications present for the consideration of the student of anatomy and surgery. The latter has probably been as carefully studied as in any other ailment afflicting the body of man; but has been mainly confined to the abnormal condition of the protruded viscera requiring surgical treatment when irreducible by taxis, and not extended to appliances for the relief of reducible hernia. To such palliative and preventive means there has been an actual indifference — submitting the precarious condition of patients to adventurous treatment, which, under the most favorable circumstances, is to compromise the life of the unfortunate individual laboring under reducible hernia. That it is of rare occurrence for persons' lives to be endangered by wearing ordinary trusses is not a mitigating argument for the indifference of surgeons when they are professionally consulted. It involves them in a responsibility for the patient's future condition. The patients, from prudential motives, having confidence in their professional attainment, intrust their lives to their protection. The surgeon is in duty bound, to the best of his ability, to tender his patient security from uncertain treatment.

The construction of a truss demands surgical skill, and a careful consideration of the indications that present in the various pathological phases of hernia.

Hernia is the result of a predisposition, in most cases, and induced most frequently from violence or a continued tendency to a protrusion of the viscera through existing apertures, abnormally distended, or, in rare instances, injury separating muscular continuity and not confined to any particular locality.

In the first condition tending to hernia, is the erect attitude of man, relaxed muscular fibre, exhausting influence — as sickness, or from long continued labor, one of the most common causes of hernia.; as gravity obtains an ascendency over the vital energies of the body, tending to hernia, prolapsus ani, procidentia uteri, retroversion, hemorrhoids and varix.

To our protection from hernia, however, there is a special provision by a tendinous expansion from the external oblique muscles that sustain the whole hypogastric region when not in a relaxed condition. This tendinous expansion, strengthened by interlacing of texture, supports immense tumors, dropsical accumulations, and other distending forces of circumscribed spherical masses. But, in the most pendent portion of the abdomen we have the passages for the spermatic cord in the male, and the round ligament in the female, the intestines and omentum extending floating masses, unfavorable circumstances so disposing, admit of portions entering these yielding apertures.

The first and most common cause tending to hernia, is a structural laxity of the visceral attachments, tending to elongate; preternaturally large openings of the abdominal rings, and a ready yielding of the margins of the apertures, often from the relaxed condition of the abdominal muscles.

The predisposition in patients exists before hernia has ensued, and presents the *first* indication to be met in the construction of a truss. The pendent portion of the abdomen should be supported and the superincumbent weight of the abdominal viscera sustained above the inguinal and femoral regions. Suffering is suggestive of means for relief; and this will be observed in the ruptured laborer when unable to get a truss; he girts his belt tightly about his loins and is thus enabled to do light work. The leather belt of the laborer is not, however, a suitable material for a truss belt. It is now the opinion of reliable authority that only suitably sized tempered steel will serve to construct the partially encircling belt of the truss, which also serves for a reliable basis of support under the varied conditions of the body; retaining its position with great certainty, being readily adapted to the body and maintaining an elastic resistance.

Figure 84 represents the forms of single, double, and child's truss.

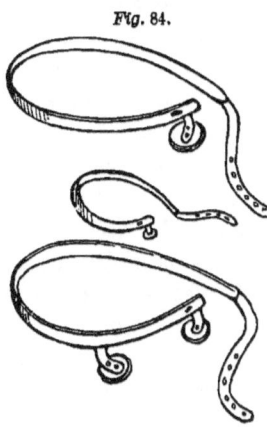

Fig. 84.

What we consider the proper form and encircling length of the spring, is to encircle three-fourths of the circumference of the pelvic portion of the body, one inch below the anterio-superior spinous process of the ilium, the posterior limb of the spring, or portion to cross the back, to have an advance of two inches beyond that of the anterior limb which should pass over and above the pubis to two or three lines beyond the internal abdominal ring, and above it. It will be observed that the spring crosses the back higher than that of ordinary trusses, hence, is less liable to be displaced by the action of the glutei muscles, and is rendered secure in position from being under the crest of the ilium in its curve round the hip, tending to the most fixed position that it is possible to place a truss on a patient. The springs are covered with some suitable material and supplied with a strap to complete the circle around the hips. This gives the desired support to the pendent portion of the abdomen and above the apertures through which the viscera escape. The *second* indication to be met is the securing of the viscera *in situ* by closing the apertures—an important and often the most difficult part of constructing a truss. The pad must be made to perfectly coaptate to the hernial parietes, which vary greatly in different persons, and in the location of the hernia. It is impossible to construct a truss with one form of pad that will in every case retain the viscera that tends to protrude even in the same form of hernia. A person greatly attenuated, and having an inguinal hernia will require a different form of pad to retain the hernia from that of a corpulent person laboring under a similar variety of hernia. The contour of the parts differ, and the floating adipose tissue intervening demands a peculiarly formed pad. Hence, the absurdity of a "radical cure truss" of fixed form. A truss that will retain the viscera most securely, is the only reliable one tending to a radical cure; and even when thus favorably conditioned, of adults there is less than two per centum cured. Of infants, nearly every case of ordinary hernia is curable when properly retained by a truss.

The idea of affecting a radical cure of hernia by pressure of concentrated force upon the inguinal canal, and inducing irritation and consequent adhesion, to the occlusion of the ring is not possible by any form of truss pad. There may be interstitial deposit of a substance analogous to that of unorganized false membrane, and it may become tolerably permanent; yet, it is always liable to absorption as is the case with all adventitious cellular tissues. And this is the rationalé of the radical cure of hernia in adults. The seat of these changes is the peritoneum, forming the neck of the sac, and the portions of the cellular tissue which naturally occupy the canal or orifice through which the sac is formed, and, at a later period, the contiguous fasciæ, or the edges of the surrounding tendons become involved, and, apparently a radical cure has been established — to the delight of the patient for the time being, but again, to their dismay, at the recurrence of their rupture in an exaggerated form,— the tissues being thinned and the aperture greatly distended from the absorption that has taken place in the abnormal occlusion of parts. The physiological condition tending to the absorption of the adventitious formation, may be attributed to the predisposing tendency to hernia, and the superincumbent pressure of the abdominal viscera upon the unsupported inguinal region. A preventive means is here suggested, viz.: that of a supporting belt to the pendent portion of the abdomen to be worn in all cases where hernia has existed and is supposed to be cured. This of course applies only to adults. The physiological tendency to absorption of abnormal formations of tissues, is witnessed in the removal of the surplus osseous deposits around a fracture and the solid deposits in the subcutaneous cellular tissues of parts, subjected to undue pressure, and this applies equally well to hard substances used in truss pads when not liberally protected with some elastic material; yet, there is no material so serviceable for a base as a soft wooden block, as it can be most easily pared into desired form, and readily attached to the truss by means of common screws — affording facility to the surgeon to remove and shape it so as to conform to the hernial parietes. In this way the surgeon is enabled to exercise his own skill in fitting his patient with a truss — the surgical instrument-maker having supplied the spring and an ordinary shaped soft wood block; having a removable cover, or sheath, well padded with blanketing.

TRUSS FOR THE RELIEF OF REDUCIBLE INGUINAL AND FEMORAL HERNIA.

Fig. 85. represents, the, truss, applied (1) — a single truss. The encircling spring is placed immediately below the crest of the ilia, and secured by the strap on a stud riveted in the spring. To this spring is attached an untempered piece of steel, three inches in length, slightly curved, and having two holes through it by which to attach the pad with common, round-headed screws. This piece of steel is so attached to the mainspring as to admit of lateral movement for adapting the pad to the hernial aperture. The pad consists of a disc of soft wood that can be shaped as desired, and then covered with soft material. For inguinal hernia a circular disc, very slightly concave so as to coaptate to the convexity of the abdomen, is required; while for femoral hernia we need an oblong convex surface to apply under Poupart's ligament. Fig. 86 represents the truss for femoral hernia (2) — location of the spring below the crest of the ilium.

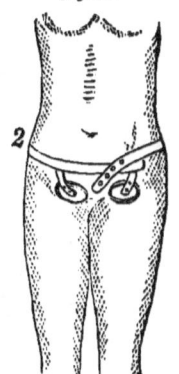

Fig. 86.

These blocks should be covered with two or three thicknesses of blanket, or other equally thick woolen fabric; over that a piece of soft leather, and on the back of the disc a circular piece of sheet brass or nickel, having holes to correspond to the holes in the piece of steel to which they are attached.

Fig. 87 represents a double inguinal truss having circular discs, the pads separated on the one encircling spring about four and a half, or five inches, for an adult. The inner pad is set about from one-eighth to one-quarter of an inch lower than the end pad, as the end of the spring dips from an encircling line about that much.

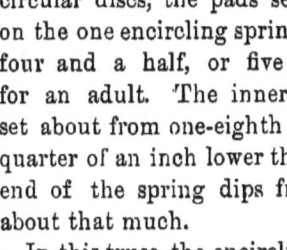

Fig. 87.

In this truss, the encircling spring crosses about three-fourths of the pendent portion of the abdomen, and, when secured by a strap, sustains the superincumbent weight of the abdominal viscera above the inguinal region — thus, in a measure, obviating the protrusion by resisting the downward force. The pads are readily adapted because of the pad strips of steel, admitting of being bent, and of a lateral movement, and the pad blocks of being shaped to the contour of the parts.

That infants are so much more readily cured of hernia than adults may be attributed to the favorable conditions of growth, and recuperative tendency of the vital forces tending to a constricting of the no longer distended aperture, and in a very limited period compared to that of an adult. Moreover they are subject to control in having the truss constantly applied day and night. If this is not attended to carefully, the children are not cured. Children cured of hernia but seldom have a relapse, because of their predisposition being curable by growth, and not existing as in the case of adults having an undue pressure of the abdominal viscera upon the inguinal region. Their ruptures are mostly produced from muscular contraction of the boundaries of the abdomen, and the open condition of the rings in the male, who is most subject to inguinal hernia, from the recent descent of the testes in which there is, if unobstructed, an active tendency to diminish and strengthen.

In exomphalos or abdominal hernia in infants, and indeed in adults, an important indication is met by encircling the abdomen with a wide belt conforming to the contour of the same — thus limiting the lateral distending tendency of the walls to the enlargement of the aperture.

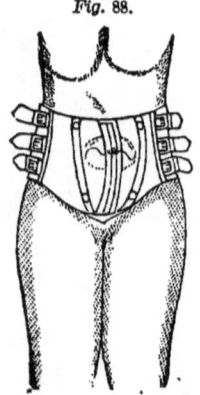

Fig. 88.

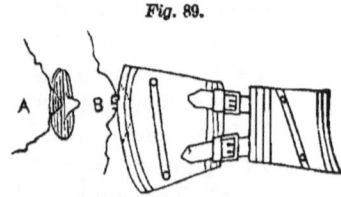

Fig. 89.

Figure 88 represents the belt applied.

Figure 89 represents the belt complete: A the pad, B the pad, tied with tapes in the belt. The small conical shaped pad as seen attached to a thin metal shield of several inches breadth and width, for adults.

This appliance cures infants in a few months and affords very great comfort to adults.

Hernias in irregular regions of the body can only be supplied with supporting appliances by ingeniously devised means.

Much relief is afforded in irreducible hernia by supporting the mass and preventing the increasing tendency to enlargement,

Fig. 90.

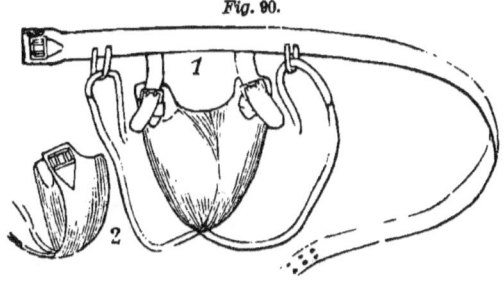

as seen in figure 90, which represents a supporting apparatus for irreducible hernia, and which may be used to much advantage in scrotal ailments, as it admits of an adjusting support when pressure is indicated, as a means of relief in circocele, and hernia humoralis. The cords attached to the pendent portion of the sac are carried backward and around each thigh and brought forward in front to be passed through the loops where, the ends being made of tape, are then tied. The straps passing through the buckles can be graduated to any desired degree.

PROCIDENTIA UTERI.

This abnormal position of the uterus is a most common ailment, and, like other organs out of their fixed locality, is relieved by surgico-mechanical appliances, for supporting the descending viscus. The pathological condition is similar to that of hernia, advancing by degrees to that of great displacement, involving the patient in serious inconvenience and suffering.

The primary cause of procidentia uteri may be considered as the result of a pathological condition of the uterine organs; and not that gravity has exceeded the resistance of vital energy in a normal condition of the person and produced the ailment. Various unfavorable influences tend to this result; and, probably, the first is aberration of function tending to an acridity of serous secretion; exciting an abnormal propulsive effort of the muscles of the pelvis and pendent portion of the abdomen. This is a reflex nervous influence from a most common condition of women; as more than a majority of them are afflicted with leucorrhœa, which, if cleanliness is not observed, becomes a decided irritant; exciting the propulsive effort, which is expressed by women as a "bearing down sensation," and, as a concomitant, pain in the lumbar region and a numbness in the thighs.

The propulsive descent of the uterus is in the direction of the axis of the pelvis; resting upon the perineum before its exit from the vagina.

In a normal condition of the uterus it is situated nearly in the centre of the pelvis; the distance of the os uteri from the *os externum* being about four inches. The os uteri is situated at the terminus of the vagina, projecting into it; the outer surface being covered by a portion of the inner membrane of the vagina. In the normal condition the distance from the opening of the os uteri to where the inner membrane of the vagina begins to be reflected over it may be nearly an inch; the distance increasing in direct ratio to the descent of the uterus — producing an inversion of the vagina. When the uterus has escaped from the os externum it usually aggravates the suffering and increases the inconvenience of the individual. Previous to its escape, its unnatural pressure upon the sacro-rectal region is a cause of hemorrhoids, and the pressure of the fundus of the uterus upon the neck of the bladder induces a constant desire to micturate — at times, without the ability; requiring surgical assistance. When in this condition it is a most difficult matter to introduce a catheter into the bladder; the point of the instrument must be be turned downward — there being also a descent of the bladder. There is an alteration of the relative situation of the parts within the pelvis and abdomen, both in regard to each other, and to the containing parts, as the parietes of the abdomen and the pelvic boundary. The bladder descends with the other displaced organs, requiring, as we have stated, in introducing the catheter that the point should be turned downward in order to have it enter the viscus — an operation frequently demanded under this condition of the patient as the descent of the bladder contorts the urethra to such a degree of imperviousness, as to prevent the passage of the urine. The rectum, instead of remaining in the curve of the sacrum, assumes an abnormal position, and with that of a portion of the abdominal viscera fills the pelvis.

The descent of the uterus is usually very gradual, being many months gliding down to the perineum where it, not unfrequently, rests, as is stated by Sir Charles Mansfield Clark, "as upon a shelf, the violence of the symptoms abating; the parts which suspend the uterus above, although much lengthened, being no longer put upon the stretch. From this circumstance it would appear that the greater number of the inconveniences attending this complaint depend less upon the pressure of the uterus in the vagina than upon the dragging of the parts above."*

Impairment of the general health tends to the descent of the uterus, and, as a sequence, the mal-positions attended with many distresses and indescribable symptoms. The most common are, pain in the side, lassitude, failure in the digestive functions, and, after a time, continued pain in the lumbar region, with inability to stand on the feet for ten minutes without having forcing pains as if to expel the uterus. The primary annoyances to the patient are constipation and leucorrhœa; then follows the disagreeable sensation of the propulsive effort, that if not relieved completes the procidentia uteri.

The first change of position is deflection; the fundus of the uterus being forced downward while the os remains in its normal position — hence, a curve in the cervix from the pressure or propulsive muscular effort brought to bear upon the pelvic viscera; the result of morbid irritation of the serous membrane of the vagina, and indicated in the leucorrhœal discharge. The next stage is retroversion, or anteversion, tending in some cases to a lateral deflection. The os tincæ being enfolded in the vagina, and pressed against the walls of the pelvis, the morbid secretion destroys the epithelial, and, finally, serous membrane, to that of an ulcerated condition of the os and cervix uteri; tending to neuralgia, congestion and inflammation. Escharotic applications under these conditions of the uterus only serve in some cases as palliative, and are often injurious in their tendency, from the fact of increasing the irritable condition of the uterus.

TREATMENT OF PROCIDENTIA UTERI.

Admitting that which we have stated in regard to the pathology of procidentia uteri, a regime tending to the restoration of the general health of the patient, manual effort at relieving the pelvic

---

* "Observations upon Diseases of Females:" 3d Ed., Lond., p. 68.

viscera by raising the fundus of the womb — thus relieving the neck of the bladder and rectum from compression — cleanliness and sedative means applied locally for relief from the irritating leucorrhœal discharge, are therapeutic means of relief; and, if successful in affording permanent relief to the patient, establish the soundness of the opinion as to the pathological condition of the patient.

The treatment of the first condition of the patient may be left to the skilful general practitioner — advising syringing of the vagina twice or three times a day with water at the temperature of the body; thus neutralizing and washing away the acrid secretions. The second, being manual effort at raising the fundus of the womb to a vertical position, requires practical dexterity that may be readily acquired from a description of the procedure.

The erect position of the patient is the most favorable for arriving at a correct diagnosis of the several mal-positions of the uterus. In the recumbent position the superincumbent weight of the abdominal viscera recedes, and the pelvis, in a measure, relieved — so much so, as to deceive the practitioner as to the actual condition of the patient; moreover, the effort at restoring the uterus to its normal position, and retaining it *in situ*, is attended with much more certainty when standing, than when in the reclining position. The effort at rising tends to a return of the abnormal position of the uterus as we have witnessed in many cases.

The procedure is as follows: The patient stands with the limbs separated, on the left side, or partly so, of the surgeon seated upon a chair. Having lubricated his fore and middle fingers with lard, or oil, he inserts one, and if it be possible without doing violence to the patient, the two fingers into the vagina, slowly exploring the internal condition. When the uterus is retroverted, the fundus may be found resting in the fossa of the sacrum, and the os tincæ pressing the neck of the bladder or the reverse — the rectum. Then by gentle elevating pressure, first with both fingers, and, when a slight yielding is obtained, with the fingers separated, one to the position of the os tincæ, making an effort to displace and bring it down. By this manipulation, and by careful perseverance, the uterus will be restored to its normal position.

Anteversion of the uterus is simply a reverse position to that of retroversion, and much more common. The os tincæ in this case will, in many instances, be found covered in a fold of the vagina, requiring dexterous manipulation to bring it forward. An effort to pass a finger behind it will prove unavailing before the fundus is raised from its position against the pubis and neck of the bladder — the latter position being the cause of the frequent inclination to urinate — while the os tincæ has become ulcerated from being so closely enfolded in the cul-de-sac of the vagina, and from the morbid secretions thus retained about it.

The uterus being replaced, the patient should be directed to get on her knees over a basin of water, and wash the vagina thoroughly with the syringe and water, then inject a small portion of the following decoction into the vagina: —

℞
    Decoct. papaver somnif.................. Oj
    Zinci sulph........................... ʒi
    M

The ablution should be practiced morning, noon and night, ending each time by using the decoction. This will relieve the irritable condition of the uterus, and heal the commonly abraded condition of the os tincæ, partially relieving the expulsive effort. As an auxiliary means of relief apply the form of belt, as shown in fig. 91, to the pendent portion of the abdomen; the modus operandi of which we believe to be an arrest by compression obliquely upward, of arresting the expulsive muscular effort of that region; as it invariably affords relief to that distressing sensation and pain in the back. These displacements of the uterus in patients in ordinary health, but seldom, if ever, fail to be relieved by the treatment here described. A repetition of the manipulation at least once a week for three or four weeks, will be required in cases of long standing, to insure success from this treatment. There is, in some cases, a lateral tendency of the uterus in its descent; but it requires no special treatment from that of the conditions described.

The uterus, in its descent, often carries before it the bladder; producing *prolapsus vesicæ*. This is one of the conditions of the retroverted viscera that is more perplexing to the practitioner than all that class of ailments. The successful treatment of procidentia or prolapsus uteri, contributes largely to its relief, and then the replacing of the bladder should follow. The patient should be made to recline on her back with her hips elevated for one or two hours, twice or three times a day, and this rest and position should be followed by an injection into the vagina of a solution of a drachm and a half of zinci sulph. to a quart of water.

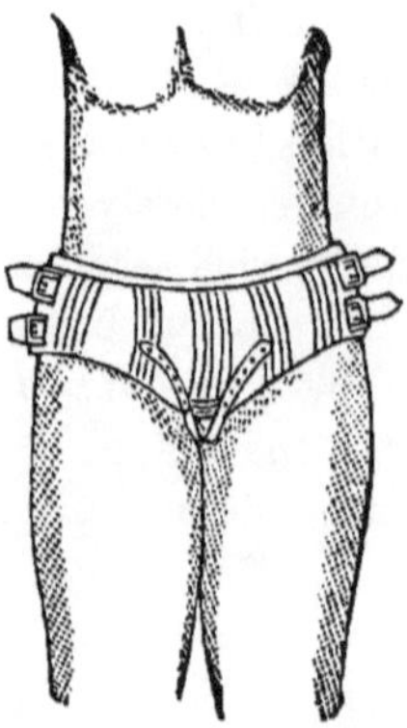

This greatly allays the irritation that tends to the expulsive effort, as also the distention of the vagina and the concomitants of the ailment. A decided auxiliary to the treatment is the abdominal support belt and perineal elevator as applied. Fig. 91.

A, the front of the belt having two buttons for attaching the cross-straps that are intended to be passed between the limbs. *B*, the back of the belt having the cross-straps *c*, attached, and the perineal elevator sheathed upon them. See Fig. 92.

Fig. 93. *A* represents the cross-strap, and *B* the elevator, of triangular form having at the base, loops of about two inches in length through which the cross-straps pass; a vertical, transverse section (*B*) giving the form — the apex to be applied to the perineum. The elevator is from an inch and a half to two inches in height, presenting an external vertical support to the uterus by elevating the perineum when firmly applied, which we believe to be more effectual in relieving procidentia or prolapsus uteri than any pessaries at present in use. Pessaries tend to excite leucorrhœal discharges, and relaxation of the vagina, and, in many cases, the patient's health is greatly impaired by their use; whilst, by the elevation of the perineum to a sustaining of the uterus, the lateral ligaments of that organ are relieved from extension, and the vagina restored to a normal condition.

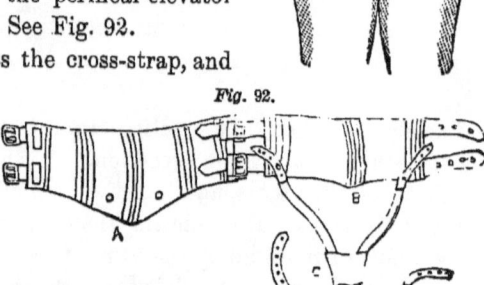

Fig. 91.

Fig. 92.

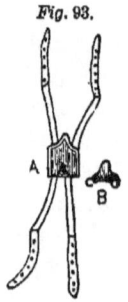

Fig. 93.

ECTROPION VESICÆ.

The most common designation of this malformation is that of *extroversion* of the bladder; as the anterior portion of this viscus is deficient, and the parietes of the abdomen that should form its normal covering are wanting, so that the posterior and lower portion protrudes under the pressure of the abdominal viscera, presenting a secreting tumor covered with a mucous membrane in which the orifices of the ureters can be seen — the urine exuding from them. The umbilicus is not distinctly marked, but usually indented by an elongated indentation. The linea alba is bifurcated at the upper angle of the umbilical cincture, and continued on either side down to the os pubis, so as to form a triangle in which the extroverted bladder lies. The pubic bones are not united by a symphysis but joined to each other by ligament. The penis is small; the corpus spongiosum is wanting; and only a remnant of the urethra remains, presenting a mere groove lined with mucous membrane on the dorsum of the rudimentary penis leading to the vesical tumor. The glans penis is full and large, and the prepuce usually of full size, but cleft above as though the operation for phymosis had been performed. The testicles are usually found in a contracted scrotum. In either groin loose folds of skin contain hernial protrusions. This applies to the male. Female subjects have been found thus conditioned; the exposed mucous membrane extending to the vagina.

This deformity is attended with much suffering; the exposed, tender surface of the protruding bladder covered with pendulous papillæ bleeds freely when, almost unavoidably, chafed by the clothing that is saturated with urine, while there is emitted a constant stench, making it altogether a hopeless, disgusting infirmity.

For the relief of this sad ailment, modern surgical operations have been unsuccessfully attempt-

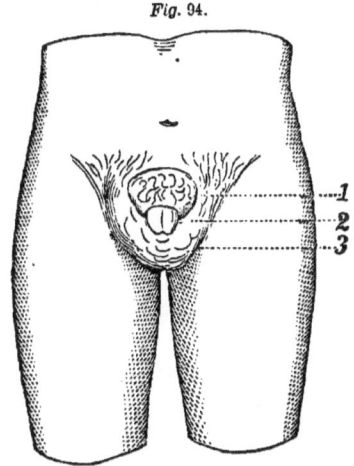

Fig. 94.

1. The secreting surface of the bladder.
2. The rudimentary penis.
3. The distended scrotum, from normal distension.

ed by the most eminent in the profession. Plastic operations have been performed, and failed in limiting the extent of the secretory surface to that of a controllable drain, by diverting the course of the urine into the rectum. Such procedures have proved most hazardous to life, with little promise of success in the future. A well-fitting female urinal serves to catch the urine and retain it, and also relieves the person from the intolerable stench, that is so offensive, arising from the saturated clothes, under ordinary circumstances. The annexed engraving represents the condition of the person. The secreting surface of the bladder, as seen above the rudimentary cleft penis, and the distended scrotum beneath and posterior to the glans penis.

RELAXED ABDOMEN.

Relaxed abdomen is an ailment that subjects both male and female to much inconvenience and suffering, tending to an increase of obesity of the abdomen far exceeding the tendency to that condition of the limbs. A belt, as described above for abdominal hernia (omitting the hernial pads) will, if applied, arrest the tendency to pendent abdomen, and afford much comfort. In women the predisposition to relaxation of the abdominal visceral supports tends to pressure upon the pelvic supports, inducing their derangements, such as procidentia uteri and vesicæ; which ailments are relieved by the belt.

The patient, under use of the belt, seldom fails to find relief from the pendent abdomen and local derangement, and, in many instances, a perfect cure, if their health is also improved by a judicious regimé. Sedative tonics, such as ext. valer. gr. 1, ferri sulph. gr. ¼; taken three times a-day, afford much relief if laboring under an irritable nervous condition. In cases of failure in the digestive functions tending to acid eructations, bismuthi subnitratis grs. 5, quiniæ sulph. gr. ½; taken morning, noon and night, is most salutary in its curative tendency.

Sea sickness is greatly mitigated by wearing this abdominal belt, as it arrests the tendency to relaxation of the abdominal walls, and arrests visceral agitation from the motion of the vessel.

Prolapsus ani and *hemorrhoids* may be considered concomitants of relaxed abdomen. The abdominal belt with the perineal support affords very efficient means of relief as an auxiliary to other treatment — affording pressure to the hypogastric region that invariably gives relief to the distressing propulsive efforts attending these ailments.

Prolapsus ani most commonly occurs in children when enfeebled and suffering from irritation of the digestive or urinary organs. In delicate adults, there is also a natural tendency to this ailment. A slight protrusion of the mucous membrane takes place during defecations, and is increased by any constitutional tendency to atony of the muscular system.

The most common causes, however, are irritation of the intestinal mucous membrane from morbid secretions in an abnormal functional condition, as habitual constipation often occurring in persons of a relaxed habit of body, a want of power in the rectum to expel its contents, requiring a constant straining at stool, and diseases of the urinary organs, as that of stricture or stone requiring effort to micturate. Mr. John E. Erichsen states: " In other cases, and indeed most usually, the prolapsus is associated with piles, the weight and dragging of the hemorrhoid drawing down the mucous membrane of it.

Diagnosis: The prolapsus, as before stated, consists in the protrusion of the mucous membrane of a red, purplish color, having a turgid, lobulated appearance varying in size. The protruded mucous membrane is continuous with that investing the sphincter, and this constitutes the mark of distinction between the ordinary prolapsus and the invagination that occasionally presents. In this condition a deep, distinct sulcus will be observed between the protrusion and the margin of the sphincter. In chronic cases of prolapsus ani the anal aperture is often permanently relaxed and widened within; externally, folds of skin are apparent, from relaxation radiating from the anal centre, having a bluish, soft, pendulous appearance and attended with intolerable itching. The protrusion, at first, occurs only after defecation, and readily retracts, or is reduced by application of steady pressure upon it. Subsequently, if not permanently relieved, it will protrude after riding, walking, or standing, and returned with much pain and difficulty.

Treatment.—In children, the cleansing and bathing the protrusion with moderately cold water, and if retraction has not ensued efforts must be made at reduction. To the accomplishing of this, a piece of worn linen folded one or two thicknesses, smeared on one side with simple cerate, should be applied to the prolapsed bowel; then find the anal orifice with the finger, passing it well into the bowel; press the protruding parts slowly into the rectum, and after a few moments withdraw the linen. Prof. Vogel, in his most excellent treatise on diseases of children, translated from the German by Prof. H. Raphael, M. D., states that "it is very useful to slide a small round piece of ice into the prolapsus before the reduction is undertaken, and then to reposite the protruded bowel with the ice." Astringent injections are reliable means of cure; sulphate of iron one to five grains in an ounce of water, injected merely within the retaining power of the sphincter, and in such small quantities as to be retained; careful attention to the maintaining of a normal condition of the bowels; a nutritious diet enforced; and a pledget of lint made wet with cold water and applied to the anal region, and sustained by the abdominal and perineal support obtains usually a cure in children.

In adults if not accompanied with hemorrhoids, the above treatment is quite efficient to their relief. In cases where hemorrhoids exist, the removal of the piles by ligature cures the prolapsus.

When prolapsus fails to be relieved by palliative treatment a portion of the protruded mucous membrane must be removed, and for the accomplishment of which we prefer the ligature, thus avoiding the tendency to exhausting hemorrhage. In case of strangulation of the parts protruded, strenuous efforts should be made for its reduction, and if not possible free incision of the sphincter may be made. If not reduced it will slough off, and thus tend to a permanent cure.

Hemorrhoids.—In this ailment much relief is afforded by having the parts regularly sponged with cold water and cloths wet in it and applied to the parts; or a solution of Barbadoes tar which is a most effectual remedy; and in cases of internal piles, injections of it are usually very efficacious. The following electuary affords very decided relief in cases thus conditioned:

℞.

  Pulv. nucis moschatæ   ℨij

  Sulph. sub.

  Conserva rosæ, āā   ℥i

          M.

  Ft. electuary.

A drachm of this taken night and morning, usually gives relief from the tenesmus and distressing pain.

In the treatment of hemorrhoids, it must be considered that the ailment is the result of some remote visceral disarrangement, requiring careful observation and a corrected régime, otherwise treatment will fail to afford permanent relief; as in all this class of ailments, the result of an enfeebled condition of the system, a well devised constitutional treatment is imperative even to the relief of the patient, then local palliative appliances will not disappoint the practitioner's expectations in treatment.

CHAPTER X.

VARICOSE VEINS—BURSÆ—GANGLION.

Persons specially amenable to Varicose Veins.—Differing indications of ailment.—Diagnosis given in Holmes's "System of Surgery."—The œdematous leg and varicose ulcer.—Women more liable than men to varicose veins.—The affection described.—Congestion may exist.—John Kent Spencer upon the pathology and treatment of ulcers and cutaneous diseases of the lower limbs. Varicose ulcers.—Description of the Syphilitic ulcer.—Treatment of ulcers of the leg.—Mechanical appliances used.—System of Measurement.—Beneficial effect of the laced stocking.—BURSÆ.—Bursæ mucosæ and bursæ synoviæ.—Distinctive symptoms described.—Mr. Paget's opinions regarding them.—The "Housemaid's Knee."—Unfavorable changes in pathological order.—Treatment of Bursæ.—Erichsen on Pathology of Chronic Enlargements of Bursæ.—GANGLION; Diseases of Sheaths of Tendons.—Diagnosis of Ganglion.—Treatment.

VARICOSE VEINS.

The pathological impairments, tending to a varicose condition of the veins of the lower extremities, have not, as yet, been satisfactorily comprehended. Various opinions have been advanced, but not admitted as the actual cause of this most common ailment, being only one-third less in frequency than abdominal hernia, with which every tenth person is afflicted; and like it, too, for relief, surgico-mechanical appliances are indicated, which but seldom afford permanent relief.

Persons suffering from varicose veins are those whose vocation requires the erect passive position of the body; as that of many trades, book-keepers, cardrivers, washerwomen, cooks, and all kinds of labor demanding a standing position for great lengths of time. These all tend to induce the ailment, and, if not relieved, in most cases, the limbs become painful, and swollen; then follow erythema, inflammation, and, finally, extensive ulceration.

Necessity compels a vast number of the indigent to labor for their subsistence, even when afflicted with varices from the unfavorable employment; and from, even, careful observation, it is not possible to decide whether there is a constitutional predisposition to the ailment, or that the pathological condition is the sequence of the unfavorable influence of position even upon all constitutions. That some yield more readily than others to the unfavorable influence, and suffer more from its encroachment must be admitted, but this is from a normal condition of the individual; being, comparatively, deficient in physical resistance to all unfavorable influences.

Persons afflicted with varicose veins, differ greatly in the indications and appearances of the limbs affected. The first and most common indication peculiar to the ailment in all those affected with it are deep, aching pains in the limb, with a sense of weight, fullness, and fatigue. These are the first indications of the congested condition of the peripheral veins, and from a slight bruise an effusion of blood in the cellular tissue ensues, leaving a dark spot; or, a very slight wound will bleed most profusely, and tend to ulcerate. This condition is common to all classes of persons subject to the ailment. However, these initiatory symptoms are the only unfavorable indications that are apparent to observation, yet the patients are subject to the ulcerative sequence. The impeded circulation is denoted by these symptoms, and limited to the deep seated consequently constricted veins, which receive all the blood from the distended peripheral veins — there being no large conduits immediately in the dermoid tissue, which becomes vitally impaired from congestion. In this class of patients the diagnosis is very obscure and fails of proper treatment, being confounded with other more serious ailments. Of this peculiarity in the ailment, writers have not made mention although a quite common condition of patients. In Holmes' "System of Surgery," a very concise description of the more conspicuous conditions of the ailment is given.*
After a time, which varies with the idiosyncrasy and occupation of the patients, small, soft, blue tumors are seen at different points in the leg; most of them will disappear on pressure, but will return when the pressure is removed, or when the patient stands. Each little tumor is caused by a superficial vein dilating to a point at which it is joined by an intra-muscular branch. Around many of the tumors a number of minute vessels are clustered, of a dark purple color; these being the small superficial veins which enter the

---

* Vol. III, p. 314.

large superficial dilating vein, and in which the blood is retarded in its passage. An increasing length of vein becomes gradually involved, and a number of irregular, knotty, convoluted tumors are developed, grouping themselves around the points at which the dilatation first began. The external and internal saphena veins are those principally affected, but long tracks of tortuous veins may extend up the leg and thigh. Dangerous and even fatal hemorrhage may ensue from the bursting of a varix through the skin. The vessels may become filled with clots, and permanently obstructed. Ulceration and disease of the skin are some of the most common pathological sequels in cases of long standing.

There are but few more painful lesions of tissue, or disenabling ailments from labor than that of the œdematous leg and varicose ulcer — the sequence of varicose veins. Persons of feeble constitutions thus affected mostly labor under superficial indolent ulceration, whilst those of a more robust condition suffer greatly from the irritable ulcer. The young rarely suffer from varicose veins. This would indicate that certain occupations eligible only to adults tended to this ailment. Women are much more liable to varicose veins than men, and when ulceration ensues, their cases become most intractable. The locality of the varicose ulcers is most commonly on the inner and lower third of the leg, the location presenting indications for a considerable time before the ulceration ensues; as that of erythema and discoloration. This discoloration is attributed to a neurose derangement of the vasa-motor nerves, and to the partial arrest of nutrition; denoting not only impeded circulation, but waste or diminution of cuticular tissue, which is often covered with dry scales that, when removed, expose a secreting surface that tends to ulceration, having a border of hardened connective tissue. This ulceration appears to be an entire destruction of the dermoid tissue, often extending, irregularly, half way around the leg.

The legs are greatly exposed to injuries, more often, in fact, than the other parts of the body, and upper extremities. Greater stress is, also, brought to bear on their vital forces from unfavorable occupations, as would naturally necessitate long continued standing in a comparatively quiescent condition. This leads to an impairment of functional nutrition of the dermal, sub-dermal and connective tissues. In the incipiency of a more serious lesion we have an erythematous irritation attended with a most intolerable itching, which, from an abrasion of the surface, secretes a morbidly acrid excretion, increasing the local irritation to the degree of a circumscribed ulcer, and a general œdematous condition of the leg below the knee, and which tends to impede the general venous circulation and increase the congestion and swelling.

This condition of the limb often exists independently of a previous varicose condition of the veins, and it, therefore, does not follow that varicose veins are the invariable indication of simple ulceration of the legs.

In the normal condition of the legs it must be admitted that the physical law of gravity is stronger than the dynamical influence governing the circulation in the ascending venous return of blood from the lower extremities; which is subject to a limit of force and is an actual cause of varicose veins — more especially in persons of feeble muscular energy, in whom ulceration of the limbs is not always a result of that abnormal condition of the veins, even when greatly distended.

Congestion of all of the soft tissues may exist, and from a similar cause to that producing varicose veins; but a varicose condition has not resulted because of a more numerous division of the superficial veins of small calibre — a peculiar condition of the limbs in some persons. In such persons numerous patches of discoloration will present, and, on examination, will apparently consist of capillary veins and exudation of venous blood under the cuticle.

From slight injury, persons thus conditioned are subject to ulcers of the leg, and no varicose veins presenting, a doubt may be entertained as to the simple character of the ulcer which is identical, in a pathological point of view, with that of the varicose ulcer, and subject to relief by similar treatment. Mr. John Kent Spencer, in his manual upon the pathology and treatment of ulcers and cutaneous diseases of the lower limbs, states that varicose and obstructed veins are the remote causes of nearly every form of non-traumatic ulcer in the lower extremities, and, that it may be said to be the most common cause of indolent ulceration of the limbs having resulted from accidental injury.

VARICOSE ULCER.

The varicose ulcer is liable to be confounded with the syphilitic ulcer, of which last we will here give a description: "Syphilitic ulcers may result from pustules, tubercles, boils, or may begin as tertiary sores; they frequently occur where the integuments are thin, or where they are moistened by the natural secretions of the parts. They are circular, with elevated edges; tend to spread in circles, with a foul greyish surface; often creeping along slowly, and destroying deeply the parts they affect; leaving cicatrices of a bluish or brown color, thin and smooth, which are apt to break open again on the application of any slight irritation." *

It is of much importance to determine the true character of the ulcers of the leg. The varicose ulcer is curable by simple local treatment, the patient being otherwise in good health, or, if feeble, rest and liberal diet is sufficient. But if the ulcer is of syphilitic origin, a general constitutional medication will be required to insure even tolerable success.

The varicose ulcer is preceded by a varicose condition of the veins, and, as a sequence, œdema, erythema, a large patch of discoloration, and lastly, ulceration; usually by dissolution of the superficial integument, leaving an irregular formed ulcer and limited to the inner, lower third of the leg — in most cases. There are exceptional cases when the ulceration exists about the malleola.

Mr. Maunder states in the London Hospital Reports, vol. ii, p. 129, that all those cases which are situated above the middle of the leg are syphilitic in their origin, and are mostly multiple, also; while the varicose ulcers are found below the middle of the leg, and are usually solitary.

TREATMENT.

In the first stage of the ailment it is necessary to consider the indications that present: Dilatation of the dermal veins, inefficiency of the valves to prevent regurgitation, œdema of the connective tissue, and passive congestion of the capillaries. These are all indications to be relieved by moderated compression, which will assist the valves in sustaining the column of blood. It will also restore the lymphatic and superficial blood vessels to their normal physiological state.

This stage of the ailment having been relieved arrests the progress to the ensuing stage, and is of the utmost importance to the patient. The most reliable means for this purpose is the non-elastic laced stocking.

---

* Erichsen's Surgery, 4th edition, p. 539.

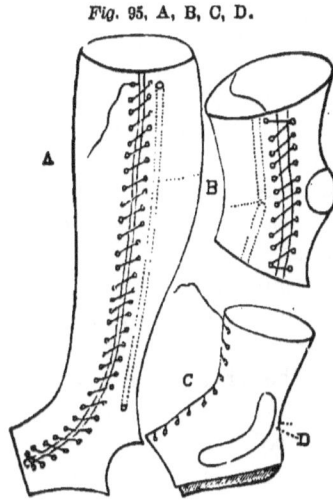

Fig. 95, A, B, C, D.

Figs. 95 represent the appliances used for compression. A, a laced stocking for the leg; the dotted lines representing whalebones encased to prevent the stocking from wrinkling. B, a knee bandage. C, an ankle bandage. D, leather stiffenings as a support under the malleoli; being thin sole leather, thinned on the edges and secured by a covering of soft leather, the leather being made wet, becomes pliable, and when the bandage is laced upon the ankle co-aptates to the form and gives an admirable support to a weak ankle.

Figs. 96 and 97 represent the proper parts for measuring, in order to obtain a suitably fitting bandage. For a knee bandage the circumference should be taken at A, B and C; and the length from A to D. For an ankle bandage or sock, the circumference at A, B, D and E; and the length from A to C. The second figure has the points of measurement for a laced stocking — circumferences at A, B, C, E and F; length from A to D. These figures delineating the measurements also represent the wire spring cloth stocking and knee-cap; the only admissible elastic material that should be used for such purposes, as it is not affected by temperature as is the case with India rubber which is not only rendered soluble by perspiration, but becomes a decided irritant; often exciting erysipelas when worn in warm rooms or during the heat of summer. Elasticity is not an essential quality in the construction of these bandages, and is objectionable because of the unavoidable thickness of the elastic material which tends to increase and maintain the existing increased temperature of the diseased limb. Light woven fabrics meet the indications more fully as, when made to fit,

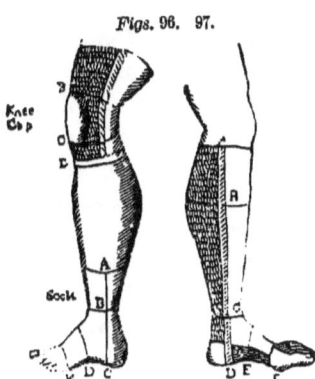

Figs. 96. 97.

and laced moderately tight, its equable pressure facilitates the return of the blood in the superficial veins, and when laced to a suitable degree, being porous, admits of exhalations passing through, greatly to the comfort and tending to the cure of the patient. The laced stocking, knee and ankle bandage, as represented in figure 96, are most simple in construction. By doubling ordinary twilled cotton cloth, and marking out with a pencil the forms as represented, obtaining the correct dimensions from proper measurement — taking one half the circumference for the measure, as the material is doubled they will, by allowing half an inch for seams, fit the limb perfectly. For the lacing, as seen, one side is to be cut and faced, all the edges to be bound with tape. Any tailor or seamstress can readily make these bandages, and thus contribute to the relief of numerous sufferers from the ailments of varicose veins, weak knees and ankles.

In the treatment of varicose ulcers, the laced stocking tends greatly to success. If inflammation and pain is excessive, and the patient's condition in regard to their subsistence will admit of their being placed at rest, it is advisable; yet relief can be given the patient even while at labor, by the application of the laced stocking, and the subjoined treatment. The following lotion and dressing should be applied at night; plumbi acet. et zinci. sulph. āā ℈i; aqua font. oj.; vin. opii. ℥ss. M. ft. sol. Several thicknesses of worn linen or cotton fabric should be wet with this lotion, or ordinary lime water, and applied to the ulcer. Over this and the exposed inflamed parts of the leg a towel, folded and made wet with water at a temperature not under 60° Fahr., should be applied. In the morning the ulcer should be dressed with fresh mutton tallow, and some lint or folded cloth to absorb the secretion—covered with oil-cloth—and the laced stocking laced firmly on the leg. If the pain is not relieved in a few days the ulcer should be penciled with a solution of nitrate of silver, four grains to the ounce of distilled water. A few applications of this solution usually affords relief. The general health of the patient, as a matter of course, claims special attention. This treatment we consider as applicable to the pathological condition of the patient, *i. e.*, congestion of the capillaries, œdema of the connective tissue, dilatation of the veins, and loss of vital energy in the superficial tissue of the parts affected—the lesion being the result of impediment from whatever cause, to the return of the venous blood in the superficial tissues of the legs.

BURSÆ.

Bursæ are an essential part of the animal economy, and like that of other functional apparatus, subject to abnormal conditions, as well as from extraordinary influences which become developed; as in the instance of bursæ on the dorsum of the foot in talipes varus, and *housemaid's knee*. Two forms of bursæ exist, and two distinct offices are performed by them. They are known as *bursæ synoviæ* and *bursæ mucosæ*. These bursæ are large, simple or irregular cavities in the subcutaneous areolar tissues, containing a clear synovial fluid, and are found in various situations; as between the integument and front of the patella, over the olecranon, the malleoli and other prominent parts. They are also found interposed between muscles or tendons as they play over projecting bony surfaces; as between the glutei muscles and surface of the great trochanter. They consist of thin walls of connective tissue partially covered by epithelium, and contain a viscid fluid. These bursæ usually communicate with the cavities of joints, as is the case with the bursæ between the tendons of the psoas and iliacus and the capsular ligament of the hip, or, the one interposed between the under surface of the subscapularis and the neck of the scapula.

Tendons passing through osseo fibrous canals are supplied with synovial sheaths. The sheaths are formed of two layers, one of which adheres to the wall of the canal, and the other is reflected upon the outer surface of the contained tendon; the space between the two free surfaces of the membrane being partially filled with synovia. These sheaths are chiefly found surrounding the tendons of the flexor muscles of the fingers and toes, as they pass through the osseofibrous canals in the hand and foot, and from unfavorable influences are liable to form a morbid condition.

Stress of pressure, and friction tends to the enlargement of these bursæ. Mr. Paget, in his Surgical Pathology, says: "Two methods obtain as regards their formation. Some — of which the best example is the bursæ over the patella and its ligament — are merely enlargements, and with various transformations of bursæ naturally existing. Not materially different from these, are the bursæ which form anew in parts subjected to occasional localized pressure, and which appear to arise essentially from the widening of spaces in areolar or fibro-cellular tissue, and the subsequent leveling or smoothing of the boundaries of these spaces; but others, such as the bursæ or ganglions which form about the sheaths of the tendons at the wrist, appear to be the cystic transformations of the cells inclosed in the fringe-like processes of the synovial membrane of the sheaths."

These bursæ undergo various changes in structure and contents — tending to the most serious consequences. Special employment is a most common cause of these morbid changes. Kneeling upon hard surfaces, as that of the floor, occasions enlargement of the bursæ patella — known as the *housemaid's knee*. Miners also are subject to an enlargement of the bursæ lying over the olecranon, commonly called *miners' elbow*, and in every situation a new bursæ may be formed by continued pressure and friction conjoined.

These unfavorable changes ensue in order somewhat as follows:

a. Continued pressure and friction tend to enlargement of the exposed bursæ which becomes filled with a clear, sero-synovial, straw-colored fluid.

b. Inflammation supervenes, tending to suppuration, when the contents change to pus and bursal secretion. The termination is then to present a point, and open, like an ordinary abscess — usually with a sloughing tendency, often terminating in a widely spreading abscess.

c. In some cases the bursæ will contain a dark fluid with a large number of small, flattened, elongated bodies of about the size of grains of rice, or melon seeds, floating in it. These bodies are of a fibroid, or fibro-plastic structure, resembling masses of imperfectly developed exudation cells, and are, apparently, portions of disintegrated lymph.

d. The tendency of the bursal tumor, in some cases, is to become perfectly solid; the walls becoming thickened by the disorganized lymph in its interior. When this condition supervenes, then a section of the bursæ presents a solidified, laminated appearance.

e. As bursæ become elastic, they have a crackling sound on being pressed, and more especially when they contain the rice-shaped bodies tending to a solid growth. The inflammation in the diseased bursæ often extends to the neighboring joint from simple continuity — there being no other connection; or inflammation and suppuration may exist within the cavity of the joint, and in the bursæ, having no connection; as in the case of strumous conditioned patients. These adventitious rice-shaped bodies often become a source of severe suffering, when formed within the capsular ligament of the knee, from thence floating within the bearing of the joint when in the erect position, as in walking, affecting the individual instantaneously, and the pain so intense as to cause them to fall as if shot.

TREATMENT OF BURSÆ.

In the first tendency to increase of development in an exposed bursæ, a removal of the cause, and gentle pressure affords permanent relief; and in many instances, in advanced cases, pressure and rest is a reliable treatment. When much inflammation has supervened, leeches, and tepid lotions containing acetate of lead will arrest the abnormal tendency.

When suppuration takes place, and they are in an indolent condition, cloths saturated with water at a temperature of 60°, holding in solution a drachm of fluid ext. belladonna and two drachms of tr. iodine to the quart, kept wet and secured with tolerable pressure, may be used with decided advantage ; and, when the patient's condition is such as to admit of saline purgatives, they should be given freely, or, calomel and jalap, of each eight grains to the dose. This treatment will often tend to a cure.

When the sac is thin, and the fluid serous from recent injury, blistering is often an effectual remedy; and especially when it occurs under the deltoid muscle. The passage of a small seton through the sac, and making a decided elastic pressure by means of raw cotton, and a laced bandage, as that of the knee bandage described in another part of this work, is quite a reliable means of relief in non-strumous or scorbutic subjects. If of strumous diathesis, violent inflammation often follows the insertion of the seton. Dynamic electricity, as an auxiliary, often contributes largely to the cure when in the suppurative condition — bearing in mind the careful enforcement of hygiene regimé, and alterative tonics ; as that of tr. cinchona and bichloride of mercury, and, if deemed advisable, iodide of potassium ; carefully regarding the constitution of the patient.

The pathology of chronic enlargements of bursæ is given by Mr. Erichson as follows : " There is an enlargement of the bursæ, and excessive secretion into its interior of simpler synovial fluid ; this, however, is discolored, probably from admixture of blood which has undergone disintegration. This fluid will be found to contain a large quantity of cholesterine, broken up blood corpuscles, and granules. The melon seed bodies are composed of largely organized fibroid matter mixed with cholesterine, and are probably separated from the serum of the bursæ.

"Solid tumors may be formed in connection with the bursæ patellæ. By many these are supposed to be the result of the deposition of a fibroid material which gradually takes the place of the fluid of an ordinary housemaid's knee, and which, instead of taking the form of melon-seed bodies, is deposited in concentric masses, and thus accumulated in the interior of the cyst. This has not been the case, however, in many instances that I have seen. In these cases I believe there is a true fibroid deposit in the bursæ from the very first; the tumor is never fluid, but hard and solid from the commencement, and continues to slowly augment in size until it occasions sufficient inconvenience to require removal. In some cases there has been a previous syphilitic taint; the patient complains of pain in the tumor like that which is experienced in nodes, and it is by no means impossible that there may be a syphilitic origin for these tumors. However that may be, in the cases that have fallen under my observation, the tumors have never been fluid, nor have they originated in pressure, but appear to have been primary deposits of fibroid matter."

The treatment for these solid tumors is plainly indicated; that of dissecting them out, carefully avoiding the opening of the capsular surroundings of the joint. Constitutional treatment is considered indispensable to their future welfare. When in the ordinary fluid condition elastic compression by the application of cotton, bathing, and the lace bandage before described, is the most safe and reliable treatment. Static electricity conjoined renders the bandaging more effectual in promoting absorption of the distending fluids.

GANGLION.

DISEASES OF SHEATHS OF TENDONS.

The sheaths of tendons are subject to two forms of disease, viz.: An accumulation of fluid within the interior of the sac, known as *cystic swelling*, or *ganglion*, and acute and chronic inflammation, or serocystitis. The common ganglion consists of a cyst containing a clear or yellowish colored serous, and in some cases, gelatinous and semi-coagulated fluid. They usually occur on the back of the wrist in the form of a distinct, round tumor, distinctly formed within the sheaths of the tendons, and are at times found upon the foot and wherever tendons are thinly covered by external integument—greatly impairing the strength of the joints over which they may be located; and, from increasing in dimensions, press upon the nerves; inducing, in some instances, considerable pain, especially when located upon the back of the wrist, by compressing some of the branches of the musculo-spiral nerve.

The palm of the hand, dorsum, sole, and inner side of the foot are subject to ganglion. In these locations they consist of a dilatation of the sheaths of the tendons, and often increase to a large size, becoming irregular in shape from the implication of several tendons in the tumor. The contained fluid is usually thinner than in simple ganglion or ganglion confined to a single tendon. The sheath itself is vascular, and lined by a red, fringed and velvety membrane; the fluid is often dark and bloody, containing masses of buff-colored fibrin, or a large number of granular bodies like those met with in certain form of enlarged bursæ. This is a most serious condition of the ailment; having a malignant appearance and being chronic in character.

Ganglion, when situated in the palm of the hand, usually extends under the annular ligament and upon the flexor tendons of the fore-arm, and is extremely hazardous in treatment; an uncontrollable inflammation being most readily induced, to the great injury of the hand. The fluid in the ganglion in this location is most readily pressed into the palm of the hand, or from the palm of the hand into the sheaths of the flexor tendons above the wrist.

TREATMENT.

The treatment of simple ganglion on the back of the wrist is most readily accomplished by a blow from the hard cover of a book. Bending the wrist so as to render the integument over the ganglion very tense, and then striking it with some force it is dispersed, and, usually, without further disturbance. It can also be disposed of by pressure from a small piece of metal placed over the ganglion and bound down as tightly as can be borne. This is a more tedious treatment, and no more safe than that of rupturing the tumor with the book. The other means of procedure are often attended with more or less serious consequences. They consist in puncturing, and the insertion of setons, thereby exciting inflammation, as in that of bursæ, the indication of each demanding similar treatment.

The ganglion located in the palm of the hand and extending under the annular ligament, and up the flexor tendons of the forearm, presents much difficulty in the treatment, and always an apprehension of serious consequences; violent inflammation being most readily excited from puncturing or the insertion of a seton, which is the ordinary treatment. Mr. Syme recommends that the cyst should be laid open and the annular ligament divided. This is certainly decided treatment, and must be attended with as much difficulty and tendency to inflammation as the practice of puncturing — if not more — and is certainly more painful to the patient, and should command much serious consideration.

CHAPTER XI.

PATHOLOGICAL CONSIDERATION OF DISEASES OF THE JOINTS.

Abnormal functional nervous energy tends to degeneration of tissues.—
Acute and chronic inflammation of the joints.— Not a local impression, but
always attended with febrile symptoms.— Richet's experiments.— Synovitis.—
Case recited by Dr. Stanley, of St. Batholomew's Hospital.— Dr. Holmes
Coote's description of a case of synovitis.— Dr. Edward Rindfleisch and Dr. R.
Volckmann on the progressive stages of synovitis.— Morbus coxarius.— Acute
and chronic forms. *First stage:* Its diagnosis and treatment.— Dr. William
Coulson on scorbutic diathesis.— Shortening of the diseased limb. *Second
stage:* Extension by weight and pulley will not prevent shortening of limb.—
Result of experience in 859 cases of hip disease.— Treatment as pursued in
the Institution for the Relief of the Ruptured and Crippled.— Hospital for sick
children in London.— Cure reported by Mr. Thomas Holmes. *Third stage:*
Illustrative cases.— Tabulated statement of results of operation.— Mr. Holmes'
analysis of the treatment at Sick Children's Hospital.— Excision of the hip.—
Barwell on diseases of the joints.— Returns from Charing Cross Hospital,
London.— Treatment.— Dr. William Coulson, surgeon to the Magdalen Hospital, London, on the pathological condition of patients suffering from this
class of ailments.— The beneficial effects of moderate exercise.— Torpidity of
the digestive processes.— Treatment in special cases.— Extension apparatus.
— Pathology of hip disease summarized. — Prognosis of unfavorable cases.
— Character of the secretion.— Treatment when unfavorable symptoms occur
after excessive spontaneous discharge.— Number of cases of morbus coxarius,
caries of the spine and synovitis, treated at Hospital for the Relief of the
Ruptured and Crippled during last nine years.— Pyemic Dyscrasia of the vital
functions a condition consequent upon complicated pathological invasions.—
The regimé imperatively necessary to be even partially successful in this hapless condition.— Diet.— Neuralgia of the diseased parts a frequent resultant
of treatment.— Remedy.— Local treatment of sloughs.— Mr. Holmes Coote's
opinion of the treatment to be pursued.— Several ailments that assimilate hip
disease.— Mr. Erichsen on chronic rheumatic arthritis.— Neuralgia of the
sacro-iliac joint.— Coxalgia.

Aberrations of nutrition tending to changes in the alimentary
materials for organized tissues must result in an anomalous order
of development. Normal maintenance is dependent upon due

functional activity in the absorption of alimentary material, circulation, respiration and excretion; and each of those functions dependent upon nervous energy. This is most readily observed in a deficiency of nerve power in a limb, or the retarded convalescence of a patient.

These considerations are of the utmost importance in the diagnosis of any apparent local disease, and not in the selection of specifics for its cure.

Abnormal, functional, nervous energy tends to degeneration of tissues, with an increasing tendency to local impairment of parts — such as that of the joints, and not merely the sequence of inflammation, the result of congestion, as formerly considered. Inflammation exists and is still continued from deficiency of nutrition and perversion of secretion in the parts presenting the following conditions:

Inflammation of the surrounding tissues — *acute* and *chronic* — extending to the cancellous tissues of the bone. Secondary effects, suppuration, necrosis, ulceration of the articular cartilages, elongation of ligaments and subluxation, as in that of the knee; and complete luxation in others, as in that of the hip.

Inflammation of a joint is not strictly a local impression, although induced by external injury; as from a blow, wound, or exposure to cold. It is invariably attended by febrile symptoms, and, in severe cases, involves the heart and pericardium. The surroundings of the joint become swelled, and the limb assumes a semi-flexed position; tending to relief from pain by the relaxing of the lateral and other ligaments relieving pressure on the inflamed tissues. Of the actual condition of the joints we have but a limited knowledge, because of the want of observation not readily attained by dissections made upon human joints at this stage of the invasion. Richet experimented upon animals by injecting irritating fluids into their articulations. The first effect was a distension of the peripheral blood vessels, a disappearance of the epithelial covering, a dull appearance of the synovial membrane, having an uneven, granular roughness of elevated, small processes — the enlarged papillæ of the synovial membrane. This vascular injection terminated at the border of the articular cartilages, for a time, and then extended to the surface; presenting there a bright vascular zone. This vascular zone gradually increased and extended upon the cartilage to the degree of congestion and the farther enlargement of the papillæ. Fibrine, then, was effused on the inner surface of the synovial membrane, tending to an increase of thickness; being of a thick, viscid consistence. It finally became turbid, and mixed with reddish serum and oil globules.

In man, this condition of synovitis is of fearful import, as it advances rapidly to the destruction of all of the implicated tissues. The synovial secretion increases largely, and where opportunity has been afforded for microscopic examination, pus corpuscles have been found in great numbers. The synovial membrane having become soft and swollen, with great distention of the joint, inducing the most excruciating pain, and compelling, as it were, the opening of the distended parts — a procedure attended with almost a certainty of the loss of the limb, and, in many instances, of the life of the patient.

Synovitis most commonly presents in a less violent form, and in young subjects. Even the acute condition subsides, in some subjects to a sub-acute, and, finally, chronic form. The sub-acute synovitis usually makes slow progress, and is attended with much sensitiveness when the limb is moved, is greatly moderated by rest, and is seldom attended with much constitutional derangement during the early progress of the ailment.

If not arrested, a recurrence of symptoms ensues with an increase of effusion and a softening of the ligaments and serious impairment of the articular cavity, with increasing distension about the joint. We then have *Hydrops Articuli*, and a chronic condition established. The hip, knee, elbow and ankle are all subject to this pathological condition; greatly distorting the parts affected from the irregular extension of the ligaments and unequal distension from effusion, even to the extent of subluxation of the joint. We are informed by such indubitable authority as that of Mr. Stanley, who had the specimen preserved in St. Bartholomew's Hospital, of a case of dislocation of the head of the femur upon the dorsum illii from excessive distension from fluid. The capsule was entire, and measured five inches, in one direction. The cavity of the acetabulum had almost disappeared; being both reduced in size and filled by fibrous tissue.

In synovitis of the knee, in its early stage, a peculiar grating sensation is perceptible when pressing and moving the patella. This arises from an abnormal secretion upon the surface of the cartilages of the joint, and not from a dryness of surface or denuded condition of the joints. The synovial membrane, after several months duration of inflammation, becomes greatly thickened, and the joint subject to an attack of pain after undue exercise or exposure to cold. Otherwise, the condition may be quite endurable, and tends, apparently, to an arrest of the ailment. This progressive condition, if not arrested, results in an effusion of lymph and fibrin within the synovial membrane; becoming greatly organized, and terminating in contracting adhesions. This, again, has its reparative tendency; in covering roughened bone or indurated cartilage on the articular extremity. If this favorable condition fails to ensue, a chronic condition becomes established. The synovial membrane is then involved to the ulceration and absorption of the cartilages, beginning at their edge and gradually extending until the denuded surfaces of the bones are exposed. The joint, when in this condition, usually contains a yellow fluid, in which floats flakes of lymph and pus, that is eventually discharged externally from draining sinuses. Various morbid changes then follow, such as infiltration and thickening of the surrounding tissues, which, in some cases, become filled with thick lardaceous deposits, softening and disorganizing the different tissues. In this condition the joints become distorted — tending to a subluxation — and the surrounding integument adherent.

Mr. Holmes Coote, in his "Treatise on Joint Disease," describes a dissection he made of a joint in this diseased condition:

"In a case examined by me, December 3, 1846, for the late Mr. Stanley, the encroachment of the synovial membrane upon the cartilage was considerable; the cartilage itself was loosened from the bone, which was more vascular than natural, but of usual firmness. Several of the glands in the popliteal space had undergone the same degeneration. I could strip off some of the pulpy mass, covered by epithelium, from the vascular layers of the synovial membrane on which it lay. The surrounding vessels, in these cases, become the seat of very active circulation."

From this we learn what an extent of induration a very slight cause may induce under a peculiar constitutional condition, tending to this result. Such cause may be a slight bruise, being then strictly local, and, under a normal condition of the system, reparative force would most readily remove the cause of irritation. Dr Edward Rindfleisch, in his commendable work on Pathological Histology, remarks: Since Von Recklinghausen found reabsorbing stigmata of the lymphatic vessels upon the serous cavity of the diaphragm, we may even advance to the opinion that the fluid in the interior of a serous cavity is subjected to a certain renewal — a change. So much the more rapidly will also an irritating body from the liquor sanguinis appear in the serous cavity. Here, however, as in the joints and endocardium, there is yet added to the — shall we say fermentative ? — irritants from the infecting body, a second auxiliary force : the movement of the opposing layers of the serous sac against each other. In virtue of this friction, the one layer rubs the infectious body straightway into the other, and I have no hesitation in perceiving an auxiliary force for the development of the inflammation. The recuperative force only exists in the healthy conditioned individual, and is of wonderful efficacy in cases of great injury done the tissues. Morbid conditions must follow disorganization from violence done tissues, tending to irritating secretions that will be thrown into serous cavities; and as Von Recklinghausen discovered reabsorbing stigmata of the lymphatic vessels upon the serous cavity, in healthy subjects it will be reabsorbed, but in enfeebled subjects, from the want of active assimilation and elaboration of nutrition, the reabsorbing stigmata are not found, and the irritating secretion remains, and, as a ferment increases, producing the anomalous condition of joint described by Mr. Coote.

R. Volkman informs us that in the forming stage of synovitis we have first an acute purulent catarrh, — a blennorrhœa of the joint. The normal epithelial cells are cast off the stratum of connective tissue, without the continuity of the surface being disturbed, producing large amounts of pus corpuscles which cloud the originally clear contents of the articular cavity and convert it, the longer it continues, more into a thickish pus, drawing out into threads because mixed with synovia. In the further cause, a turn for the worse sets in, especially by this, that the articular cartilage is irritated by the stagnant pus, thereby undergoing a kind of decomposition, and is excited to a kind of superficial ulceration which destroys the cartilage, layer by layer, and after this, may pass over to the bone itself. In this case the cartilage perishes by a process which begins with cellular division and ends with the complete dissolution of the cells as also of the inter-cellular substance. If we make a vertical section through the cartilage we discover the first layer and cellular divisions, perhaps, in the tenth to the twelfth layer, counting from above. These, primarily, go the way of simple hyperplasia of cartilage; we see groups of four to ten cells, which yet distinctly bear the character of cartilage cells, in a common capsule. Farther toward the surface, the ordinary pus corpuscle takes the place of the cartilage cells, together with very considerable dilatation of the cartilage cavities, and the capsules gradually become indistinct. The basis substance has, meanwhile become finely granular, cloudy; toward the surface it more and more disappears, and, finally, liquefies in the contents of the articular cavity. Simultaneously with this complete liquefaction of the basis substance is produced the opening of the most superficial of the cartilage cavities. The pus corpuscles which have mostly been converted by retrogrenine metamorphosis into fatty or fatty granular detritus, mingle with the pus in the articular cavity; a semi-circular erosion of the free edge remains behind yet for a time, but with it vanishes the last trace of the cartilaginous structure.

The suppuration, however, like every excessive increase of new formation, requires space — under certain circumstances, very much space — and the mechanical force that is developed in the continued division and multiplication of cells is one so considerable that against it the integrity and turgidity of vessels cannot maintain themselves if both are to exist, side by side, in a given space, and are not capable of further expansion. In this manner, by the compression and rupture of vessels, suppuration gives occasion to a series of the most manifold and profound disturbances of nutrition, as far as they affect the osseous tissue, fall entirely into the province of necrosis and caries.

MORBUS COXARIUS.

Morbus coxarius has its acute and chronic forms, and is susceptible of being divided into three distinct stages. That it occurs in scrofulous children not a doubt can be entertained; but it is often observed in individuals who do not present a single indication of a strumous diathesis — if we confine the strumous diathesis to tuberculous deposits involving the glandular and parenchymal tissues. In many patients laboring under morbus coxarius the scorbutic diathesis is decidedly indicated by spongy gums, mouth bleeding at night, and aphthous condition of the mucous membrane and no enlargement of the glands, nor the most remote tendency to phthisis pulmonalis in after-life. This condition may be considered as the " strumous inflammation without tubercle" of some authors.

First stage. — The first indication of this ailment is stiffness of the limb, most apparent after remaining quiet in a sitting posture for some duration of time, or, on leaving the bed in the morning. After a time, there is impediment in stooping and a noticeable pain when drawing on the stocking; also, a sense of fatigue is experienced after slight exertion. Occasional pains are now felt down the thigh, and, more especially, in the knee — often exclusively there; misleading the family practitioner as to the actual condition of the patient. The patient refers to the knee in more than a majority of cases, and the knee is in some instances found to be swollen slightly, yet the seat of the disease is in the hip.

Various opinions are advanced to account for its occurrence. By some it is believed that the pain is conveyed by the branches of the anterior crural nerve down the thigh. By others it is asserted that the obturator nerve, in passing through the thyroid foramen is affected, and in supplying the muscles of the thigh sends a branch to the inner side of the knee where the pain is most commonly located. The pain is often found along the middle and on the outer part of the thigh, where the obturator nerve is distributed to the muscles of the limb.

In the incipiency of this stage there is no apparent indication from superficial examination — taking a mere view of the limb — but upon pressing the fingers behind the trochanter, an elastic fulness will be detected, indicating an increased quantity of synovial fluid within the capsular ligament, and often tenderness, — firm pressure increasing the pain in the knee. The patient being directed to walk, the afflicted limb will incline the foot inward or outward more than its fellow, and will be restricted in movement, having a stiffened appearance while, in some cases, slight pain is induced upon attempting to separate the limbs or close them, laterally. These are indications of importance but commonly overlooked by the parents or friends until actual limping is apparent; and even then they are attributed to "growing pains" or a strain of the joints.

There is no definite period for this stage to continue; months, and even years may pass, the patient only complaining at times and mostly at night. It is in the advancement of this stage that there is a perceptible change in the contour of the nates. On the affected side a flattening and elongation will be observed, and an increased breadth of the lateral folding under of the nates. See Fig. 98.

The foot of that side will be carried more directly forward than its fellow; and in some cases incline inward in walking.

Fig. 98.

The general health in this stage is usually, to all appearances, unaffected; but upon careful observation indications of fatigue upon slight exertion, an indisposition to join in active sports — probably because of tenderness, if not actual pain, and, toward evening, indications of fever may be observed; the latter so slight, however, as to be scarcely noticed or appreciable by the thermometer.

In this stage the constitutional diathesis, scrofulous or scorbutic, tends to a retarded or more active progress of the ailment. Scrofulous subjects having a tolerable condition of health will continue in this incipient stage of hip disease for a year or eighteen months, as in other scrofulous diseases. The indolent condition renders the first attack obscure — often without pain — with intervals of apparent arrest of the disease, and after months of advance a diffused swelling will be observed about the part affected. A slow indolent abscess forms, to the astonishment of the attendant, there having been so limited an expression of pain by the patient compared to the advancement of the disease. The child, probably of rotund form and apparent good health, only occasionally complaining of pain in the hip for a few hours during the day, elicits no special attention; but at night its rest is usually disturbed; it will scream out as though frightened or in seeming agony, even bounding up in the bed, and when spoken to will not give a satisfactory answer — lying down again and sleeping quite composedly, probably, to again cry out in a few hours time. This is the most decided indication of the advancing hip disease.

In the scorbutic diathesis, or inflammation of the joints without tuberculous deposits, the constitutional symptoms are much more severe. Mr. William Coulson, Fellow of the Royal Medico-chirurgical Society of London, gives the following statement: "If the complaint occurs in a scrofulous subject, the inflammatory action will be very much modified by this state of the constitution. * * * On the contrary, in persons who are not of a scrofulous habit, the local and constitutional symptoms are much more severe."* When patients are thus conditioned they suffer intense pain, and the hip is so sensitive that the vibration produced from persons walking briskly over the floor increases the morbidly painful condition of the patient. After this painful condition has subsided, a limping gait is induced, and greatly increased from walking any considerable distance, and much to their injury. When walking over an irregular surface, they are liable to fall, as the muscles of the leg and thigh will have lost some voluntary power, and have diminished in size, very considerably.

In the cases of strumous diathesis, if the surgeon makes pressure behind the trochanter major or upon the psoas magnus and iliacus internus, or rotates the head of the femur against the acetabulum, the patient, in most cases, will complain of pain, indicating the seat of the disease, while if much motion is made it serves greatly to aggravate, and not a doubt can be entertained, but that it more fully establishes the disease. Patients that have been able to walk about with tolerable comfort, after a severe examination at the hands of an inconsiderate surgeon, dates his first confinement to the house from an inability to walk, because of an increase of pain and loss of power in the limb, from that time.

In cases of scorbutic diathesis the severe pain and tenderness about the hip and thigh, causing the patient to flex the thigh upon the pelvis and rest on the opposite hip, determines the case without further examination; and yet these cases are more promising of cure than those of torpid disposition. The inflammation having been subdued and the patient's general health improved, they are in many cases cured, perfectly, in three or four months; after having suffered for a month from most severe pain, night and day, and requiring the constant attendance of nurses during that period — even opiates failing to afford relief.

It is in this stage of the ailment that the lengthening of the diseased limb is observed in some patients, and from the obscurity in which it is involved, the cause has been greatly discussed, but without arriving at any positive determination. The patient being laid upon a bench or table, because of the firmness when thus

---

* Coulson on Diseases of Hip Joint: Lond. Ed., p. 5.

placed, and the limbs straightened as much as possible, the trochanter patella and malleoli will appear lower in the diseased than in the sound limb. The patient when attempting to walk extends the sound limb, and is compelled to flex the other by advancing the knee, the foot being, most commonly, everted. Some contend that in reality there is no lengthening, and others insist that it is actually longer. The opinion of those who consider the limb to be lengthened is that the head of the bone is enlarged or there is a large increase of synovial fluid pressing the head of the bone forward and downward; while their opponents in theory, hold that the apparent elongation is produced by position of the pelvis — being depressed upon the affected side. The patient being disposed to avail himself of the most easy position relaxes all the muscles of that side of the body and limbs, and it is not improbable that the muscles of the affected side have been impaired in tone, tending to the apparent elongation of the limb over which the patient has no control. This is our opinion; and, also, that the limb is, in reality, not lengthened, as the patient when cured of the disease in the hip, has limbs of equal length.

The most common condition of the diseased limb is to find it shortened. The muscles contract from irritation and the head of the bone is drawn firmly against the upper edge of the acetabulum, and the knee slightly advanced, is the cause of shortening. This shortening in this stage of the ailment never impedes motion, as in the third stage. It is in the advancement of this stage that a change of position of the trochanter will be observed, being advanced more forward, and the nates of the affected side flaccid, lengthened, and flattened. The contraction is mainly in the front of the thigh and the limb somewhat adducted, nutrition being partially impeded, to the lessening of the size of the limb below the hip, compared to that of its fellow. The pain now becomes more steady, and an increase of temperature about the hip ensues. The patient is at times pale and then flushed, the skin being mostly moist, the tongue white, and the strength declining. The pain at the knee, which was before intermittent, now becomes steady, the bowels constipated, and the ailment passes into the second stage.

Second stage: The second stage of the ailment is merely a prolongation of the symptoms as stated, tending to an apparent torpidity of mental and physical status of the individual, and to the increase of suffering in the diseased limb; yet, at times there will be an intermission of suffering, even to the entertaining of hopes of recovery; but a powerless condition of the limb continues, and more decidedly if the weight and pulley has been applied early to the relief of pain : a deceptive hope, as the nerves of sensation have been partially paralyzed from continued extension, and the disease not arrested. It is in this stage that the elongated limb now gradually shortens — in some suddenly, and to even two or more inches ; and the patient, when in the erect position, can only touch the toes to the floor, having a disposition to place the foot of the affected limb on the dorsum of the other foot when standing,

Fig. 99.

(see Fig. 99), and even the continued extension by weight and pulley will not save the patient from this shortening, as nothing is more fallacious than to expect a cure to be made by a continued extension of the limb, as in that of the weight and pulley, which continues to extend to the extent of yielding, and, physiologically considered, to the impairment of the tone of the muscles. In our experience it is only palliative to pain, and not curative, but actually injurious as a treatment. This conclusion has been arrived at from the treatment of thirteen hundred and twenty-six cases of hip disease in eleven years, in the Institution for the Relief of the Ruptured and Crippled, where more than a majority were cases that had been under treatment before being sent to this institution, and invariably by the weight and pulleys;
the treatment having failed to arrest the disease in its progress or the limbs from shortening ; whilst many of the cases that were received in the incipient stage, and others advanced and not previously treated were more readily relieved than those having been treated by the weight and pulley. No treatment of this kind is permitted in this institution, believing it to be injurious by impairing the vital energy of the limbs, although an exceedingly popular remedy. These remarks are made from a serious impression resulting from experience in its not having arrested disease in the first stage of so large a number of cases, — convincing us most fully that it is not a reliable treatment under the most favorable condition of the patient — a conclusion determined from long experience. Therefore we feel justified in thus expressing our objections, but with all due deference to the opinion of others, where they have

had as extensive a practical experience in the treatment of this class of ailments as we have had for the past thirty-five years. A few cases and an occasional success will not determine the reliability of any treatment; it is only a large experience with close observation that is to be relied upon.

It is in this stage that much tumefaction ensues; or, in other words, bursal developments, or abscesses, as they are commonly termed, are formed, which may be dispersed in many instances by elastic pressure, and the patient experience a perfect recovery; and recovery may be expected from the advanced conditions of the several stages of this ailment. The ligamentum teres may have disappeared, and bands of fibrous tissue formed partial anchylosis, to the restoration of the limb to nearly a normal condition. We have had cases under treatment in which the head of the femur could be dislocated from the articulation upon the dorsum ilii, and there distinctly felt, and as readily reduced. And cases of advanced hip disease, where from accident — a fall on an unfavorable position of the limb — a dislocation and shortening was produced, that were reduced and the patients fully restored to the use of the limbs.

Mr. Thomas Holmes states, in his work on the surgical treatment of children (p. 438), that a cure was made under his care at the Hospital for Sick Children, in London, 1865. He says: "The patient had not suffered from any congenital affection of the hip. There was no formation of matter; there was no grating of the bones on each other. But by a slight manipulation the head of the femur could be dislocated on to the dorsum ilii, as proved by the sensation of the head slipping out of the socket, which could be plainly perceived; and the characteristic shortening of the limb was then immediately produced, and the head could be felt on the dorsum ilii. It was equally easy to reduce the bone into its natural position." Mr. Barwell, in the "Diseases of the Joints," (p. 297), says, that "when the subsynovial tissues in which ligaments are placed inflame, the ligaments themselves suffer, soften and become thickened or absorbed, as the case may tend." The inflammation having subsided leaves the limb conditioned as we have described it, — subject to dislocation from slight causes. This apparent laxity of the surrounding tissues of the joint does not preclude the possibility of recovery, and to a substantial condition, as we have witnessed in a number of cases.

It is in this stage that the muscles become more or less atrophied, and the breadth and flattening of the nates on the diseased side becomes more apparent. (See Fig. 100.)

Some noted authorities on this subject express their opinion that a failure in nutrition is the cause of the first impairment to the limb, preceding even inflammation about the joint, and hence the early flattening of nates or relaxed condition of the muscles, tending to that appearance, the thigh being flexed upon the pelvis. Abscess now presents, and most frequently about the upper third of the thigh, and upon the outer border of the rectus femoris, commencing in some cases much higher but gradually gravitating to the lower position. After the formation of the abscess the patient is usually relieved from pain and presents an interim, as it were, from progress in the ailment — a favorable tendency to an arrest of the disease; many being perfectly relieved in the second stage. The abscess, from persevering and judicious treatment, often entirely disappears, and the limb is restored to a nearly normal condition of usefulness but not of size. If not arrested at this stage of the ailment it portends an unfavorable prognosis for the future.

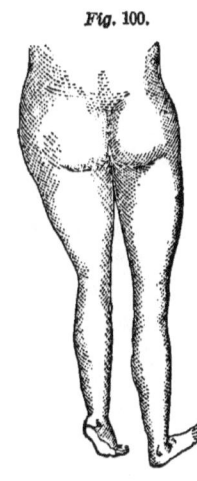

Fig. 100.

Third stage: The pathological condition in this stage of the ailment is most unfavorable to the patient; caries of the bone has now ensued after a failure to arrest the two previous stages, and Mr. Thomas Holmes asks the question: "Can any thing be done to rescue the child from impending death?" This must be considered a truly critical condition of the patient. This eminent authority states: "If the pelvis is much diseased, if sinuses are numerous and extensive, and if the internal organs (chiefly the lungs and liver) give clear symptoms of degeneration, the result of the disease, if left to itself, will usually be fatal." And further: "I have seen patients recover even from such a condition after excision of the hip. Whether they would have recovered without operation is more than I can say; but I think not." * This very candid writer impresses us most unfavorably in regard to the results of exsection of the hip joint, and his conclusions confined to the third stage of morbus

---

* T. Thomas Holmes, M. A., Cantab., on "Diseases of Infancy and Children." Lond., 1868.

coxarius are most reasonably considered, tending to confirm our objection to the hazardous treatment of the patient when in this extreme condition. Mr. Holmes claims to have performed the operation of exsection of the head of the thigh bone more extensively than any other surgeon, and gives the results of nineteen cases of which he had preserved notes out of a still greater number of cases. And it is reasonable to suppose them to have been the most favorably disposed cases. However, he premises his statement of those reported by stating: "The results will always vary according to the kind of cases operated on. I dare say this seems a truism, but it is at any rate, one which is very little dwelt upon in treating of the subject; I mean this: If a surgeon restricts any operation, say that of an exsection of the hip, to the best or most curable cases of confirmed disease, he will obtain a good percentage of success; but the question will remain whether the same success might not have been obtained by the expectant method. If, on the other hand, he restricts himself to cases in which, according to all reasonable probability, spontaneous cure is impossible, and operates upon every case in which the patient is at all in a condition to allow of his surviving the operation, then this tale of success will be much less; but then, also, all the successes must be considered as clear gain." . . . . "I think a fair classification for practical purposes might be made by separating the deaths into those who die from the direct results of the operation, and those who sink from constitutional causes; and the recoveries into those in whom the wound entirely heals and the limb is perfectly useful; those in whom the limb is useful, but the wound remains open for an indefinite period; and those in whom the patient recovers, but with a more or less useless limb and open wound — in fact relapses into much the same state as we usually find in chronic hip disease."

Here we have a synopsis of the results — three conditions: *First*, restored to a useful limb; *second*, the wound remains open for an indefinite period; *third*, having a useless limb and open wound — in fact relapses into much the same state as we usually find in chronic hip disease. Is it possible that a wise, deliberate consideration of these unfavorable results from excision of the head of the thigh bone should be accepted as admissible surgery, when it can be shown, from careful record, that not more than five per centum die from the sequellæ of the ailment in the third stage, under expectant treatment: that is, a judiciously prescribed régime of constitutional treatment. This author then states, in regard to his "fair classification for practical purposes:" "Taken in this way my nineteen cases will show, in the first place, seven deaths, in six of which I should refer the fatal issue to the direct effects of the operation, five of them dying of pyæmia, and one of gangrene of the wound. The other died of causes that had been acting, I believe, before the operation, which had, in fact, been put off till the patient was in a dying condition. In one of these cases which died of pyæmia, the consequence of acute osteomyelitis of the femur, I amputated the limb at the hip, with the desire, if possible, of removing the cause of pyæmia, but unsuccessfully, inasmuch as deposit in the lungs had already occurred, as was shown by post mortem examination. In another case I amputated with success, the operation having been followed by chronic osteomyelitis of the femur. Rapid recovery ensued, but the patient had had cerebral symptoms before the operation, and he died of abscesses of the brain some months after amputation.

"Two other cases have died since the operation, but at periods of time very remote from that of the excision, and from causes quite unconnected with it. In one of these cases (Isaac Richards) disease showed itself in the opposite hip to the one excised, and soon went on to abscess. The boy lingered for a long while in an asylum for incurables, where he ultimately died. In another case (Margaret Horing) though she recovered from the operation, never had any use of the limb, which remained in a chronic condition of suppuration. This leaves nine cases, one of which (Alfred Davis) is, I think, in an incurable condition, and will probably ultimately die. One who has been twice operated on (William Morgan) I have not seen for a long while. When last seen he was improving in general health and in flesh, but the limb was much shortened and distorted, and there were still open wounds leading to softened bone. In two others (Lydia Smith and George Punter) I think ultimate success is likely to be obtained, though in the former certainly with much deformity of the limb. In two other children (James Tapson and Lydia Bygrave) the sores are nearly healed, and the limb is very useful. Success is nearly certain, I should hope, in those cases." [The writer states that two recovered.] "Two others are walking about with useful limbs, the wounds being perfectly sound (William Watts and Mary Ann Hall). In Margaret Kirby's case the result was equally good; but the child died some time after recovery from an accidental attack of pneumonia, not tubercular. Thus, out of nineteen cases,

"6 died from direct effects of the operation (in one case after amputation).

"1 died after the operation from the previous effects of the disease.

"1 died of independent disease some time after recovery from amputation.

"2 recovered from the operation, but not from the disease, and died a long while afterward.

"2 were little, if at all, benefited.

"1 (twice excised) was doubtful.

"3 have useful limbs, but with sinuses.

"3 recovered completely."

This author continues his remarks on excision of the hip, and asserts that Dr. Hodges' work on this subject is the most trustworthy which has yet appeared, and it is stated that out of 111 cases, 53 terminated in death and two in amputation, while 56 recovered with more or less useful limbs; but it appears that the evidence of the power of walking was obtained in only 34 of these. I have already pointed out elsewhere the fallacious nature of what are called the statistics of most of these surgical operations, and have shown that there is every reason to believe that the success of the excision of the knee joint has been much exaggerated by the enthusiastic partisans of that operation, and that it has really been, on the whole, far less successful than the amputations of the thigh performed in similar cases, viz.: in chronic disease of the knee. But I do not on that account dissuade excision of the knee when applied to appropriate cases; nor, if the mortality after excision of the hip could be shown to be even higher than fifty per cent, should I admit that fact as a valid argument against operating in any given instance." *

In this reasonable conclusion of the writer, we fully concur. There are cases that present of sad deformity from a dislocation of the head of the thigh bone resting firmly in the ischiatic notch, the result of unfavorable position of the limb during the destructive processes of the tissues about the hip joint, and the disease having subsided leaving the patient in a tolerable condition of health. Excision of the head of the thigh bone tending to the redressing

---

* A paper by T. Holmes — London Lancet, October 29, 1864.

of the limb to a vertical position with the body, would most certainly justify the operation. In such cases the prospect of relief without any great risk of life to the person, and a prospect of affording an incalculable relief to the decrepit patient, and no other relief can be afforded when the limb is thus situated. The thigh when flexed anteriorly, and the head of the bone thrown back in the fossæ of the ilium, and the inflammation having subsided, and before firm ligamental attachment has ensued, can be readily redressed to a vertical bearing with the body, and the limb maintain its length within an inch, by the application of the apparatus represented in Fig. 105.

Another very eminent authority is Richard C. Barwell, F.R.C.S., assistant surgeon to Charing Cross Hospital. In a treatise on "Diseases of the Joints," page 431, he says: "Since 1848 the operation has been increasing in credit, both in England and in Germany (White operated in 1822, Hewson, of Dublin, in 1828), and we are now able to give a succinct account of its effects. I can thus gather altogether 104 cases. Twelve times the operation was performed for injury (eleven times gunshot injury, once for fracture of the neck of the thigh and descending ramus pubis.) Of these twelve cases but one recovered. Of 92 cases in which joint diseases was the cause of operation, we find that 56 recovered, 32 are dead, 4 remain uncertain. Therefore, in 88 cases 56 recover. * * * It must be, nevertheless, acknowledged, that several of the patients, after having lived and even walked about for some months, or even more, ultimately succumbed to internal disease, generally to tuberculosis.

Concerning the power, or use of the limb afterward, it is necessary to speak with the greatest caution. Very many of the cases, after having been reported as cured, with perfect use of the limb, have been lost sight of just when the critical time for testing the use of the member has arrived. Many of these are, I believe, dead; others have not so much use of the limb as the first result of the operation might lead us to expect. We may tabulate the only attainable numbers thus; but the quality of "useful limbs" is very much too high. Of the 56 recoveries I get no reliable information in 14 — in 6 the limb is useless, in 36 the limb is reported as useful."

In a foot note of this reliable authority he remarks: "I am greatly assisted by a valuable paper by Dr. C. Fock, of Magdeburg, in the Archiv der Chirurgie. The cases which he has gathered together are 90; 46 English, 7 American, 34 German, 2 French, 1 Belgian. * * * These numbers are thus got together: Dr. Fock gives of all his 90 cases, 78 as being performed for joint disease; of these 38 recovered, 26 died, 14 are doubtful. He obtains authentication of a useful limb in 22 cases only. I obtained, through the kind replies of the profession to my inquiries, authentication of useful limb in 6 out of the 14 cases he was obliged to leave uncertain, and of 8 out of the 14 which I have added to this table."

We have presented this extended evidence of the unfavorable results of excision of the head of the thigh bone to sustain our position in refusing to sanction it as a favorable means of relief in the advanced stage of morbus coxarius; and more especially when it is known to general practitioners, that not five per cent die from even the sequence of the ailment, under an ordinary careful regime, as that afforded in a comfortable home, and judicious medical advice. The patients themselves are evidence, having recovered with shortened legs and numerous cicatrices — true indications of the loss of bone tissue, and the former existence of extensive sinuses, and drain of serum attended with all its depressing influence upon the system.

Then, under this ordinary favorable condition of patients thus afflicted, is it not possible in a well regulated sanitarium to not only mitigate a vast amount of suffering, but shorten its duration, and in ratio to the stamina of the patient, — thus presenting the most feasible and favorable means of saving the patient's life, by expectant treatment under such favorable circumstances?

Treatment. — Mr. Thomas Holmes, on the Surgical Treatment of the Diseases of Children, p. 444, (Lond. Ed.), says: "There is hardly an early case of hip disease which is not curable if the patient can obtain careful nursing, prolonged repose, plenty of fresh air, good diet, and an appropriate medical treatment." One of the essentials is fresh air, which cannot be made available when the weights and pulleys are applied for treatment; nor can there be a desire for a "good diet" if the quiescent condition of confinement to bed is enforced. To have a desire for nutritious food an expenditure of vital force is imperative. Muscular motion and fresh air are the excitants to appetite for nutrition — the latter being essential to the accumulation of vital force and recuperative tendency. Enforced eating of nutritious diet engenders pathological conditions of the digestive organs decidedly inimical to the restorative powers of the system, and more especially when there is no expenditure of muscular energy. Hence, we are opposed to confining patients to bed in any chronic ailment, and especially diseases of the joints.

The disease being located in *any* of the joints, we know but little difference in regard to treatment. It is an ailment impairing and destroying the integrity of a joint, or joints, peculiar in character, and the result of a specific condition of the system tending to that end. If this were not the case, *all* bruises and injuries would produce synovitis, and in the joint sustaining the violence; but it is well known that such is not the case. Children receive great injuries from falls and bruises that do not result in hip diseases or caries of the spine. Admitting this fact we must consider the patient as laboring under a peculiar organic dyscrasia that determines the chronic character of the ailment — and failure of assimilation, to a normal condition of the person. Confinement to bed cannot improve this condition; and upon this ground we oppose its adoption in the treatment of this disease.

To assure the reader that we are sustained in our conclusions in regard to this matter, we refer to a treatise on hip joint disease written by William Coulson, Surgeon to the Magdalen Hospital, Consulting Surgeon to the City of London Lying-in Hospital, Fellow of the Royal Medico-chirurgical Society of London, Corresponding Member of the Medico-chirurgical Society of Berlin, published in 1841, — about the time we commenced a careful investigation of the pathological condition of patients laboring under this class of ailments. This eminent practitioner in surgery — and so considered by James Jonsen, M. D., editor of the Medico-chirurgical Review of that day, as will be seen — entertained the same views, and long experience on our part has fully confirmed our opinion in their favor. He says:

"In this class of patients rest should not be so strictly prescribed as to endanger the health of the patient. To obviate, in some degree, the ill consequences of want of exercise, the patient should be taken as much as possible into the open air, which acts as a stimulus to the vital powers; and gentle exercise, provided pain in the joint does not follow, may be allowed. The difficulty, in these cases, is to know at what precise periods of the disease does more than increased synovial secretion take place, and the irritation which attends it, and at what period does organic change or injury of the synovial membrane supervene. In fact, I firmly believe that the doctrine of rest is carried to too great an extreme, and that modified exercise is of vast importance in this disease." Lugol seems to entertain the same opinion.* "I may venture," says Dr. L., "to solicit the notice of practitioners to the results of my general experience in which I never observed any accident or inconvenience to result from this innovation (the employment of exercise). Of seventy-six scrofulous patients at present in my wards, there are thirty-two who, if treated according to the too general custom, would be restricted to absolute confinement to bed, under my direction, walk daily in the hospital promenade in the same manner as the different individuals afflicted with other forms of the malady. The study of scrofula, as regards its causes and diagnosis, denotes that this disease has, for its general character, an original weakness, which arrests the development of organs, but which renders them, subsequently, subject to a sudden and exaggerated increase. Rest has even been regarded as a debilitating agent; it is the ordinary associate of all antiphlogistic systems of treatment. The most vigorous and robust constitution would inevitably be weakened and brought to a state of etiolation by long-continued repose. If rest thus debilitates the vigorous, still more should an invalid of primary weak constitution be enfeebled by its operation, and his malady proportionately increased. But, the matter is not one of argument alone; visit those patients confined to bed for six months, and on a debilitating regimen; they are pale, emaciated, weak and depressed. I admit that the motion of a diseased joint is attended with some inconvenience, but the advantages derived from it are very great — beyond all proportion. In fine, for three years that I have followed this method, I have never been induced to change it, or even modify it but for a transitory period in some unusual cases."

From this inactivity of the system the digestive processes are rendered torpid, and a failure to supply nutrition ensues, which results in the tendency of impairing all the normal functions of the entire organization. This may be placing the subject of dieting in a strong light, but it is none the less true. If patients are not thus carefully regulated for treatment, their ailment will fail to be relieved under the most persistent and careful local treatment. It must be borne in mind that the local ailment is a manifestation of a dyscrasia, or constitutional degradation from a normal condition,

---

*On Scrofula, p. 148; trans. by W. B. O'Shaughnessy, M. D.

DISEASES OF THE JOINTS. 273

and requiring restoration before the local ailment can be cured; as it is sustained by all of the morbid secretions of deficient assimilation. Hence an approach to a healthy condition of the patient must be attained to even favor an arrest of the local disease. Palliatives may relieve extreme suffering, and are of great value to the patient during the preparatory efforts to progress in cure, which is dependent upon the normal stamina of the individual when in that critical condition. Hereditary influence is here to be considered, and abuse of constitutional powers, which, under a careful regime, are subject to much improvement — tending to the cure of other local ailments.

As we have before stated, the muscles and blood-vessels, and also that condition of the nervous centres, resulting from the use of tonics, by which they are enabled to exercise their power more energetically, and give greater force to the arrest of disturbing influences. Such influence is exhibited in the control evinced by some of the tonics over various diseases; as chorea, epilepsy, and neuralgia.

SPECIAL TREATMENT.

It is in the second stage — having failed to arrest the disease in the first — that treatment yet gives promise of saving the patient, and even from much deformity of the limb. Hygienic treatment has been, and is yet, of much importance, including that of alterative tonics: hydrarg. bichloridum, tr. cinchonæ, potass. iodidi, ol. morrhuæ, and the several preparations of iron. The roller applied to the hips over cotton batting is as important as in the first stage of the disease. The manner of applying the roller is seen in Fig. 93. Fresh air, moderated exercises, nutritious diet and cheerful entertainment tends the most largely to the cure. Nothing is more prejudicial than monotonous confinement; as that of confining the patient to bed. In more than a majority of cases thus treated (as we have witnessed), the progress of the disease has not been arrested, but, in our opinion, rendered obscure because of the relief from pain obtained by the sedative effect of the continued extension made by means of weight and pulleys, to the impair-

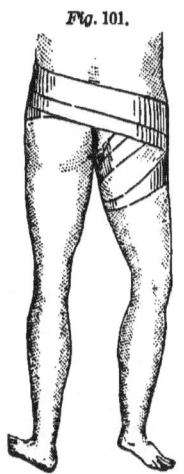

Fig. 101.

ment of nutrition and normal excitability of the limb — this being in accordance with admitted physiological laws; that is, that free contraction and extension of the muscles is essential to the maintenance of the normal increase from assimilation in the animal system. The most nutritious diet, and a desire for it, affords no strength if there is not an expenditure of muscular energy. Witness the idle gourmand, and compare him with the laborer. The system becomes torpid from confinement, both physically and mentally, tending to a concentration of excitement in the diseased hip, or other parts similarly affected and sustained at the expense of the general system. Hence, the demand for general and local treatment; general, in order to obtain and maintain an equilibrium of excitement in all parts of the system, and local treatment to remove the cause of excitement in the part affected. The primary local invasion being probably a constriction of the peripheral to the engorgement of the larger blood-vessels with an excess of serum, and, from distension, pain arises, indicating the relief that should be afforded to the arrest of the tendency to disorganization of the locally distended tissues. Now, whatever will tend to this relief is applicable in the first stage as in the second, and in the first gives greater promise of permanent relief. In the second stage the relief may not be as permanent because of the impairment of the implicated tissues — they having become weakened from the duration of distension. This distension consisting largely of serum, an effort must be made to lessen the excess by derivatives, which have a curative tendency by relieving the distended vessels; leeching, cupping and, indirectly, cathartics tend decidedly to this relief, but greatly reduce the patient's strength. Vesication is not so objectionable when freely poulticed and treated as a derivative, and not as an irritant by dressings of stimulant ointments. Vesication is an invaluable therapeutic agent for the purpose intended, as irritants in the region of the lesion are inimical because of their increasing the morbidly localizing tendency in the part affected. The curative tendency is to lessen the engorgement. This relieves pain without impairing normal sensibility, and the vital forces; and to maintain the relief we apply compression, and all therapeutic agents having a tendency to promote absorption or dispersion of accumulated serum in the congested vessels. Iodine and belladonna are reliable, either alone or in combination, also elastic compression, such as cotton batting applied with the roller after having reduced temperature, which should be carefully determined by the thermometer of Dr. Seguin, constructed for the purpose of determining comparative temperature in different parts of the person, and the excess of heat, when localized. Cold water frequently applied by means of a folded towel or something sufficiently large to retain moisture.

℞.
  Ammon. chlor.......... ℥ i
  Plumbi acet............ ℥ ii
  Tr. opii................ ℥ ii
  Acid acetic.
  Aq. font............ āā ℥ viii.
  M.
  Ft. loti................

This lotion freely applied affords much relief from active inflammation and pain, obviating the necessity in some cases for vesication and poulticing. It is, however, the safest practice to vesicate freely while pain and swelling continue even to the extent of eight or ten repetitions — healing having been permitted between each renewal of vesication.

It is in the second stage that the limb becomes powerless and presents an apparent diminution in circumference compared with the fellow limb, while the pain at the knee becomes more severe and the tendency to shorten increases. An increase of temperature, exceeding that of the body several degrees, will now be observable, and as the disease advances the pulse becomes accelerated, the face alternately pale and flushed, the skin moist and clammy, the tongue white, and the person emaciated. The patient starts and screams out during sleep from distress occasioned by involuntary contraction of the irritated muscles. Through all the increase of suffering, in many cases, there will not be any appearance of abscess forming, while the leg is shortened two or three inches. In some cases a deceptive interim ensues — an entire remission of all the symptoms, at least the painful ones, and yet the disease advances to the third stage. When indolent abscesses form we do not open them but apply the roller firmly, or have a lacing applied, as it is more permanent than the roller, and by this means often dispel even abscesses of large size.

The second having advanced to the third stage, a most formidable set of symptoms supervene. The elongated limb shortens so, that the extremity of the metatarsals and toes, only, touch the ground, and when in the erect position the patient places the foot of the shortened limb on the dorsum of the foot of the sound side, but if attempting to walk everts the foot of the impaired limb. It is the exception to see them turn the toes inward. The stick will now no longer serve the patient in walking, and the crutch must be resorted to as a means of avoiding pain when compelled to move about. When in the house the safest means of support for the patient is, to push a chair before him, especially if he be a child, as the crutch is liable to slip and let the patient fall, to the great injury, and often to the dislocation of the thigh bone upon the dorsum of the ilium. See Fig. 102.

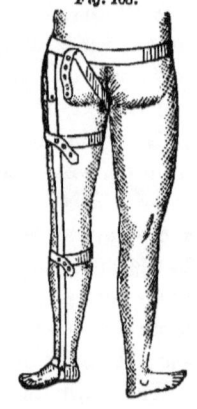

Fig. 102.

This unfortunate occurrence demands prompt attention. An anæsthetic should be resorted to, and the dislocation reduced. The muscles being impaired in tone do not offer the resistance of muscles when in a normal condition, but there is a decided tendency of the head of the bone toward escaping again, spontaneously, from the acetabulum. To arrest this tendency, before or after, we apply an extension support, used in other conditions of the diseased limb — that is, when inflammation is subdued — in hip disease and in cases of dental paralysis.

Fig. 103 represents an extension apparatus. Two fixed points of extension are observable, a cushioned strap crossing the perineum and attached to a steel belt encircling the hips. Two vertical bars of steel having encircling steel belts to inclose the limb, being secured by leather straps; the foot resting on a plate of steel having an encircling cup for the heel, and the foot held firmly by a leather band crossing the instep. The inner vertical bar extends upward to within two inches of the perineum, the outer bar having a joint about two inches from its attachment to the hip-band corresponding to the hip joint, or means of obviating an anchylosed condition of the joint, movement being made available, to the head of the thigh bone maintained *in situ* by the appliance. The method of applying the support may be thus described: The hips being firmly encircled

Fig. 103.

with a roller, as seen in Fig. 101, the leg is then placed in the frame, and the foot banded to the foot-plate. The band encircling the hips is then secured and the perineal strap tightened to the extent of the patient's endurance. By this means the limb is supported, which support should be continued until the limb is firmly attached. From the use of this appliance a favorable result is often obtained without anchylosis — restoration from the disease having ensued.

Fig. 104 represents the apparatus unapplied, with or without a joint at the knee (A). B, the form of the joint having a slide to fix the joint, if desired. C, the instep strap to secure the foot for extension at the hip.

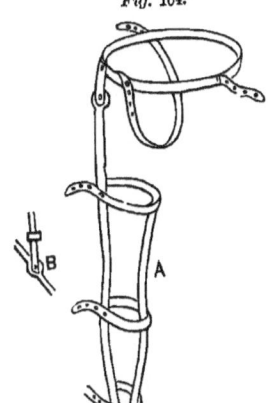

Fig. 104.

This we devised for this purpose, and applied it to many cases during the past twenty years, in all cases after inflammation had been subdued, and mobility in the limb sustained, thus enabling the patient to walk without the support of a crutch. The use of the crutch is objectionable because of its tendency to shorten the limb by the adaptation of the spine to the obliquity of the pelvis — the diseased hip being elevated — hence the necessity for enabling the patient to place the entire foot upon the ground, to stand, and then to push a chair before him and walk.

In regard to shortening of the limb, much difference of opinion has been expressed by authorities upon the subject. It is in the third stage of the ailment that we have the bones actually diseased. In the first and second stages, the disease is mainly limited to the surrounding tissues, tending to indolent abscess, that by improvement of the physical condition of the patient, and the application of elastic compression, even when as large as half a goose egg, often disappear and the patient relieved — without a limp. In the third stage, however, the shortening of the limb becomes most decided, and there is apparent dislocation of the head of the thigh bone upon the dorsum ilii and, as we have seen cases, into the ischiatic notch; this last probably arising from accidental cause. When dislocated upward and backward, Mr. Nelaton gives a mode of determining the condition: "When in the normal position, if a string is stretched from the anterior or superior spinous process of the ilium to the lower edge of the tuberosity of the ischium it will touch the upper margin of the trochanter major of the thigh bone. If dislocated upon the dorsum of the ilium the line will be entirely below the trochanter." This learned gentleman states that this will determine the question as to the dislocation of the head of the thigh bone. In advanced hip-disease we have distinctly felt the head of the bone thrown back in the fossa of the ilium, two or more inches from the acetabulum, and the psoas magnus and the iliacus intermus, tensely contracted — flexing the thigh upon the pelvis, and shortening the leg from four to five inches. In the early condition of this shortening much improvement can be made in the straightening of the limb by lateral elastic force obtained from the bedframe represented in Fig. 57; the patient being exercised upon it for about fifteen minutes, twice a day. Then, to maintain the improved condition of the contracted muscles, and admit of limited motion in the joint, we apply the bracing support represented.

Fig. 105. A, a steel frame encircling the body and having lacings in front, and straps to pass over the shoulders, and button on the back. B, steel springs having metal discs padded and arranged so as to co-aptate to the posterior part of the head of the thigh bone, and maintained in position by means of a leather strap attached to the center of the lower encircling band of the body brace, and from thence to the pendent bars, of which one or both may be used.

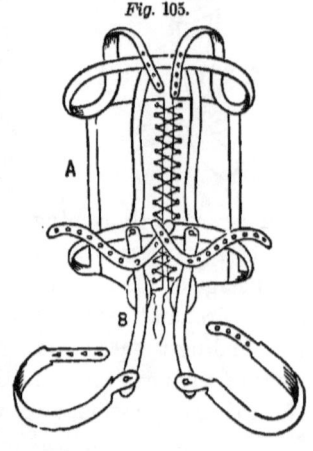

Fig. 105.

The distal end of the spring is confined to the leg with leather straps, giving force by means of the attached springs, the bone being carried forward and the femur brought in a vertical line with the body. By this means the shortening will be limited to an inch or an inch and a half at most. The origin of the glutei muscles sustain the head of the bone partly on the upper edge of the acetabulum, a firm resistance being maintained when in this position, and tolerable movement admitted by the sustaining attachments, which health and time ensures. To relieve the deficiency in length of limb, a wedge-shaped piece of cork should be fitted within the gaiter (see Fig. 106), and which relieve the patient from any apparent limp.

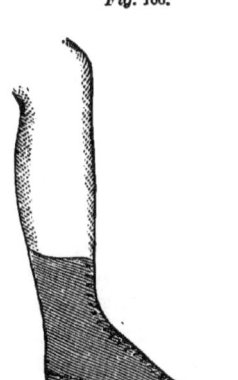

Fig. 106.

The dotted line above the sole of the gaiter shows the shape and position of the cork wedge.

The redressing treatment must be commenced as early as the diseased condition of the limb will admit, as the more decided relief will be obtained; although cases of a year or two's standing have been much improved by this treatment.

The limit of our work will not admit of a presentation of the various opinions expressed, or theories advanced in regard to the different pathological conditions attending hip disease; nor is it in accordance with the design of our clinical remarks, which is intended only to present the result of our own experience in the treatment of diseases as stated, and also of the means of relief, when curable, and in cases of confirmed displacement and shortening of the limb as in the case of the loss of the toes or permanent shortening of the limb.

Fig. 107 A, represents an appliance to be used when the toes have been amputated. A steel plate to the sole of the foot extended to the length of its fellow, and having cork shaped to the form of the missing toes attached. A metal cup to secure the heel, covered, and laced upon the instep; upright bars having metal encircling bands about the leg, and at the ankle a limited joint to resist the tendency of the gastrocnemius muscles from shorten-

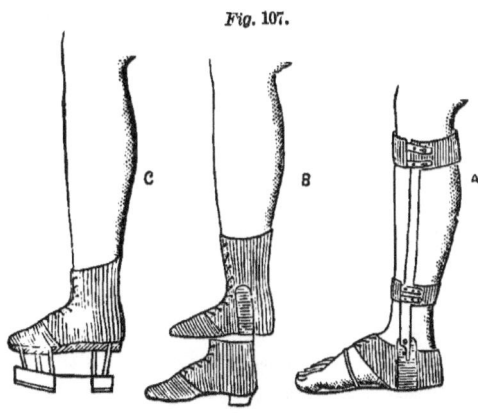

Fig. 107.

ing, having been relieved by the severing of the flexor tendons. This simple appliance we devised some twenty years since, and have found it to answer the purpose intended, namely, that of a resistance to the shortening of gastrocnemius muscle, and retaining the foot at a right angle with the leg; thus enabling the patient to walk, after some practice, without a limp.

B, is an artificial foot attached to an ordinary gaiter having a steel plate fitted to the sole, and upright pieces to support the ankle. This was devised mainly for women, and some time previous to the appliance C.

C, is for a similar purpose, being more convenient for men, and, we believe preferable to the ordinary cork sole. This having steel wire studs supporting two steel plates, one attached to the shoe and the other covered with leather, having a toe-piece and heel of sole leather.

PROGNOSIS OF UNFAVORABLE CASES.

Having considered the cases that are subject to remedy, we will now notice cases that tend to a fatal termination, yet, with careful treatment, many recover to a tolerable condition of health, and strength of the diseased limb.

It is in this advanced stage of the ailment that we observe the constitutional tendencies to destructive disease, being at first confined apparently to the hip, but now extended to the involving of the entire system. Inflammation has extended its destructive influence to all the surrounding tissues of the hip, involving the bones of the pelvis. The nates have become tumefied, the surrounding cellular tissue inflamed and the skin reticulated with distended veins; presenting reddened spots on the nates or outer side of the thigh, indicating the opening of sinuses of great depth and excessive drainage.

Before opening, there is much pain and hectic symptoms; the glands in the groin becoming enlarged and painful. In some cases the ulceration extends through the acetabulum, forms a sac that unites with the intestine in the pelvis; and by fistulous openings the matter passes into the rectum. The flatus passing through the sinus indicates the extent and parts affected. We have a case now under treatment in which fœcal matter passes outward into the sac of an abscess on the front of the thigh and escaped through an opening three inches below the groin. The matter from the abscess is now passed per anus, and the sinus on the thigh nearly closed.

The character of the secretion thrown off varies much in quality; in some it is healthy pus; in others, fetid sanies containing small grains of spiculæ of bone. The quantity and duration of the discharge also varies greatly; continuing in some cases for considerable duration of time, and in others ceasing, and returning again, and at times with hemorrhage, to the degree of endangering the life of the patient. In those cases where the discharge is well-formed pus, a favorable prognosis can be made; but if the discharge assumes the thin, dark, fetid character, hectic fever ensues, and the strength of the patient fails, but little hope for recovery remains.

Unfavorable symptoms ensue after spontaneous discharge, if of large quantity of pus and serum, and the patient fails in strength; but as a general result, less constitutional derangement follows than in the early puncturing of these torpid abscesses. Time, in many instances, tends to the restoration of the general health of the patient, if so situated as to have the benefit of a proper regimé; when, under these circumstances, as before stated, large abscesses, in many cases disappear, and the patient recovers from all indications of the disease. This we consider to be sufficient to justify a reasonable delay in opening them. In cases where there is much suffering from distension, and the abscess tends to a point that would soon open, a small opening affords relief; and a considerable thickness of wet cloths secured by the roller, will, in many instances, relieve the patient without shock to the nervous system. We consider poultices to be injurious from their tendency to encourage excessive draining when the abscess is open,—*and before!* A large soft cloth folded and kept wet is much more cleanly and quite as efficient in reducing the heat and pain in the tumefied part. A lotion of poppy-head, hop˙tea, stramonium leaves, or, that of muriate of ammonia, acetate of lead and opium, before mentioned, usually affords much relief.

Most unfavorable symptoms of general prostration follow, in some cases, a spontaneous opening of these large abscesses, even under the most careful treatment, such as excessive perspiration and colliquative diarrhœa. And, what is remarkable, patients will keep their beds for months, wholly unable to help themselves, exceedingly irritable and sensitive when handled for change of position to their protection, comfort and preservation from bed-sores; indicating carious bones and indurated tissues surrounding them. And yet, as unpromising as these cases may appear, the patients, in many instances, recover; and more often than in cases of excision of the head of the thigh bone, and, in cases where nutrition can be maintained, much more successfully. Therefore, we do not consider excision of the head of the thigh bone a justifiable treatment even in extreme cases.

Out of nineteen cases of excision, only three recovered completely. The patients unfavorably conditioned as we suppose them to have been far advanced in caries and necrosis of the bones. Of such, and incipient cases, we have restored seventy-five per cent to an ability to labor — the disease having been arrested, with more or less shortening. 1326 cases of morbus coxarius, 1679 cases of caries of the spine, and 798 cases of synovitis of the knee, ankle, shoulder and elbow, have received treatment within the past eleven years in the Hospital for the Relief of the Ruptured and Crippled. Of this number, the most unfavorable cases relieved of morbus coxarius were only about two and a half inches shortened by the disease, and the most of them not over an inch; the thigh bone having been brought to a vertical bearing with the body. This we accomplish after the active condition of the disease has been subdued; it being always advisable to desist from posterior extension when pain is induced, and to delay the effort from time to time, until the patient can endure the treatment without suffering, or risk of exciting inflammation tending to a return of the former morbid condition.

We have been considering the several conditions of hip disease to the most perplexing advancement where complicated pathological invasions supervene, to the retarding of the recuperative tendency. In this advanced stage we have a pyemic dyscrasia of the vital functions and advancing to an incurable condition of the patient. And this dyscrasia arising from the contents of abscess, the morbid condition of which results from impaired nutrition, inducing a cachectic condition of the system. To give the least promise of success when in this condition, a proper regimé is of primary importance. Pure air, the temperature not such as to chill the body, pleasant mental excitement tending to muscular exertion, the skin carefully cleansed and gently excited by the warm hand, or flannel, avoiding alcohol or any lotion that evaporates readily, as they diminish the enfeebled vital forces by reducing temperature, and are only applicable to the reduction of local excitement and increased temperature in the part. This tends to a demand for nutriment, that must be supplied, of easy digestion and in moderate quantity, and at frequent, stated intervals. The system responding to this treatment will require but a very limited quantity of medicine of careful selection. Alterative excitants are admissible, as that of hydrarg. bichlor., in tr. cinchon. com., ferri iod., potass. iod., and, in cases of excessive discharge from abscesses, cod-liver oil, in quantity limited to the power of digestion. If not assimilated, because of excessive quantity or enfeebled digestive powers, as before stated, it is inimical to the digestive functions, and decidedly injurious to the patient. Milk, yet warm from the cow, beef tea made from fresh beef of the best quality — avoiding the fat — and palatably salted, properly fermented bread, having been baked forty-eight hours, may be considered as proper diet. mucilage made of this bread is a valuable nutrient in cases where the patient cannot relish the bread, and is prepared as follows: four ounces of bread, eight ounces of boiling water poured upon it and set aside to cool. When cool, strain, and add one ounce of best cognac brandy and two ounces of white sugar. The condition of the patient must determine the quantity of this nutrient to be taken in twenty-four hours. Milk, and milk punch is a favorite nutrient with many practitioners. Experience has rendered it objectionable in some cases, for it is liable to coagulate and form a mass of curd in the stomach, which greatly distresses the patient, tending to serious consequences if not ejected by an emetic, as we have witnessed quantities of curd thrown off, to the great relief of the patient. In ordinary cases the amount of mucilage must be given carefully at stated periods, and containing more brandy in cases of diarrhœa, excessive discharge from abscess, or hemorrhage caused from spiculæ presenting in sinuses. In cases of colliquative diarrhœa, argenti nitras in one-fourth of a grain doses, every four hours, has been a valuable remedy in our practice, not omitting a liberal quantity of opiates. The subnitrate of bismuth and gum acacia affords most decided relief in mild cases.

Patients that have been so much relieved by this treatment as to give promise of recovery, in many cases suffer from attacks of neuralgia in and about the diseased parts. For their relief we have found the solution of potass. arsenitis the most reliable remedy; and also in the several stages of the ailment, increasing the quantity to nauseating the patient, if required, for relief. In all cases where the stomach will bear a dose of the decoction, that of cortex cinch., rad. gentian, rad. serpentaria virg. and sem. cardamom, may be administered, and will relieve the neuralgic pains, increase the appetite, and give strength to the patient. One ounce of this should be given for an hour after meals.

The treatment of the discharging abscesses and sinuses demands careful consideration at all stages, but more especially when sphacelus or slough ensues, leaving extensive ulcers to treat. To arrest sloughing, carbo ligni, cinchona, and gum myrrhæ, added to poultices of linseed meal, is a most reliable dressing when changed every three or four hours. The slough being detached, frequent pencilling of the ulcer with a solution of argenti nitrat., ten grains to the ounce, of water, greatly strengthens the rising granulations. For the dressing, carbolic acid, and oil; fresh mutton tallow, or Turner's cerate. If painful, a lotion consisting of the following may be applied as a reliable means of relief, cloths to be kept saturated with it:

℞
    Zinci sulph......................
    Plumbi acet. āā grs.............. vi
    Vin. opii....................... ℨi
    Aq. font....................... ℨvi
                                         M
    Ft. lotio ....

This, however, gives only temporary relief, and the cause must be sought and a more permanent relief given. A very common excitant is constipation, and torpidity of the liver, when the tongue will be coated in the centre and the edges papilated, and more than ordinarily red. A decided purgative is then indicated that will act upon the entire alimentary canal, producing watery evacuations. The patient will be greatly relieved by taking the following:

℞
    Hydrarg. chlor. mit...... ....
    Pulv. jalapæ āā grs.............. viii
                                         M
    Ft. pulv.
    and in eight or ten hours a portion of rochelle salts.

The local treatment of these sloughs demands most careful consideration. The patient's strength must be maintained by every effectual means. A purgative, as given above, tends largely to increase the appetite, enabling him to digest a greater quantity of nutritious diet, which, if persevered in, without the purgative, would only over-burden the stomach and impair the appetite, inducing a disrelish and actual nausea at the sight of food.

In cases where we have extensive sloughing of large sacs, to all appearances of the cuticle alone; exposing the muscular sheathing, and where the edges are abrupt and seemingly not adherent, with sinuses extending in different directions to a very considerable extent, these extensive secreting surfaces tend greatly to the exhaustion of the patient, and must be remedied by whatever means possible. Adhesive strips approximating large openings, and elastic compression, has been tried with the most favorable results; ravelings from worn linen or cotton is the material that gives the most comfort to the patient, as it is soft, light, and not as compact as oakum lint or raw cotton, for dressing. A firm bandage, laced, is preferable to the roller, and should be applied as tightly as the patient can bear. In removing the dressing, great care must be taken not to elevate the edges of integument surrounding the ulcer. Cleanse it with warm water and a large camel's hair brush, in preference to the sponge, and if the discharge is fetid, apply a soft piece of linen wet in Labarraque's solution properly diluted, carbolic acid and oil, or a solution of ferri sulph., two drachms to the pint of water. By perseverance in the treatment of these advanced cases when under favorable hygienic influence, it is only remarkable that so few die from the ailment, not one-tenth of the number that die after the exsection of the head of the thigh bone. Of cases in their incipiency to these advanced cases, we find, in our records, as before stated, seventy-five per cent have been restored to self-sustaining ability. Of these cases two-thirds were in the second and third stages of the disease, having been subjected to the weight and pulley treatment; and yet the limbs flexed more or less upon the pelvis. One case, inviting special attention, was a girl of eleven years of age, having both hips diseased and the limbs so flexed that it was impossible for her to stand alone. This child had had the heads of the extensor femoris muscles divided, and the weights and pulley applied steadily for nine months, yet was left in a most decrepit condition, and the disease not arrested. In eighteen months this child was enabled to walk, like that of a case of congenital dislocation of the hips by the treatment herein described; that is extension produced by posterior force after the inflammation was reduced.

Mr. Holmes Coote, F. R. C. S., surgeon to St. Bartholomew's Hospital, London,* says: "I do no approve of any attempt to straighten the limb until all morbid action ceases and the joint is free from pain; and then extension should not be so rapid and forcible by which a great amount of injury can be effected, but it should be slow and gradual and effected by means of proper apparatus." Our experience fully agrees with this learned gentleman's views of extension in the treatment of hip disease; that it should not be applied before all inflammation and tenderness is subdued, and then with much close observation to guard against the tendency to a renewal of the inflammation.

There are several ailments that assimilate hip-disease and are noticed by authors as that of psoas abscess, diseased condition of the glands in the groin. Close investigation of the early condition of the patient would most readily determine a correct diagnosis.

Mr. Erichsen, in his work on the Science and Art of Surgery, 1864, p. 746, Lond. Ed. 4, says: " Chronic Rheumatic Arthritis: It commonly affects the hip. I have met with cases of disease of this joint presenting all the character of this affection during life, though, as there has been no opportunity of examining the state of the parts after death, it is impossible to speak positively as to the true nature of the disease. Chronic rheumatic arthritis is an active disease of the bones and fibrous expansions about the joints; it is especially characterized by considerable increase in the size, and by alteration, of the shape of the osseous structure, which becomes porous in some parts, porcelaneous in others, by thickening of the fibrous capsule of the joint with deposition of masses or plates of bone in it, and ultimate destruction of the cartilages and synovial membranes. The suffering is considerable; the disease greatly cripples the utility of the joint, at last produces incomplete anchylosis of it, and is incurable." This greatly simulates hip disease of the ordinary kind in its incipient stage, except that it is a dull, constant, aching pain, attacking persons about the age of puberty, and that have been subject to rheumatic seizures during changes of weather, and not the result of a fall or bruise about the hips. From a careful investigation of the history of the case a true diagnosis is obtained.

---

* On Joint Diseases, 1867, p. 124, Lond. Ed.

Neuralgia of the sacro-iliac joint — This disease was first noticed by Bayer and Chelius, and since by Nelaton and Erichsen. The latter writer remarks that "the disease may be confounded with some of the varieties of coxalgia, and that it is essentially a very chronic affection, lasting for months and years. The disease appears to be strumous in its origin, partaking of the nature and ordinary character of 'white swelling.'" I have never seen it in young children, and in all the cases which form the basis of these observations it has occurred in young adults from 14 to 30 years old. The exciting causes of this disease are obscure. I have not been able to trace it to a blow or injury in any of the cases that have been under my care, although there can be very little doubt that such causes might excite it. We have had cases of this ailment to treat, as we believe from the indications, and in one case, a boy of eleven years of age who died after two years of great suffering, and the only relief afforded him was from increased doses of morphine from time to time. His parents refused an examination after his death. The appearance of the hip in the early stage was of a more than ordinary flattening or laxity of the glutei muscles, and the child would place his finger upon the sacro-iliac symphysis when asked to touch the part that pained him. The limb was longer than its fellow when first examined, and became attenuated to mere skin and bone. The hip was not so largely swelled in the anterior region as in that of ordinary hip disease. There were no perceptible changes of the trochanter major throughout the progress of the disease, yet there was a decided rotation of the toes inward. An abscess opened upon the ilium near the upper junction of the sacro iliac symphysis, and discharged a most fetid sanies and purulent matter, but not an excessive quantity. And, in all the cases that we have diagnosed as this ailment, the patients have never located the pain at the knee. Pain in the limb, thigh, and below the knee at times, and not of any great duration were complained of, but the pain was, mainly, confined to the location of the disease.

The difference of indications tending to a correct diagnosis of each ailment are as follows:

a. In coxalgia the patient suffers most severely when pressure is made posteriorly and above the trochanter, and in the depression beneath the prominence, or compression against the anterior part of the hip joint through the pectineus muscle.

In the sacro-iliac disease no pain is experienced on pressure in these situations.

b. In coxalgia, abduction and rotation outwards, and pressure of the head of the thigh bone into the acetabulum gives pain and suffering to the patient.

In sacro-iliac diseases these movements give no pain if the pelvis is supported by an assistant.

c. In coxalgia, in the advanced stages, a decided shortening of the limb exists.

In sacro-iliac disease no shortening ensues throughout the ailment.

These indications will designate coxalgia from all simulating ailments, if carefully observed.

CHAPTER XII.

DISEASES OF THE BONES — NECROSIS.

Various causes of Necrosis. — Traumatic Necrosis. — Diagnostic-history. — Expectant course of treatment. — Periostitis. — Instances among patients in Hospital for Relief of the Ruptured and Crippled. — Young more susceptible than adults. —Seldom confined to one bone. — Violence done to the recuperative processes by sawing, etc., often arrest the normal tendency to cure. — Acute subperiosteal inflammation.—Seven years' experience in Baltimore General Dispensary. — Objection offered to expectant treatment.— SYNOVITIS OF THE KNEE JOINT.— " White Swelling." — Premonitory symptoms. -- Inflammation of the two bursae. — But two structures in composition of joint subject to inflammation. — Fully 80 per cent of cases synovitis affect the knee-joint. — Acute and chronic synovitis. — Formation of abscesses. — Erichsen on synovitis. — Treatment.— Apparatus.— ANCHYLOSIS, IN ITS VARIOUS CONDITIONS.— Proper diagnosis of vital importance, to distinguish between true and false anchylosis. — Distinguishing features in true anchylosis. — False anchylosis frequently the result of disease of the articulation. — The Gonyometer. — Sub-luxation of the Joint. — Treatment. — The Roller. — Holmes Coote and Dr. Hodges on the restoration of the limb to its perpendicular condition. — Resection. — Illustrative Cases. — Result of 1,326 cases treated in Hospital for Relief of Ruptured and Crippled. — ANCHYLOSIS OF JOINTS AND TREATMENT. — Locomotive apparatus invented by Stephen W. Smith's Sons, of New York City. — Swing manufactured by E. L. Horman. — FLOATING CARTILAGES IN JOINTS. — Erichsen's method of treatment. — The Knee Bandage. — CARIES OF THE VERTEBRÆ. — Diagnosis. — Dyspnœa an early symptom. — KYPHOSIS, OR POSTERIOR CURVATURE OF THE SPINE. — Diagnosis of the several stages. — Symptoms of adults vary from those of children. — Abscesses arising from Caries of the Spine. — Prognosis more favorable in children than in adults.— Caries of the cervical vertebræ. -- Causes and seat of the disease. — Motor paralysis a concomitant of caries of the cervical vertebræ. -- TREATMENT OF CARIES OF THE SPINE. — Social enjoyment, and amusement, a potent remedial influence, as exemplified at the Hospital for Relief of Ruptured and Crippled. — Children should not be confined to bed. -- General Treatment. — Diet. — MECHANICAL APPLIANCES. — Holmes Coote on Remedies for Caries of the Spine. -- Medical Treatment. — Preventive means to be adopted. — Unskilful treatment induces paralysis. — Exposure of charlatans. — Advanced stages of caries of the spine the least amenable to cure of all bone diseases. — Physical signs of formation of spinal abscesses. — Objection to removal by use of knife, trocar or caustic. — Drs. Holmes and Erichsen on Treatment. — Our method of treatment.

Various causes may tend to the arrest of nutrition in a limited portion of bone, which, it may be said, destroys the life of that affected portion. When, as a foreign substance, the indurated bone becomes an irritant, it induces an inflammation that results in an exfoliation of the dead bone, and an effort, as it were, to cast it off, which, for a time, is impossible without surgical interference. If an injury is done the bone, as is most commonly the case, and purulent periostitis ensues, which is in proportion to the depravity of the constitution, we then have *traumatic necrosis*—pus having formed between the periosteum and the bone. Under this condition, the nutrient vessels are either ruptured or compressed to the arrest of supply, and the result is necrosis. In more healthy subjects, necrosis will not always follow purulent periostitis; the reparative force being equal to the emergency in removing the irritant cause, and the restoration of the nutrient function that had been partially arrested by the violence done the parts. But, if this favorable condition does not exist, sequestering inflammation progresses to the conversion of the periosteum into a pyogenic membrane, that protects the organism against its own secreting product. In the bone, there develops a rarefying ostitis, whose office it is to protect the organism by a layer of granulation tissue surrounding the dead part.

This rarefying ostitis is nourished from the medulla, and reaches as far into the compact substance as its vascular tracks have remained open and in circulation. There are cases where it develops upon the outer surface, where the Haversian canals dilate, and the tissues of granulation spring up in numberless vascular villosities and unite into a continuous layer which then connects with the elevated periosteum into an abscess membrane, inclosed all around. Most frequently, however, the outer layers of the compact substance having been too long deprived of circulation, and the exchange of material, that their revitalization could be possible by the medulla; and the rarefying ostitis appears as a sequestering inflammation which separates the lamellæ that have perished, and ends in pus which fills the abscess cavity. The dead bone is denominated *sequestrum;* and the rarefying ostitis that separates it, *demarcation.* The demarcation may occupy months and years.

During this time the suppuration continues uninterruptedly; the elevated periosteum, however, again returns to its ossifying capacity, and forms under the pyogenic surface a layer of new osseous tissue which, in time, may attain a very considerable thickness. The bony capsule (called coffin), thus produced, lodges more or less loosened sequestrum in its interior. The adjacent attached periosteum also participates by ossifying periostitis. For a distance of several inches upwards and downwards, osteophytes and exostoses arise in varying form and number. All these phenomena of inflammation extending to some distance, disappear as soon as the sequestrum is successfully removed. Even the bony capsule diminishes, and, by the obliteration of the cavity of the abscess, becomes attached to the surface of the bone, the exostosis disappears, and the bone assumes its normal form.* The health of the patient being vigorous, or improved by favorable hygienic influence, tends to the disposition of the sequestrum; if in large portions — and more especially in young subjects — it is reduced by the process of absorption to spiculæ that become displaced, and their thinned points penetrate the surrounding soft tissues, exciting in them inflammation and suppuration that results in an opening from which the irritant spicula escapes, or can be readily removed. This is known as the expectant course of restoration to a normal condition, and usually results favorably even in delicate constitutions when a carefully enforced regime is made available by treatment, as in a hospital especially designed for the treatment of this class of patients.

Periostitis in children, tending to necrosis, involves, more or less, all the long bones; as we have now in the Hospital for the Relief of the Ruptured and Crippled a girl of fourteen years of age having necrosis of the tibia, brachii, and both clavicles, and an out patient, a boy of about the same age, having necrosis of the femur, ulna, and one clavicle — and others, in progress of cure. This pathological condition is one of the constitutional ailments presenting pathognomic indications of scorbutus; there being no evidence of tubercular deposits, such as enlarged glands, or, primarily, enlargement of cancellated bone.

This is an ailment observable in young persons from the age of eleven to fourteen, and never involving so many of the bones at the same time in adults. When acute it runs its course most rapidly, and is mostly induced by injury in a scorbutic diathesis.

The inception of the ailment is not always from accidental injury, but an idiopathic indication tending to spontaneous, slowly increasing, periosteal inflammation apparent over the shafts of bones

* "Text-book of Pathological Histology, by Dr. Edward Rindfleisch, p. 581—1872.

and attended at times with much suffering, more especially in the early stage of the ailment; the patient being much reduced in strength, with decided constitutional cachexia, indicating serious encroachment of disease. It is but seldom confined to one bone, and ordinarily limited to the diaphysis, involving the joint only when extending the entire length of the bone, and, even then, but seldom impairs the integrity of the joint. Inflammation usually extends along the entire shaft of the affected bone, tending to extensive tumefaction, and induration of the surrounding tissues. After a time, small abscesses form; and, when open, expose extensive sinuses and denuded bone. But the less probing and meddling with the diseased bone, other than when a spicula presents free and projecting, to remove it, the greater the success that will be attained in saving the life and limb of the patient.

Much experience in the treatment of this ailment will determine the expectant treatment to be the most successful. The greatest effort should be made to improve the health by the enforcement of careful hygiene and alterative tonics, careful attention to the digestive functions, and patient waiting for the liberated spiculæ in small pieces to be thrown off from time to time. This will result most favorably to the patient. The violence done to the recuperative processes by splitting, sawing, and chiseling, for the purpose of removing the sequestrum of decaying bone, often arrests the normal tendency to a cure, and impairs vitality to the extent of a necessity for amputation which seldom saves the patient's life. Much probing and handling tends to like results when the patient is in the condition of health that might result favorably under the expectant treatment. This conclusion has not been arrived at without experience, having spent seven years in the Baltimore General Dispensary, practicing and witnessing the practice of others in the removal of necrosed bone, in many cases by efforts made when the sequestra appeared to be free of attachment, to the loss of the limb in some, and of the life of the patient in others. Like cases were, from the expectant treatment, as subsequently practiced, attended with the most favorable results.

In the acute subperiosteal inflammation and necrosis of the bone, this being more of a local lesion and active destruction of the limb, indications of pyæmia supervening, demands more decided relief, and amputation as an imperative treatment. These cases are mostly attended with decided constitutional derangement, as that of prostration, great pain, rapid pulse, foul tongue, frequent rigors, and delirium.

The objection offered to the expectant treatment, in the chronic condition of the patient, is that of long confinement, and as by some, to the bed. Now the latter is objectionable and not required. Judicious treatment consists in having the patient exposed to fresh air, social entertainment, and, at the same time, agreeable employment. And of children, the attainment of an education, subject to careful limitation, thus avoiding the tedium of confinement. These are the rational means of restoring a recuperative condition of the system, tending to the exfoliation and absorption of the non-vitalized bone, resulting in disintegration, to that of spicula that can be removed without injury to the ossifying process — the surrounding tissues being readily restored.

SYNOVITIS OF THE KNEE JOINT.

This disease of the knee is commonly known as " white swelling." In this, as in all chronic inflammations tending to the destruction of joints (and all joints are subject to the invasion, we believe) when there is a constitutional predisposition. Bruises, lacerations, and incised wounds, heal with limited impairment of the joint in normally conditioned persons, whilst the slightest bruise will induce a slow, progressive induration of a joint in those predisposed to this tendency; and the invasion not being superficial, is unattended with redness, as in ordinary inflammations; there appearing to be, rather, a torpidity of the superficial tissues that in time becomes implicated and attended with redness and dissolution in parts, permitting the morbid matter to escape by sinuses from the depth of the diseased joint.

Deep seated pain in the knee attended with slight chill and an increase of the circulation, indicating one or two degrees increase in temperature above that of the other limb, demands prompt treatment, which will, in a majority of cases, arrest the supervening tendency; as that of change of form; a uniform surface quite tense and tumaceous; the patella seeming to float in the great increase of synovial fluid, which can be made to fluctuate from side to side of the knee. This enlargement indicates fluid passing up between the femur and extensor muscles, tending to a fulness of the popliteal space, and, conjoined with this apparent change, is semiflexion of about 120°.

Inflammation of the two bursæ somewhat simulates synovitis of the knee-joint; the one below the patella, and a small one, when in a normal condition, above. When these are inflamed they are circumscribed and prominent, rising above the patella, and not involving the entire region of the knee, nor is the disease as painful as synovitis of the knee-joint. These are the most commonly affected. Those under the quadriceps extensor cruris and the tendon of the semi-membranosus only become involved when the surrounding tissues are decidedly diseased. These bursæ, when in a normal condition, supply the joint with lubricating fluid, communicating directly with the joint, and it is only remarkable that they do not become largely implicated in the incipient stage of synovitis of the knee joints.

The knee has, of all the joints, the most extensive synovial membrane, and from its exposed position, is the most subject to violence, the injurious effects of cold, and constitutional irritation. There are but two structures entering into the composition of the joint subject to primary inflammations: the synovial membrane and the cancellous tissue of the articular extremities of the bones. Of synovitis it is stated that 80 per cent of all cases affect the knee joint.* It is subject to acute and chronic inflammations of the synovial membrane. In the former the symptoms are more severe and attended with more general disturbance of the system. In the chronic form there is less disturbance, the joint less distended, and less pain. Cases often present for treatment in which the patient states that there is no pain; yet, the leg is flexed and the synovial membrane thickened, indicating the want of treatment.

Abscesses forming about the knee, may be recognized by their form and redness, and limited fluctuation. They should be opened early, as they might involve the interior of the joint. Being relieved of pus, they are most readily subdued; having, however, a tendency to repetition, if attention is not given the general health of the patient. The opening of these abscesses demands special discrimination, as the joint may be penetrated, to the great injury of the patient; there being an irregular distending point of the diseased tissues of the joint that may be mistaken for a circumscribed abscess.

The external tissues surrounding the knee joint in this ailment usually become infiltrated with fatty and plastic matter, tending to gelatinous infiltration and suppuration. The cartilages become

---

* Holmes Coote, on Joint Disease, p. 129.

softened and changed in appearance, the synovial membrane is removed in parts and replaced by large quantities of semi-transparent gelatinous fatty deposit, or pulpy grey and brownish fibrocellular material. The ligaments become inflamed, softened, and destroyed, being converted into somewhat similar materials; and the interior of the joint is filled with a purulent looking synovial fluid, thin and yellow, containing a quantity of fatty matter. Important changes take place in the articulations of the bones; they become enlarged, the cancellated cells filled with bloody, fatty, serous fluid, and the bone so softened as to be readily cut with a knife, and being so filled with fat, presents, when cut, a homogeneous surface resembling that of healthy bone. Mr. Erichsen, in remarking on this subject, says: "From this it would appear that the principal changes that take place in a joint affected with white-swelling consists in a kind of fatty degeneration of the tissues that enter into the formation of the articulation associated with an unhealthy strumous inflammation of the parts, and in the consequent deposition of considerable quantities of semi-transparent and lowly organized plastic matter, which in its turn has a tendency to undergo the same structural change, or to run into unhealthy suppuration."

Treatment of synovitis in the knee joint applies equally to all other joints similarly affected. The first object is, if possible, to arrest the tendency to suppuration. Alterative tonics, and an enforced hygiene, blistering and poulticing; and on the denuded surface sprinkling a small quantity of starch and morphia, keeping the knee perfectly supported, and the patient out of bed and agreeably entertained. The last is of much importance. Brooding over his condition greatly impairs the general health of the patient, to the increase of the local derangement.

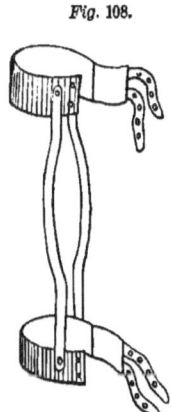

Fig. 108.

The knee can be most readily supported with the light steel frame which we devised for the purpose several years since, and known as the knee support.

Two laterally curved steel bars secured to broad belts of sheet steel by means of rivets, admitting of free movement for adaptation to the limb. The belts are suitably padded, and broad straps are affixed to encircle the limb above and below the joint, and with a pliant roller carefully applied. This secures rest, and admits of the dressings.

After blistering and poulticing, to the relief of severe pain and tenderness, cloths saturated with the lotion of ammon. chlor. and plumbi acetas described in the treatment of hip disease should be applied and kept constantly wet with the solution until the temperature is reduced to a normal condition and so maintained with water dressings. The following liniment may then be applied, and if a more extended support is desired, the following form (Fig. 109) may be made available as answering a better purpose for walking when the patient can be permitted to do so.

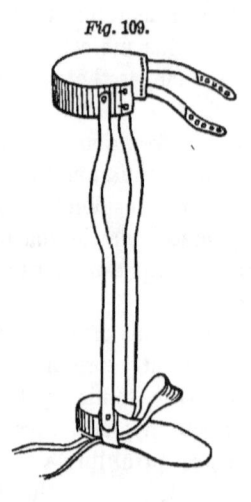

Fig. 109.

℞.
    Lin. sapo. camph....... ℨ vi
    Tr. belladonnæ........
    " iodini, fort. āā...... ℨ i
    Ol. cajeput........... ℨ i
                   M.
    Ft. emb.

This embrocation should be applied once in twenty-four hours, and the knee firmly bandaged with a flannel roller, or a thick application of cotton wadding, and over this a lacing of firm woven fabric in form of the knee bandage. The bowels should be kept quite free with an occasional dose of calomel and jalap or scammony, comp. ext. of jalap or colocynth. When not under the influence of the purgative, and reduced in strength and general development, give an ordinary teaspoonful of cod-liver oil an hour after meals and from eight to ten minims of syr. ferri iod., night and morning. To patients, rotund in form and seeming full in habit, give the following:

℞.
    Hydrarg. bichlor.............. grs. ii
    Tr. cinchon. comp............ ℨ vi
                     M.
    Ft. mist.

Of this give a teaspoonful three times a day; the iron and oil being inadmissible to such patients. The inflammation having been subdued, and the patient disposed to take exercise with a crutch or chair (the latter preferable), the straight upright brace (Fig. 109.) should be secured to the limb by the roller. See Fig. 110.

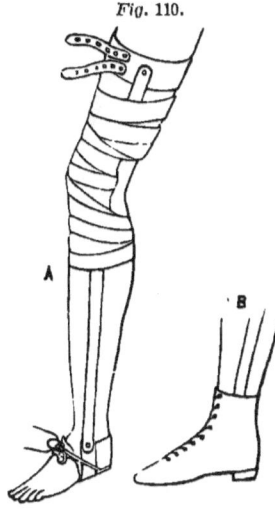

Fig. 110.

Fig. 110, *A*, the roller, a few turns or so having been taken within the bars and over the knee, thus tending to bring the knee within the bars extending the limb, and then encircling the limb. *B*, a gaiter applied over the brace, for walking.

This serves as a protection against unfavorable position, in case of slipping or getting a fall, to the injury of the knee.

In this stage of the ailment there is thickening and stiffness of the joint, and an effort must be made to restore flexibility of the articulation by gentle manipulation, always avoiding painful extension by applying the roller too firmly about the joint. The forcible extension of the joint has in many instances, after much suffering, compromised the patient's life.

Mr. Erichsen truly states, in his work on Surgery, that "the swelling and puffiness that are left, together with the debility dependent upon relaxation of the ligaments, are perhaps best remedied by the use of Scott's strapping; but pressure should not be applied so long as there is evidence of active inflammation going on in the articulation, which it would certainly increase."

Fig. 111.

The limb, even in the early stage of synovitic invasion of the knee, is in some cases flexed to a right angle, and the tendons under the knee very tense. This we never attempt to control, when such attempts give the slightest pain. Support is essential to avoid undue motion, and if tenseness exists, the knee should be bathed for fifteen or twenty minutes in warm water holding in solution half a drachm of Potass. Carb. to the quart, then dried, and the following lotion applied night and morning, and a brace as shown in Fig. 111.

Fig. 112.

℞.
Tr. bellad. ..... ʒi
Ol. olivæ. .... ʒvi
M.

Ft. embro.

The pressure should be, at each dressing, carefully governed by the pain induced.

This brace serves an admirable purpose where extension is applicable to the overcoming of angular contraction of the knee or elbow in cases relieved of all inflammation and in a flexed condition, as seen in Figs. 104 and 105.

Fig. 113.

ANCHYLOSIS, IN ITS VARIOUS CONDITIONS.

Of all the joints in the human frame, the knee is the most often found to be anchylosed in various degrees. The knee consists largely of fibrous tissue thinly covered compared to other joints, and more exposed to injury. Inflammation readily follows the injuries in degree to the force of violence, and a scrofulous diathesis of the patient, which results in contraction and deformity, if not relieved in the incipient stage of the increasing excitement. When not arrested, deep-seated pain and great distension of the joint follows without redness, but with decided increase of temperature. After much suffering, anchylosis supervenes by osseous deposit, partial or complete; or, we may have what is most common, partial or membranous stiffening of the knee.

A proper diagnosis of the condition of the knee, under these circumstances, is of great importance — to distinguish true from false anchylosis. Cases frequently present after eminent practitioners in surgery have given their opinion, and as to what they have stated, having knowledge of the history of the ailment, we are often not correctly informed. Our judgment in such cases is alone to be founded on our experience from careful observation in practice. To attempt a long, tedious course of treatment and fail to straighten the contracted leg (as all cases of anchylosis are slow in being redressed), the surgeon seriously compromises his professional status by incurring the displeasure of the patient, and friends who had been advised of the incurable condition of the limb, but ventured upon another's supposed superior judgment. This impairs confidence in others who might be readily cured by skilful treatment.

The straight condition of the leg from anchylosis of the knee, presents serious inconvenience to the patient, inducing him to seek advice from the most skilful surgeon, and if discouraged as to the possibility of obtaining relief, intrust their limb to adventurous treatment in hope of finding relief, and often pass through the ordeal of much suffering without the least possible relief; yet many of these cases thus conditioned are, by practical skill, curable. False anchylosis is exceedingly deceptive, and cases often pronounced incurable by eminent surgeons in general practice, are subsequently cured by those having more practical skill in the manipulation of anchylosed limbs.

The leg, when straight from anchylosis, commonly presents slight rotation outwards — the toes everted, which, in the act of walking, the foot more easily avoids inequalities on the surface of the ground. This condition of the knee is often the result of imperfect treatment of anchylosis, violent or gradual extension having been accomplished without careful attention to that of giving free motion to the joint. The limb may appear very rigid, and yet only by the attachment of the fibrous tissues, being in a curable condition; and most commonly so when carefully straightened. Under these circumstances a favorable prognosis may be given — the limb having yielded to extension that placed it in the straightened position without having excited inflammation to any great extent, only sufficient to give a fixedness to the joint, which is in quite a favorable condition for the restoration of flexion and voluntary movement.

In all cases it is necessary to consider carefully the probable nature of the impediment to motion in the joint, by ascertaining whether it arises from adhesions resulting from effusion of lymph in the articular tissues, or bands of adventitious matter, the product of cicatricial sinuses, or from membranous and calcerous adhesions of the articular surfaces. To determine the existence of mobility careful investigation is essential, as if very slight movement and positively determined — the patient's health being good — a favorable prognosis can be given, and if successful in obtaining mobility to the extent of a few degrees it will greatly favor the patient and compensate the surgeon for his efforts. Before, however, attempting to relieve the fixed condition of the joint, ascertain the patient's condition of health, and, if possible, the extent of inflammation that has existed, as also its termination, by resolution or destruction of the articular tissues within the joint. The history of the case enlightens the surgeon considerably as to the extent of relief to be afforded the patient. This is not always readily obtained; the parents of the child are very desirous of obtaining a favorable opinion of the case, or do not recollect all the suffering the patient has endured in months or years past. The surgeon will often be informed some weeks after he has been treating the patient, of some serious incident that occurred in the early invasion of the disease, impressing him with doubts as to a favorable result in treatment. Hence, a necessity for questioning the attendants as to the most unfavorable circumstances that the patient could have labored under.

False anchylosis is frequently the result of disease of the articulation, presenting numerous grades of severity which, from previous disease, have sustained a very limited invasion of structure and function, or a great amount of disorganization by ulceration, adhesion, cicatrices, and sub-luxation which in many cases are amenable to relief — the contraction being from a few degrees to that of closely approximating the heel to the nates. The angle may be determined by the *gonyometer*, as represented in Fig. 114.

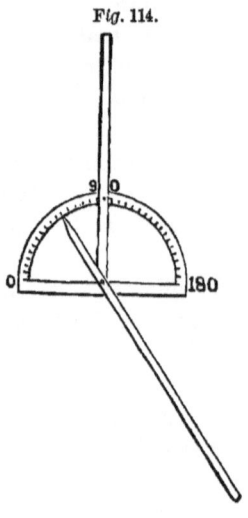

Fig. 114.

An arc of a circle divided into 180°, having two arms of about eighteen inches in length; one arm attached to the zenith of the arc, the other being free and having an extension point that determines the angle of flexion in the limb.

As before stated, the history of the invasion is of the utmost importance in determining the anchylosed condition of the joint. The true or osseous anchylosis but seldom occurs as a sequence of scrofulous or scorbutic inflammation in persons of tolerable constitution. In incomplete, or false anchylosis, as the latter is usually termed, and in which there is always some degree of mobility that may be only perceptible on a very careful examination in a limb flexed to about 35°. When ossified union has ensued, there is not any yielding of the joint; and this may be most readily determined by placing the patient, if a case of anchylosis of the knee, on a firm seat, as that of a bench, or narrow seat, elevated a suitable distance from the floor, and firmly strapping the thigh to the seat, then attaching to the ankle a small plummet having a graduated scale placed vertically to the plummet line; the point of the plummet, in ten or fifteen minutes, will determine the extent of yielding. The anchylosis of the thigh may be determined by a similar procedure, fixing the body and pelvis firmly to the bench. Other proceedings than this are much less reliable. The applying of the extension splint, as represented in Fig. 111, is another tolerable means of determining the yielding of the knee joint, when applied firmly upon the flexed limb to the inducing of pain, and continued for six or eight minutes; then, if pain subsides, and the bandage has slackened, a yielding of the joint is indicated to the encouragement of further efforts. This is our most common practice in the hospital.

If anchylosed joints yield to effort, which a seeming elasticity will indicate, a favorable prognosis may be given as to their extension. And, in some cases, in a few months; in others, from careful perseverance, after several years of effort, a restoration, even of free motion, may be calculated upon. This we have witnessed in many cases. The greatest difficulty is to keep the patient satisfied that they will eventually reap the benefit of perseverance in efforts to this end. In cases where there has been serious lesion of all the tissues about the joint, plastic attachment of the soft tissues ensues, often binding the joint most firmly, and without completing the anchylosis. This is the condition in two-thirds of the cases that recover from the ailment.

There is a great tendency in synovitis of the knee to subluxation of the joint; usually rotation of the head of the tibia outwards and backwards, more or less, into the popliteal space; see fig. 115; and in some cases, inwards, as in genu valgum, greatly everting the front of the foot. In nearly all cases there is more or less flexion of the leg, as in that position movement outward is most readily attained. Rotation outward is effected by the limited action of the biceps muscle attached to the fibula assisted by the tensor vaginæ femoris which acts on a strong band of the fascia of the thigh and extending downwards to the leg. Flexion of the knee relaxes the ligaments, especially the lateral, and relieves the joint which is sensitive to pressure of the articular surfaces when self effort is made to straighten the limb. From this we are impressed with what must result from painful extension force, whilst inflammation exists in the joint. It is only when this tenderness has subsided that it is safe to apply extension. This aberration of the knee joint can be redressed by means of the apparatus represented in fig. 116, by applying the roller above the knee, around the leg and outer bar, then reversing and passing it below the knee and around the leg and inner bar, applying force in opposite directions; thus making counter-pressure, and completing the whole by encircling the limb with the roller; a few turns being made within the bars and over the knee, half encircling it. This meets all the indications presenting in the aberration — flexion and subluxation. Subluxation directly backwards results from synovitis of the knee joint, and can be as readily reduced as that of lateral subluxation, and with the same apparatus, half encircling the leg above the knee by passing the roller within the bars and over the front of the limb above the knee, then reversing the application of the roller below the knee, half encircling the calf of the leg.

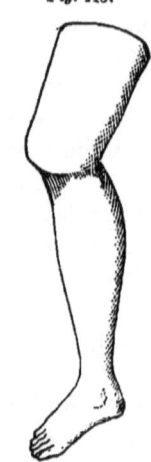

Fig. 115.

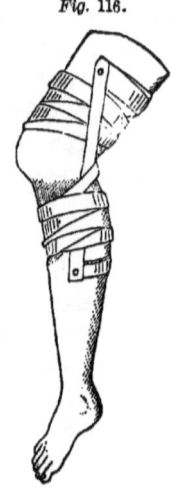

Fig. 116.

The dexterous application of the roller serves more effectually the desired intentions to be met in most of the aberrations of the limbs than any therapeutic means in our possession. The steel frame serves as a basis of support and resistance, but their use is made available and complete only by the skilful application of the roller, by which every contortion of the arm, hand, leg, or foot can be resisted with a curative tendency, because of attaining the most complete adjustment of the means to the purpose with less fixed resistance, than in the use of the more complex apparatus worked by screws, they being unyielding tend to concentrated force upon the salient points to the great injury of the soft tissues; and which by their fixed resistance to motion, impair muscular tone and motion in joints essential to the obtaining of a normal condition of the distorted limb. Elasticity, as a resisting force to the redressing of distortion, is in consonance with the laws of physiology, while fixed force for that purpose, is in opposition to the vital forces and injurious in its tendency, in direct ratio to the extent and continuation of the application.

As to re-section of the knee joint for the restoration of a perpendicular condition of the limb from flexion and destruction of the articular surfaces, we are not in favor of the practice, because of having witnessed many failures; and we find that we are not alone in our opposition, for Dr. Holmes Coote of Bartholomew's Hospital, London, says, in his work on Joint Diseases, p. 139: "I am no strong advocate for the operation of resection of the knee joint under any circumstances, and disapprove of it almost unconditionally in children, in whom the limb becomes, year after year, weaker, more withered, and less equal to the opposite member. Dr. Hodges considers that the small degree of success following the few cases of excision performed for traumatic cases does not warrant interference favorable to its future adoption as a substitute for amputation. In cases of disease, one death occurs in every $3\frac{7}{15}$ operations. Therefore, although occasionally yielding brilliant results, it is an operation to be practiced with great reserve. Partial operation upon the knee, as in other ginglymoid joints, adds, naturally, to the number of unfavorable cases. Inasmuch as the non-removal of the patella converts the excision into a partial one, the bone should never be allowed to remain.

"Re-section among the young has its unsatisfactory aspects. On February 14th, 1863, one of my colleagues amputated, at St. Bartholomew's Hospital, the left thigh of a boy, aged twelve, who had undergone the operation of excision of the knee joint in another hospital, in consequence of scrofulous disease of the articulation, about nine months previously. The condition of the boy before the second operation was as follows: He was not much emaciated, nor had his face lost all trace of the usual florid color. The left lower extremity was an inch and a half shorter than the opposite; the thigh was hot and swelled; sinuses led from the incisions about the knee to dead or denuded bone. There was some mobility at the knee, the limb was utterly useless, and the opinion was general that the sooner it was removed by amputation the better. Upon the usual incisions it was found that the tissues were much infiltrated and some considerable sinuses extended upwards among the muscles of the thigh. The hemorrhage was troublesome, in consequence of the generally increased vascularity of the limb. The wound ultimately cicatrized, and the boy left the hospital well and able to walk about on an artificial limb. The examination of the amputated parts, including the seat of the resection, was full of interest. In the first place, very little of the bony surface of the tibia or femur had been cut away at the time of the resection; their usual outline remained; therefore, although the epiphyses had been left, arrest of development was still a consequence of the operation. Secondly, the greater part of the synovial membrane was left behind in a thickened, discolored, and pulpy state, keeping up the effect of disease in full activity. Thirdly, there were numerous scrofulous abscesses in the popliteal space. Fourthly, there were two or more points where the surface of the tibia and femur were incomplete, being partly fibrous and partly osseous; but the bone was dark-colored, soft, sodden, and spongy, and unsuited to effect firm union. Sixthly, the opposed surfaces were not in rectilinear apposition.

"Could any result have been less satisfactory? But, it may be said, such cases are exceptional. I am not prepared to acquiesce in that statement. In December, 1862, another surgeon amputated the left lower extremity of an infant on whom excision of the knee joint had been performed a year before in another hospital, and a third case occurred in my own practice, in which the parents declined any further operation, although the limb was short, withered and useless, and the child could walk only with the aid of crutches.

"We are told that excision of the knee joint is a less fatal operation than that of amputation of the thigh; but Dr. Hodges observes in his essay on the excision of joints, that: 'Out of a considerable number of cases, one-third died, more than one-third are known to have failed, and there is no direct evidence of success in more than one-third of the cases, even accepting the statement of those who furnished the notes.'"

The experience of these very eminent surgeons fully sustains us in our objection to the excision of the knee joint. In our experience but a very limited number recover even under favorable circumstances. Out of 798 cases that received expectant treatment in the Institution for the Relief of the Ruptured and Crippled, in eleven years, only four per cent failed in being relieved of the disease; some having limited motion, and others perfectly free joints; and of those having failed to be relieved under treatment in the institution, a few of them left us to receive other treatment, after having been a long time under our care without permanent relief — at times being much relieved, and again relapsing, and of these we have had favorable report. The régime of the house having improved their health tending to their cure, even when subsequently under a less favorable condition than that which had contributed largely to their recuperative tendency, and final recovery. This establishes an important consideration with regard to hygienic influence upon this class of patients. The physical condition of the patient being improved, tends to the cure of the local disease when subsequently under more unfavorable circumstances, and no special medical treatment.

TREATMENT OF ANCHYLOSIS.

Inflammation having seriously impaired the integrity of a joint, three conditions usually present:

First — A weakened condition subject to relief from supporting bandages.

Second — *Fibro-cellular Anchylosis*, limited motion in degree to that of the injury sustained from destructive inflammation, as that of thickening and induration of its fibrous capsule, or formation of fibroid bands within the joint, or in consequence of the cartilages and synovial membrane being in part or wholly removed, and their place supplied by a fibroid or fibro-cellular tissue, by which the articular ends are partially united. Muscular contraction often contributes most largely to the limiting of motion in the joint. In some instances the disuse of the joint tends to its becoming fixed in position. These several impairments of free motion in the joints are all favorable conditions to restoration from judicious treatment.

Third — *Osseous Anchylosis:* This consists of the indurated osseous surfaces having coalesced into direct bony union, and is most common in the spine, hip, knee and elbow.

The two first conditions of the partially anchylosed joints will be considered, and the means of relief. When thus conditioned, there have been fibro-cellular deposits, or degeneration of the joint; and the bones united by bridges of osseous matter thrown out externally to the articulation extending across from side to side, being accidental formations occurring from strumous articular inflammation of considerable duration. When in this condition some degree of mobility, often very slight and only discovered by experienced manipulation, may be detected, tending to a favorable diagnosis.

In efforts to restore mobility to partially anchylosed joints, great care must be observed as to the impression made, tending to a relapse of the inflammation. Carefully applied pressure from the roller, as described in a previous chapter, upon a straight splint, serves as a valuable means of restoring normal form and motion in cases of recent formation. Gentle manipulation of efforts at flexion and extension is a most successful treatment, and the hand the most reliable means of accomplishing the desired object of restoring motion to the joints in cases of moderate invasion. Some degree of motion is always perceptible in such cases, and in many instances if left to the ordinary exercise of the patient, in months after the inflammation has been subdued motion is gradually restored. It is in these cases that the bonesetter obtains flattering encomiums for his successful treatment, in having restored motion to the stiffened limb. In more obstinate cases, and where the immobility appears to depend, in some degree, on fibrous bands, they may be divided subcutaneously and to the immediate relief of the patient. Flexion of the limb, with apparent anchylosis, is often the result of continued unfavorable position when in a painful condition, and the restoration by means of extension in some instances most tedious and seemingly impossible. In such cases the. division of the tendons subcutaneously assist greatly in the restoration of motion to the limb.

In the case of children, they being most liable to invasions of the joints from inflammation, when in a condition to safely bear treatment tending to restoration of mobility to the joints, mechanical devices for flexing and extending the limbs are most valuable means when the apparatus is so constructed as to afford pleasurable exercise, as all means of extension is attended with some expense of pain, thus serving the desirable purpose of restoring motion to the joint and not being objectionable to the child. This is of valuable consideration, as it tends to invite a desire rather than resistance to the exercise, as in that of a fixed frame affording monotonous movement. The value of conjoined mental and physical exercise to a recuperative tendency, as we have stated in a previous chapter and here present as a practical point in the redressing of partially anchylosed joints. The mechanical device we use in restoring motion to the hip and knee joints is known as the Locomotive Cantering Horse.* (See fig. 117.)

Fig. 117.

This locomotive apparatus consists of a wooden horse placed upon three wheels, one behind and two in front; the hind wheel admitting of a vibratory movement, and controlled by two wires passing up to the horse's mouth and attached to a free bit of a bridle. The rider governs the desired direction of the horse by means of the bridle, as in that of riding the ordinary horse. Two *shafts* and the horse's hind legs are attached to the axle of this single wheel, and extend forward to a crank shaft, and connected close to the axle of the front wheels, where there is an elevation or depression in the progressive movement. The crank shaft, consisting of from three to five inches elevation from the axle, the horse's fore legs being attached to the elevated portion of the crank shaft, admits of alternate depression and elevation of the horse's head, when power is applied to propel the horse on wheels. The rider being mounted, and a cripple, because of a stiff knee or hip (even partially anchylosed), has his feet strapped to the pedals on the shafts where they are kept stationary; the crank being elevated, the rider's weight of body is thrown forward, depressing the horse's head and crank, and the movement continued by the rider making effort to elevate the horse's head. By these alternate efforts, the apparatus is moved and kept moving, and directed by the bit. From this description of the efforts to travel with the wooden horse, it will be seen that the legs are alternately contracted and extended, to the overcoming of the anchylosed condition of the joint. Paralytic patients derive much benefit from this exercise; and when one arm is paralyzed, a heavy wire must take the place of the bridle

---

* These Cantering Horses are manufactured by Stephen W. Smith's Sons, New York city, who also are the inventors of the Invalid Traveling Chairs.

reins, so that one hand can direct the horse, and the paralyzed arm exercised by being fastened to the wire rein.

In this exercise the patient requires careful supervision in limiting the efforts at riding to the condition of the patient ; for, if permitted, he will indulge to excess, tending to a serious relapse of inflammation in the joints. Patients will indulge in this exercise even at the expense of severe pain, as we have witnessed little patients shedding tears and yet desire to ride. Any level surface, floor or pavement, can be made available for this exercise.

These horses run rapidly over any good road, sidewalk, or park grounds, propelled by the weight and strength of the rider, flexing and extending the legs, arms and body. They may be made available for healthful out-door exercise for both adults and children when unable to walk.

Fig. 118.

Another most useful means of affording exercise in cases of paraplegia, or other inability to walk, as that of rheumatic seizures, is that of the swing.* (See Fig. 118.)

The seat is suspended by ropes attached to the door or other upright frame having a cap of some eight inches in breadth from one side; two pendent ropes suspend the seat; on them is sheathed handles to glide readily, and attached to two shorter ropes, secured to the opposite side of the door frame. The patient taking a seat in the swing grasps the handles and makes efforts at drawing them down. To assist in starting the swing, a gentle push may be given, when the patient will be enabled by the strength of his arms to increase the move-

---

* Manufactured by E. L. Horsman, New York city.

ment, which is a pleasant exercise, requiring considerable muscular effort of both body and limbs.

Similar devices may be made available, thus tending to pleasurable entertainment both physically and mentally, and contributing largely as an auxiliary means to the relief of patients.

FLOATING CARTILAGES IN JOINTS.

Slight injuries of the knee joints inducing inflammation, that subsides in many cases without inviting any special attention, often tend, however, to the formation of these adventitious substances, to the great annoyance of the person thus injured. In time they induce sudden pain with great uncertainty of walking, and are met with in the different articulations.

These condensed, spherical substances floating within the capsules, formed and lined with the synovial membrane about joints, are not always found to be cartilaginous formations, but condensed, indurated, fibroid tissue simulating a warty excrescence of the synovial membrane, usually found in such cases; some of these formations being pedunculated and attached, others floating about; indicating different stages of growth, and that their origin was a warty pathological condition of the synovial membrane. They vary in size, from that of a grain of wheat to the dimensions of a chestnut; the larger being flattened, smooth, shiny, and usually of a yellowish or greyish white color. They are found within the surroundings of the knee, most commonly, but occur in the elbow, the lower jaw and the shoulder joints, being in some cases quite numerous. When attached to the synovial membrane they produce but little inconvenience; when loose they are liable to be caught between the opposite articular surfaces and produce such sudden and violent pain as to arrest all motion in the joint, and the patient, if walking at the time, will drop as if shot — faintness and sickness following. Richet thinks it may be owing to the synovial membrane being pinched between the foreign body and one of the articular surfaces. Mr. Erichsen is of the opinion that "it is most probably due to the foreign body being drawn in between the opposite surfaces of the joint when these are separated anteriorly in the act of flexion of the knee, and then, when the limb is extended acting as a wedge between these, tending to keep them separate and interfering with the complete straightening of the limb. In consequence of this wedge-like action of the loose cartilage the ligaments are violently stretched, and the sickening pain consequent upon this act is experienced, followed, as happens, in a violent strain by rapid synovial effusion."

As regards treatment, this very skilled surgeon says: "But it must be borne in mind that whilst the loose cartilage is at most an inconvenience, though a serious one, any operation for its removal by which the joint is opened becomes a source of actual danger to the limb, and even to life. It is far less dangerous, in fact, to leave the foreign body than to perform the operation necessary for its extraction." Our own experience impresses us with the same precaution in the removal of these substances from within the joint, by operation; the violent symptoms that have followed the procedure endangering the loss of the limb. Perfect relief can be afforded by wearing the *knee bandage*, made of strong, woven fabric — other elasticity being rather a disadvantage than conducive of any good purpose. When elasticity is supposed to be required, it should be of coiled wire sheathed in chamois leather or soft kid, and forming only one part of the bandage — one-half of the circumference of the knee. It is, however, very clumsy, and by no means as pleasant to wear as the plain, single fabric. By this pressure, in time, they disappear, and the wearer of the knee bandage has perfect immunity from the ailment.

Synovitic inflammations in the several joints, being pathologically alike, require similar treatment; as they differ in no essential manifestation demanding a special treatment we leave the subject of inflammation and its results in the joints of the limbs — only considering the weakness of the ankle joint as the sequence of strains — and refer to a bandage already referred to, which affords quite as much relief to the ankle as that of the knee bandage to the knee when impaired by inflammation or these adventitious deposits. The ankle bandage is made of similar material as that for the knee, but stiffened under one, or other, or both malleoli; this affords most decided support to the weak ankle, as well as protection from the injuries it is constantly liable to sustain when weak, more especially than when in a normal condition.

CARIES OF THE VERTEBRÆ.

This is a pathological condition of the vertebræ resulting, as we believe, from a dyscrasia of the system induced most commonly by zymotic disease of virulence and long continuance, inducing functional impairment and a latent disposition to synovitic disease tending to involve the bones. A bruise or blow is the exciting cause, and if occurring in the region of a joint synovitis tending to caries is the result; if over the shaft of a bone, periostitis and necrosis; or, if in the soft tissues alone, interstitial abscess. We believe this predisposition may also exist constitutionally from hereditary influence. Synovitis and caries of the spine evidently present in scrofulous and scorbutic diathesis which we believe may be induced in children from inefficient diet, and bad ventilation, or whatever may tend to impair nutrition in childhood. Many practitioners in medicine whose position in their profession entitles them to special regard, we are obliged to differ from in opinion (with all due deference) as to the etiology of caries of the vertebræ; being decidedly impressed with the belief that there is a predisposition to these lesions, of not only the articulation of the limbs but of the spine, and that the invasion is, in many cases, the result of some external violence. Children often receive severe injuries from falls or blows, even to the fracturing of their bones, and recover without a symptom of synovitis or caries of the vertebræ. Then why are they not affected with one or the other of these ailments, if they are entirely of traumatic origin? If patients were not predisposed to these ailments they would be as readily cured of contusions, wounds, and fractures, as healthy children.

Caries of the vertebræ is most commonly attributed to an injury sustained at some anterior period, and by the parents of the child, to some accident, often to carelessness of the nurse who is then discharged, bearing the disgrace of having so severely injured the child, when, in all probability, a very slight injury may have inadvertently been the cause, which, under ordinary circumstances, would not have injured the child and thus brought unjust censure upon the nurse. The true history of the case invariably reveals the actual circumstances that have preceded the development of the disease and usually in an advanced stage when advice is sought. The parents, one or both, being of feeble constitution, or the mother, from various causes, fails to afford the child the essential nourishment for a healthy stamina.

DIAGNOSIS.

Caries of the vertebræ is one of the most insidious ailments of childhood. At first, slow of progress, and attended with very decided indications unobserved by most of parents, or those having charge of the afflicted child, as before stated, treatment is sought mainly in the advanced stage — a slight projection beyond the limits of the other spinous processes. This is, in some cases, months after the invasion or incipient stage of the ailment has passed. Inflammation has preceded absorption of the interstitial substance tending to the angular projection from an approximation of the bones of the vertebræ.

From close observing intelligent parents, we learn the first indication of a tendency to caries of the spine — an apparent lassitude, and disposition of the child to recline the breast upon the seat of a chair or other convenience for resting. If in active play, after a time the child will assume a semi-reclining position — the shoulders advanced, with a slight flexion of the legs and run to a chair to assume the position described, and frequently complain of pains in the stomach or bowels, while, to an ordinary observer, the child is in apparent good health. After reclining for some time an effort will

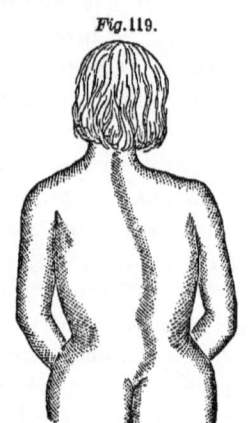

Fig. 119.

be made, slowly (for a child in good health), by the arms to gently raise the body, and steps are at first softly taken, when the play is again resumed with the usual vivacity. This incipient stage may exist without increase, apparently, for several months in some cases; in others, but a few weeks. On examination of the spinal column it is but seldom that a tender point is found. In most cases a slight, limited *incurvation* in some portion of the spinal column will be observed, and that may arise from a swelling of the intervertebral substance from inflammation and congestion. This indication will be observed by close examination in a majority of the cases of incipient caries of the spine, and an apparent disposition to steady the spine by limiting its normal elasticity in conformity to the movements in walking, which stiffness increases as the disease advances. Dyspnœa, or interrupted breathing, is an early symptom, increasing as the ailment progresses and becoming eventually, if not a serious concomitant, a great annoyance.

KYPHOSIS, OR POSTERIOR CURVATURE OF THE SPINE.

After these first indications have passed, it is then that the surgeon is consulted, because of the discovery of a small projection of the spine, after months have passed, which is the most decided symptom of the advanced ailment. The family physician has already been consulted and failed to discover the true condition of the patient, because of his not having had a case within his sphere of practice for several years, although his practice may be extensive. This truly unfortunate circumstance is much to be deplored, as it in many cases precludes the possibility of arresting the deformity. The cancellated portion of the vertebræ being in a state of dissolution and absorption, clinical teaching is not readily attained, nor are persons possessed of information in special regard to the ailment, the treatment of which has mainly been left to adventurous practice, regardless of the actual pathological condition of the patient; and, when in advanced stages an exclusive reliance upon some *patented spinal supporter* is all that is tendered to the unfortunate patient, when the most skilfully constructed spinal supporter is only auxiliary to skilful treatment. When the caries is in the sixth and the seventh dorsal vertebræ, and only a single, limited projection, the form of the person is but slightly impaired, as in Fig. 120. Or, when affecting the lumbar vertebræ, as seen in Fig. 121.

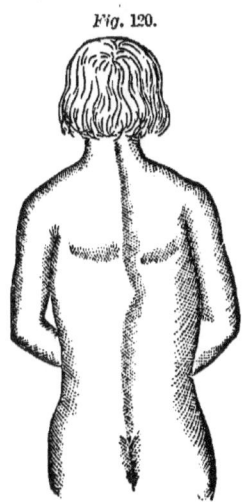

Fig. 120.

Fig. 121.

But when the caries advances, and more especially in the lumbar vertebræ, the body is thrown forwards, and a disposition made to support it by grasping the legs above the knees to support the body when walking.

Paralysis but seldom ensues when the caries affects these portions of the spine, although the projection is very large and limiting the stature, as seen in the last figure, tending to an increase of anterio-posterior deflection. Even at this stage the patient is amenable to relief, both as to form and condition of health; not, however, to the restoration of the normal form of the spinal column. Any effort at reducing the angular projection is attended with the most serious consequences, even to the compromising the life of the patient.

The progress of the disease or pathological condition from incipiency to the absorption of the bodies of the vertebræ, demands our especial attention to prevent the sad deformity that ensues from inefficient treatment.

Inflammation having invaded the intervertebral cartilages, soon involves the cancellated structure of the bone, and as ulceration proceeds the angle increases, and becomes more or less acute, according to the number of vertebræ affected and the loss of substance. This impairs the physical integrity of the spinal column, rendering it no longer a center for sustaining the weight of the head and shoulders; and if not supported with a properly constructed brace the flexion and attrition become a mechanical irritant. As the disease progresses pain increases upon motion, but more especially from that of a sudden step or jolting in a traveling conveyance. The general health suffers from constant irritation, the appetite fails, the bowels become irregular, the secretions morbidly conditioned, tending to hectic fever and abscess. The anterior surfaces of the vertebræ being the primary seat of the invasion, extending in some cases to one side more than to the other, the body inclines greatly to the side, yielding from the loss of substance, as seen in the last engraving. Fig. 122.

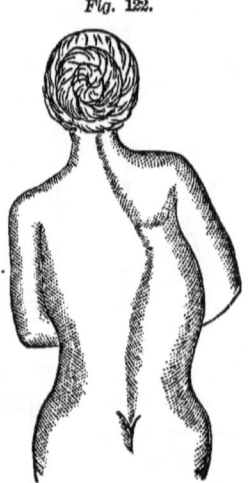

Fig. 122.

At this stage of the ailment, in many cases, abscess is formed, and indicated by increased disturbance in the general health and by an enlargement on either side of the projection or in the lower parts of the body.

On examination of the diseased vertebræ the indications of pre-existing inflammation are readily discovered by the textural disintegration, ulceration, abscess and caries.

The abscess is imbedded in the substance of the bone, or found near the surface. They vary both in form and size; some contain pus, others pus and blood, and still others contain pus surrounding the diseased bone. The matter in many cases becomes inspissated by absorption of its watery constituents, when it acquires a yellow color and thicker consistency.*

In cases of progressive disease the disintegration of bone proceeds until great loss of substance ensues, the vertebral column yields, and the spine projects to an extraordinary degree; the patients recovering after the invasion of from two to six vertebræ, and retaining their power of locomotion. It is usually in these sad cases that internal abscesses form and descend, often of very great size, containing a large quantity of watery matter, passing down the course of the psoas muscle and presenting in the thigh, constituting "psoas abscess;" or, forming behind the tendinous origin of the transversalis muscle and presenting in the loins, constituting a lumbar abscess, and, in some cases, in the region of the diseased vertebræ. They present internally in some cases, and enter the bronchi or colon, by no means observing any regularity in their tendency. This sad occurrence is but rarely met with.

In adults the symptoms vary considerably from those of childhood. If the disease is located in the cervical vertebræ, as the bodies are comparatively thin, caries rapidly penetrates to the spinal canal, involving the cord, endangering thereby the life of the patient. Where the dorsal or lumbar vertebræ become diseased the lesion is not so serious in regard to life, but if not arrested in the early stage of the ailment great deformity ensues, to the compression of the thoracic and abdominal viscera, inducing difficult breathing and severe attacks of asthma. We have, in some cases, caries without angular curvature impairing the flexibility and mobility of the spine, and thus compelling the patient to retain the body in a fixed position when walking, and when lying upon the back unable to raise from the position without help. The lower limbs now become weak, the patient walks unsteadily, and the health partly fails because of mental depression on contemplating such deplorable prospect from the progressing inability to walk. The weakness is especially marked in going up stairs, as they labor under an inability to stand on one leg and raise the other. Spasm of the muscles of the lower extremities, with a tendency to relaxation of the

---

* Holmes Coote on Joint Diseases, p. 159.

sphincter ani and retention of urine are accompanying symptoms. In some cases abscess forms before deformity is apparent. Mr. Stanley remarks that pain and irritation of the spinal cord are usually lessened for a time after abscess has formed. Mr. Erichsen remarks, in his work on the Science and Art of Surgery, p. 788: "It must not, however, be supposed that abscess necessarily forms in all cases; indeed, the formation of matter will, I believe, chiefly depend upon whether the disease of the vertebræ be tuberculous or not. Simple congestive or inflammatory caries of the spine may take place to a very considerable extent, and yet no suppuration occur; the bodies of the vertebræ undergoing erosion and absorption, and coalescing so as to become fused together into one soft and friable mass of bone across which bridges of osseous tissue are sometimes thrown out so as to strengthen the otherwise weakened spine. In these cases, masses of porcelaneous deposits will not unfrequently be found intermingled with and adhered to the carious bone. Indeed, this anchylosis and fusion of the bodies of the diseased vertebræ may be looked upon as the natural mode of cure of angular curvature of the spine; the only way in which it can take place when once the disease has advanced to any considerable extent." This is the most probable rationale of the process of cure in cases of caries of the spine.

Diagnosis of abscess arising from caries of the spine is in some cases very difficult to determine; the only guidance and tolerable certainty being in the history of the case, as before stated; its course to external development is not limited, often presenting in localities subject to abscess; as in the groin, where we have lymphatic collections in the subcutaneous and cellular tissues; from the cellular tissues about the kidneys, pericœcal abscess; abscess in the iliac fascia, from hip disease; glandular abscess, from empyema perforating the pleura and passing down behind the diaphragm. In all these cases there is an absence of pain in the dorsal region and no spinal projection, which is a most common attendant upon psoas abscess. If the abscess is perinephritic, there would be symptoms of renal disease. In other rare cases the attendant symptoms usually determine the ailment with tolerable certainty.

Prognosis in spinal caries is much more favorable in children than in adults. Children, even after complete paraplegia and anæsthesia, will recover from both in ratio to the reduction of inflammation. The occurrence of abscess renders the prognosis less favorable on account of its influence upon the vital energy of the system. It is remarkable that psoas abscess, in some cases, is attended with less implication of the spinal cord than in those where no suppuration presents; but where these are indicative of active inflammation and softening of the spinal marrow, marked by prolonged nervous symptoms and total paralysis of the sphincter ani, the prognosis becomes decidely unfavorable.

Children recover after several years progress of the disease, having accessions of various intercurrent disorders; irritable and tumid bowels, hectic fever, vesical irritation, discharging abscess, and bed-sores. If nutrition can be sustained after the acute symptoms have passed, there is a tolerable prospect for recovery; that is, to an arrest of inflammation, consequent pain, and dissolution of tissues, minus the parts and processes that have been wasted in morbid secretions. The nervous tissues maintain their integrity to a greater extent than any other tissue.

CARIES OF THE CERVICAL VERTEBRÆ.

The tendency to caries of the cervical vertebræ in children claims special consideration, as in its progress peculiar impairment to the physical condition of the system presents; as that of most painful neuralgia, in some cases followed by paraplegia and partial paralysis of the pelvic viscera, which but seldom supervenes in caries of the lower portion of the spinal column.

When the disease presents in the first and second articulations of the cervical vertebræ, it constitutes one of the most serious forms of vertebral caries. The movement of the head is limited from pain and swelling attended with the formation of abscess that tends, in some cases, externally, and in others into the cellular tissue behind the pharynx and extending along its posterior wall against the nasal aperture, even to the extruding of the tongue, occasioning much suffering from a difficulty of breathing. In these cases the sterno-mastoid muscles are made tense and prominent, and the neck perfectly rigid, with a disposition to support the head with both hands under the chin. This, in some cases, has proved fatal from luxation of the vertebræ forwards, causing compression of the cord, in others from hectic exhaustion. The patient having recovered from this fearful condition is usually left with a peculiar position of the head; the occiput elevated, and chin depressed; inability to turn the head, and a round prominence upon the upper posterior portion of the neck.

The next, and much more common seat of the disease in the cervical vertebræ, is inflammation involving the last cervical and first dorsal, producing a very peculiar series of painful, consecutive implications. The first indication is that of a set position of the head and neck, attended in some cases with most excruciating pain on motion, and, frequently, when in a state of rest. The sensations extend over the cervical fascia. This severe neuralgic seizure is apparently the only ailment of the child, in some cases for several months, when progressive paraplegia will be observed, with inability to control the sphincter ani; the paralysis supervening in nearly every case where the disease is not arrested in its incipiency. The paralysis, however, in these cases, being strictly from local cause, is recovered from in time, as the active stage of the disease subsides.

It is remarkable that motor paralysis should be a concomitant of the inflammation of the last cervical and first dorsal vertebræ; and even as caries increases in development, recovery from the paralysis is progressing, as we have witnessed in many cases. From this we must not infer that it is caused by compression from angular curvature, for the paralysis precedes the curvature; and, yet, the patient is more readily relieved from spinal support by suspending the head, than by any other treatment. Paralysis seldom results from incipient invasion of the dorsal or lumbar vertebræ, unless it is of traumatic origin and actual fracture of the vertebræ.

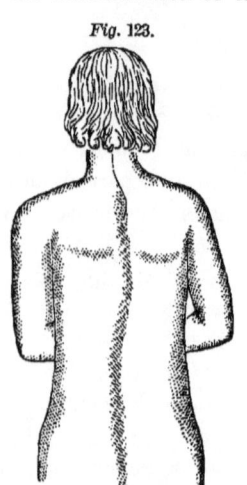

Fig. 123.

The incipient or active inflammatory stage having subsided, and absorption of the intervertebral substance and bodies of the bone advanced, a small knuckle-like projection will be observed, and the patients restored to the use of their limbs. See Fig. 123.

This condition of the patient is, in many instances, maintained, the patient possessing a tolerable stamina of constitution, although a paralysis may have supervened and recovery eventuated as the inflammatory stage subsided; but most commonly caries of the vertebral bones progresses to deformity, as seen in Fig. 124, and

DISEASES OF THE BONES. 319

Fig. 124.

the patient eventually enjoys a tolerable condition of health. This favorable result is not always obtained, and we have abscess in those cases where caries has continued to the destruction of the bodies of the diseased vertebræ, in some cases, months after it is supposed the disease has been arrested. As the ulceration affects the anterior surfaces of the bodies of the vertebræ, it is in this situation the abscess is formed. Matter having formed in this location has passed the entire length of the body and presented in the iliac region, and in front of the thigh, and, in some cases, has perforated the rectum, and the patient recovered after months of suffering.

TREATMENT OF CARIES OF THE SPINE.

In the treatment of incipient or advanced caries of the spine favorable hygiene is required, as success depends mainly upon the enforcement of a suitable regime and proper sanitary relations thereto, and which is imperative to the making of nutrition available to the recuperative powers essential to an arrest of the disease. This favorable condition can most readily be obtained in a well designed hospital. Patients at home meet with many interruptions to progress in cure from unfortunate indulgences, to their serious injury. This is the experience of every surgeon in attending to private practice. It is possible to obtain a prescribed diet, and medicines administered at stated periods, but it is impossible to have an observation of all the circumstantial incidents to the surrounding and varied condition of the patient, that would govern the surgeon in determining upon a suitable diet and medication. Proper ventilation is not readily obtained in a private chamber; monotonous confinement tends to torpidity of the vital energies and to the increase of local excitement from concentration; and the want of diffusion from pleasurable mental and muscular exercise, which excites assimilation and a demand for nutrition. Hence, a hospital especially designed for this class of chronic ailments contributes most largely to their cure, from the advantages accruing from persistent treatment.

Children laboring under chronic disease of the joints or osseous tissue, confining them for a great length of time to the bed or room, become greatly depressed, even to despondency, and form a dislike to any interruption from contemplating their increasing deplorable condition; the mind becoming morbidly sensitive to any interference, by treatment, that would tend to their benefit, while their only desire is to be undisturbed. Then parents and friends, expressing their sympathy, contribute greatly to increase this tendency. And, in many instances, these promiscuous influences are unavoidable at their homes, even in the most opulent families. Social enjoyment is only attainable among a number of joyous children, where a selection of suitable companions can be found, as may be daily observed in the Hospital for the Relief of the Ruptured and Crippled, where children from comfortable homes and those from no homes are associated. The little patient, on entering, is quite homesick for a day or two in some instances, but by no means are all so affected, as pleasurable entertainment is made one of the essentials to the regime of the house, and in which the child becomes interested. Some children take part in these amusements at once, while others remain in their acquired quiescent condition for days, so depressing has been the surroundings at home, and excessive the drains from large abscesses. The child having failed to take a nourishing diet, because of frequent persuasion at irregular periods, is supposed by the parent to be in a starving condition from its refusal to eat a reasonable portion of nourishing food; forgetting that the child had been hourly tempted to eat something pleasing to the taste, however unsuitable as a diet in its enfeebled condition. A child in this condition having entered the hospital is offered a suitable diet at stated periods; if it is refused, no attention is made apparent until the next period, and this course is continued. From time to time the child takes a small portion of the food, and within a week will, in most instances, astonish its parents by the excellent appetite it will display, and the quantity it will eat at a meal. This is entirely contingent upon taking food at stated periods, and upon the excited energy induced by freely associating and joining in with the other children in their amusements; even when confined to bed, amusements are offered the child and it is induced to join in them. Music contributes usually most largely to the pleasure of the child, and if it can be induced to join in singing it is one of the most salutary exercises that can be offered the feeble little patient, even though not able to sing above a whisper. Whether it is the increased quantity of oxygen inhaled, or the pleasurable excitement to the nervous system, or both, we are not prepared to say, but state the result as contributing largely to an increase of appetite and a solace to their sufferings; for while singing they cease complaining for the time, and for some time after.

Children disposed to move about without being much inconvenienced from pain, should not be confined to bed, as it tends greatly to torpidity of the physical power and depression of the mental faculties. The physical powers are accelerated by pleasurable reflections which tend to a demand for nutrition and a consequent increase of vitality — the only power of resistance to disorganization, or, in other words, disease, under whatever form it may present. Temperature and ventilation claim special consideration, and can only be made available in apartments designed for the purpose, and having capacity and construction to supply each individual with not less than nine hundred cubic feet of air, and a change every ten seconds; the air passing off through registers in the floor, beneath which the entire space should constitute an air chamber connected with a heated conduit of some eighty feet in height, kept at a temperature of 120°, thus obviating the pernicious influence of inhaling the diffused gases arising from purulent fœtid matter emanating from discharging abscesses and foul breath. It is pure air that should be supplied the patient, protecting him from miasma or pyæmic influence; the most common cause of failure when otherwise skilfully treated. Proper ventilation is of primary importance in the preliminaries of preparatory arrangements for treatment. In an ordinary dwelling, there being a fire-place, a tolerably pure condition of atmosphere can be obtained by having a fire in the fire place, or a jet of gas burning therein, and this should be made available at all seasons of the year. When damp easterly winds prevail in midsummer, or close rainy weather, the room where there is a patient laboring under any of these chronic ailments, there should be, invariably, heat in the open fire-place. This would give a more rapid change of air, essential to the escape of any morbid exhalations emanating from open disease or body. Frequent sponging and cleansing the body and limbs, keeping the skin in a moist, pliant condition, is a most salutary means of improving the condition of the patient. To free the emunctories, and promote excreta from the cuticular surfaces is of nearly as much importance as that of the regular evacuations of the internal organs, and, when carefully incited to a normal condition, tend largely to the cure of the local ailment. As it is invariably to be observed in these chronic conditions, the skin is torpid in its secretory functions, there being a deficiency of vital energy; a shriveled, dry, inelastic condition as in the case of an excessive internal irritation. This is in the incipiency of the ailments, and for a time during their susceptibility of cure. When the system becomes prostrate, as in chronic diarrhœa, or draining abscess, the emunctories of the skin become patulous from an extreme loss of vitality, and the sweating stage ensues.

The diet of the patient laboring under these chronic ailments is of the utmost importance, and is dependent upon the sanitary relations having prepared the assimilative functions for the reception of nutriment. When this has been accomplished, suitable nutrients are not only to be selected but carefully prescribed in suitable quantity, quality, and at stated periods. This requires close observation of the patient's condition and habits. An excess in quantity of solids or fluids in diet, is as inimical to health as a pernicious selection, and if taken at irregular periods disposes the digestive function to a morbid tendency. If the food is not properly digested, the quality of the blood is impaired, the general function of nutrition fails to sustain an extraordinary demand — that of the normal supply and the drain of the abscess or ulcer. The disease thus obtains an ascendency to the increasing depression of the whole organization, arresting in degree the recuperative tendency. By moderately exciting the digestive forces we tend to restore the due qualities to the blood, and the energy of nutrition — the stomach recovers its powers when properly nourished, and a normal condition is restored — tending to recuperative energy and ability of vital force to relieve the system, even of local irritation and destruction. The restoration of normal power in the stomach is of primary importance, inviting very special consideration. An excess of functional effort exerted upon the most valued nutrients will prove inimical, and tend to the destruction of the patient. Careful observation and experience in selection and quantity, determining the power of the stomach to digest in every case that presents for treatment, is that which will contribute most largely to successful practice.

MECHANICAL APPLIANCES.

Holmes Coote on Joint Diseases, published 1867, p. 164, says: "A carious spine is unfit to support the superincumbent weight; it is not suited to the movements necessary in progression; no apparatus, however skilfully made, is fitted for a case of actual disease, and yet the patient's recovery depends on the function of nutrition being well carried on, a process to which exercise is most necessary."

This eminent writer believes exercise to be imperative to the recovery from caries of the spine, but objects to spinal supports. This last is because of their imperfect construction, and impinging upon the diseased spine they become a source of irritation. A skilfully devised brace or supporter is of inestimable benefit, because of its enabling the patient to take exercise with impunity, and thereby obtain a very necessary part of the treatment. The spine supporter we have applied invariably affords great relief to those laboring under caries of the spine in all stages. If applied in the incipient stage it arrests aberration of the spinal column and sternum, that is, if the patient's stamina is susceptible of improvement from a strictly enjoined regime and alterative remedies. The spine supporter we now use we devised over twenty years since.

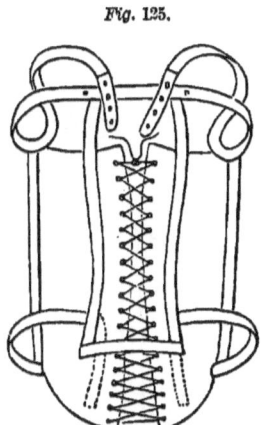

Fig. 125.

It will be observed that it is a light steel frame encircling two-thirds of the body. A back view is presented. The lower bars rest upon the crest of the ilium, and the two longitudinal bars extend below, thus giving a lengthened support, and obtaining greater firmness, so that if the caries should be in the lower lumbar regions no injury could be sustained from pressure. Shoulder straps, well-padded, pass from the front over the shoulders and button on the cross-bar, passing over the scapulæ and under the axillæ. Strong woven fabric incloses the front by lacing.

It will be observed that this brace gives lateral support and not extension; giving a fixed, non-elastic support, thus arresting motion, and consequently attrition in the articulations of the diseased vertebræ — being the first intention to be met with a therapeutic agent (in our opinion), and confirmed by the benefit they have afforded many hundreds of persons afflicted, from two to thirty years of age; enabling them to take exercise with impunity. Certain modifications are required for the several points of the spinal column that may be diseased. The two vertical bars to the spinal column have to be carried above the scapular cross-bar when the caries is in the upper dorsal region; and when in the lower dorsal and lumbar regions the tendency of the body to incline to one side is corrected by extending a small bar of steel from the lower cross band, having a pad attached and intended to press upon the hip of the side inclined to, and thus the body is sustained in the erect attitude. See Fig. 126, back view.

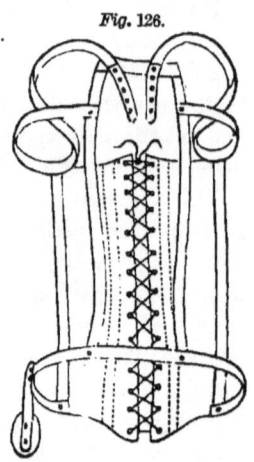

Fig. 126.

This same form of brace is made available to support the head; an essential addition in cases of caries of the cervical vertebræ, and more especially in cases of disease located in the last cervical and first dorsal vertebræ; in all cases relieving the paraplegia that subsequently ensues from caries in this region, as well as the distressing neuralgic suffering attendant upon the same. This form of apparatus has, in many cases, relieved torticollis by having a very firm upright bar carried over the head, with a swivel piece to which is attached the straps to be passed under the occiput and chin; the swivel permitting the rotation of the head. See Fig. 127.

Fig. 127. A, the body of the brace; B, the upright bar curved to pass over the head, and sustained in a strong leather sheath attached to the upper and lower cross bands of the body brace. A back view is given of the entire apparatus.

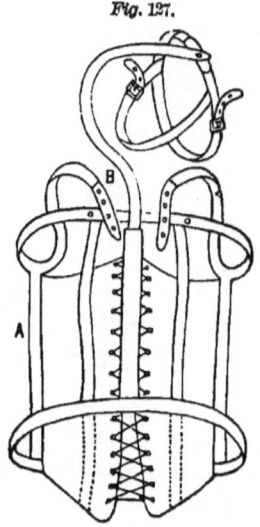

Fig. 127.

These spinal supports should, under all circumstances, be worn night and day — thus to obviate pressure upon the spinal cord.

In the treatment of the extraordinary painful condition, often preceding caries of the last cervical and first dorsal vertebræ, the spine supporter (Fig. 126) affords much relief by steadying the head from the bases of support upon the ilium when the patient is in the erect position; thus enabling him to have the benefit of open air exposure, and by no means interfering with the aplication of other means of relief, as the bar supporting the head is sustained in position some distance from the spine.

℞.
    Ext. bellad.................... ʒj.
    Aq. font..................... ℥viij
    M.
Ft. lotio.

Cloths wet in the above lotion and applied to the affected parts (extending a few inches on either side), with the support applied, often afford immediate relief for the time being, and must consequently be frequently saturated with the lotion. Cantharidal vesication, followed by poulticing for forty-eight hours, the support removed, and the patient reclining for a time, is the most valuable means of relief in obstinate cases as it relieves the congested vessels of their serum. Even this decided treatment, however, will not, in some cases, afford immediate relief; the liq. potass. arsenitis, is then the most reliable remedy in doses of four drops three times a day for a child of four years of age, and to older children in proportion. For several weeks after a patient has been relieved of the painful symptoms, a slight jar will induce a recurrence of the suffering, tending frequently to a more persistent condition; leaving him in a state of such extreme sensitiveness that great difficulty is experienced in the removal of the clothing. The early application of the brace and head support protects the patient from much suffering, especially if the child is to be conveyed any distance in a railroad car, otherwise much suffering would be induced. With the brace, however, the little patient can ride with comparative comfort. Another peculiarity in the ailment is that of an obscure condition, in some cases, for several months, rendering the diagnosis doubtful. Several cases of more than ordinary interest have been found in this condition of the ailment after having been carefully examined by experienced surgeons without a discovery of the actual condition of the patients having the premonitory symptoms of caries in the last cervical and first dorsal vertebræ.

Maggie ———, a little girl, four years of age, in charge of her grandmother,— the mother being dead — the family possessed of considerable wealth, had been presented for advice before the first talent in the profession, without obtaining a reliable diagnosis as to the pathological condition of the patient, or of finding any relief for the suffering girl. The grandmother's statement was to the effect that some three months previously the child complained of the nurse hurting her while being dressed — giving her severe pain in the back of the neck. After this complaint had been made some two or three weeks, the grandmother tapped the child gently upon the shoulder, at which she cried most pitifully from the suffering induced by this playful intention. From that time, it was observed that to lift the child under the arms would give her great pain and suffering for an hour or two; and she appeared to be subject to periodical seizures of this suffering without any apparent cause. The suffering had become so extreme that anæsthetics had been resorted to for relief on occasions of examining her condition. By the advice of their family physician the child was placed under our care for professional treatment. From having seen several other cases quite similar in indication, we placed the child under treatment for caries of the last cervical and first dorsal vertebræ. The grandmother objected most strenuously to the application of the brace, and was sustained in her objections by several of the medical gentlemen who had previously examined the child. The father of the child, however, insisted upon the treatment being tried, as all others had failed, and contended in favor of the brace that it would steady the child's head and neck — as he had observed that, in certain movements of the head this suffering was induced. The child was of lymphatic temperament; blue eyes and light hair; suffering at times from aphthous ulceration of the fauces and bleeding from the gums. The teeth were short and discolored. Even after all her suffering, the child was rotund in form; being evidently of a scorbutic diathesis, or strumous inflammation without tubercle.

Treatment: emplastrum lyttae vesicatorum, 4×5 inches was applied to the cervical region extending down the back two inches at ten o'clock at night, and to be carefully dressed in the morning with a large flaxseed poultice; the dressing being changed every four hours for forty-eight hours. Of "Fowler's solution" of arsenic, four drops were given morning, noon and night; to be continued for ten days unless nausea was induced. After three days the brace was applied, and without giving pain to the child. The nurse was cautioned to keep the brace on the child day and night, and not to permit her to receive a jar by getting down from chairs, or in any other manner. After ten days' treatment, the child was yet subject to suffering from slight causes, but was much relieved from her former condition by the support given the head, so much so as to cause her to tell the nurse that she must not take the brace off, for it relieved her pain. It was then thought advisable to blister again, and to increase the dose of arsenical solution to five drops three times a day, when, on the second day, much nausea was induced and the remedy discontinued. To the great alarm of the family, the little patient was now becoming partially paralyzed, and in one month from that time she had complete paraplegia ; the neuralgia entirely subdued, and, upon removing the head brace, the chin and head would rest upon the sternum; but upon reapplying the support, no unusual posterior projection of the spine could be discovered. The father of the child wished me to continue my attendance, as he was assured that others had recovered the power in their limbs when similarly conditioned; that the paralysis was from the advanced condition of the caries and impairment sustained from the long continued inflammation that was now subdued. The following mixture was then ordered, and administered teaspoonful doses, night and morning:

R.
  Hydrarg. bichlorid................ grs. ij.
  Tr. cinchon. comp................ ℥ vj.
              M.
  Ft. mist.

Plain nutritious diet was given at stated periods, and fruit in the forenoon, two hours before taking a regular meal. Her legs were manipulated at least twice a day by firmly grasping and pressing the muscles, then the body sponged with water at the temperature of 80°, and well rubbed off with a coarse towel. The child was lifted to her feet at frequent intervals during the day and encouraged to make efforts to stand, and on spastic trism being observed permitted to rest upon her limbs as much as possible. Out door exposure on all favorable occasions was made available. Three months passed without improvement in walking; the sustaining bar of the head brace was then shortened half an inch, that the bodies of the vertebræ might be brought into closer approximation (an important point to be observed in spinal supporters. If much of the cancellated structure of the bodies of the vertebræ is absorbed anchylosis cannot readily take place to close up the intervening space which is the case in advanced cases of caries of the spine; a projection being unavoidable to a cure, and, as far as they have taken place, cannot be interfered with without endangering the patient's power of locomotion. And, what is remarkable, this holds good for any portion of the spinal column; that the projection cannot be interfered with to straighten with impunity). The little patient then commenced to improve by bearing weight upon her legs when in the erect position, and in fifteen months was restored to the full and free use of her limbs; a slight angular projection having resulted from the lowering of the head brace.

In all cases tending to *caries* of the spine, the first indication to be met is an arrest of motion by lateral support, which tends to the arrest of inflammation; and if it has tended to absorption of the intervertebral substance, angular projection will be observed. But a more favorable diagnosis presents in the early stage of the ailment if the patient when lifted from under the arms, the projection disappears, there is a prospect of cure without any apparent projection remaining. This, however, is only in cases where the inflammation has been of short duration and a very limited portion of the bone removed, admitting of being bridged over.

All spinal braces admitting of motion to the spine, evidently permit attrition of the diseased vertebræ upon each other, and cannot be considered as complete in rendering relief, as those arresting all motion by lateral support. Such extension as is given by crutches to elevate the body from pressure under the arms should be avoided. This means of elevating the body, in advanced cases, separates the sound portions of the bodies of the vertebræ, and prevents anchylosis from completing a cure; as, in cases when anchylosis has made some progress, and angular projection is very apparent, attempts to overcome the projection by efforts at extension of the spine have, in many instances, produced paralysis.

Yet, there are practitioners who boast of straightening the spine in such cases, and who we know have produced paralysis by their efforts — the injured subjects now living. The patient having been

DISEASES OF THE BONES. 329

placed in such position as to give the deformity the greatest prominence, a plaster cast is taken of the deformed portion of the body. This serves to determine the improvement that may be made in a very brief period of time. The next step in the procedure is to place the patient, face downward, on a sofa or firm mattress; then reduce, if possible, the projecting portion of the spine, and apply, most commonly, a patented brace. After a few days, the patient being placed in the most favorable position to represent their improved condition, another cast is taken. These casts are preserved for the deception of others that may be so unfortunate as to fall into their hands for treatment. The paralyzed patients are not noticed in the history of the extraordinary successful treatment.

Indifference in the profession to the treatment of ailments that they are not familiar with, favors these charlatans in their nefarious conduct, resulting in great injury to the decrepit, who are, in many instances, advised by their family physician to apply to these creatures for treatment. But most commonly, in distant parts of the country, the poor cripple having excited the sympathy of the postmaster, who is supplied with circulars filled with wood-cuts and descriptions of extraordinary cures, receives one of these from him, in common with every other cripple in the neighborhood. This is a most profitable source of income to these wretches, flourishing upon human affliction, which they only serve to increase. The sad history of these impositions are almost daily related by the afflicted when making application to the Hospital for the Relief of the Ruptured and Crippled for treatment for themselves or their children; having paid all they could obtain from their friends, in addition to that of their own hard earnings, for a few weeks' treatment, when they were coolly informed that if they could remain longer they would be cured — after having been informed in the first place that they could be cured in the brief period for which they had paid, when most of them had chronic ailments that would require years of skilful treatment — as that of dental paralysis, extraordinary cases of diseased joints, lateral curvature and caries of the spine.

The advanced stage of caries of the spine presents, probably, the least assurance of successful treatment of any of the diseases of the bones, as removal of the diseased bone cannot be made available to the cure. Large abscesses, as before stated, form, and travel in various directions, some much more serious than others, and all of doubtful character, as to an arrest of the exhausting influences upon the vital energies of the system, which are seriously compromised under the most favorable circumstances — the obscurity and almost unlimited mass of matter accumulated within the body impoverishing the system and making an unfavorable impression upon the assimilating process of nutrition, thus impairing the very sources upon which reliance may be placed for recuperative power. Hectic symptoms, even, are noticed in many cases, before the opening of those monster abscesses, commonly known as psoas and spinal abscesses — having descended over the surface of the psoas muscles and presented in the groin, often passing down the thigh, enlarging it to a size nearly equaling two of the sound limb. These abscesses present at different parts of the body, and, in some cases, in the immediate region of the caries, almost always tending externally, and progressing in size very slowly, with little or no pain in most instances. In some cases abscess is formed before any of the other signs, except weakness of the spine, and before deformity has taken place. In cases when much pain has attended caries, as is remarked by Mr. Stanley, the pain and irritation cease when the abscess forms.

Any thing like a satisfactory prognosis is impossible under such circumstances. *Physical signs* excite suspicion that an abscess is forming; the pulse more tense and quick, an increase of temperature attended with occasional chills, hectic flush of the cheek, and, on making slight exertion, a peculiar whiteness about the nose and mouth, with great increase of the interruption in breathing; a disposition, at night, to fold the thighs upon the abdomen, with the face brought down almost to the knees, thus relaxing the abdominal muscles and obviating pressure upon the viscera that would press the abscess from ordinary tenseness of the muscles; an unusual full appearance of one inguinal region, and a disposition to flex the thigh in walking, as in advanced hip disease — all indicate the approaching development of the abscess in the groin.

The patient being assisted to the erect position when these indications present, by careful observation from percussion, fluctuation will be discovered, and, in a few days, the abscess development will be observed, when a roller should be applied about the hip so as to give support and pressure — limiting, if possible, the increasing development of the sac. By this means a farther development is, in many cases, arrested, and even a diminishing of the contents occasioned, to the degree of entire absorption. The child's physical condition having been favorably maintained — a precaution that should be most carefully enforced by the advice of the surgeon, from the early premonitory indications of disease of the spine. It is a wasting disease, and requires preparatory measures to sustain the exhausting influence upon the system — observable by the experienced practitioner in medicine when there is not an external sign, as that of projecting spinous process, or developing abscess. In caries, alone, when abscess is not induced, prescribed regime and an arrest of motion in the diseased locality is of vital importance in the arrest of caries in its incipiency: thus saving the patient from sad deformity and general impairment of health.

To the opening with knife, trocar, or caustic, we are decidedly opposed at any stage of these spinal abscesses, and we are not alone in our opposition to the interference with the immense collection of serum which is not inimical to the health of the patient, if absorbed. Mr. T. Holmes, in his work on the "Surgical Treatment of the Diseases of Infancy and Childhood," p. 543, says: "With regard to the opening abscesses connected with spinal disease, my own experience leads me to dissuade it. However affected, and with whatever precaution, I think it generally does more harm than good. There are, of course, some cases where the rapid increase of the quantity of fluid and the pain which it causes compel the surgeon to interfere; and then, the abscess should be tapped with a trocar, the opening being closed, or Mr. Seister's method of dressing the opening with carbolic acid should be used. If the surgeon prefer to use Thompson's canula, by which the abscess is opened under water, there can be no objection. I have, myself, little confidence in any of these plans, and greatly prefer, when possible, to leave the matter to find its own way to the surface." Mr. Erichsen remarks: "When abscess has formed the surgeon should be in no hurry to open it; but, in accordance with the principles laid down when treating of this affection, he should delay doing so, lest injurious, fatal constitutional irritation be set up." Mr. Holmes Coote, in his work on joint diseases (p. 159), says: "It is in such cases that large abscesses are so apt to form, either pursuing the course of the psoas muscle and presenting in the thigh, constituting a 'psoas abscess,' or forming behind the tendinous origin of the transversalis muscle and presenting in the loins, constituting a lumbar abscess.

"When there is an apparent necessity for opening the abscess, let not the surgeon act hastily! That pus may become inspissated by absorption of its watery elements is acknowledged by the latest pathologists; that it may be, and often is removed by the absorbents is my firm belief. The gradual disappearance of a large collection of pus is an event by no means rare in surgery."

When the prognosis is clear that spinal abscess has ensued, from apparent development, or otherwise, every possible means must be made available toward invigorating the patient; and it is quite possible in the majority of cases to greatly increase their vital energies. To the accomplishment of this very desirable object, pleasing entertainment at home or abroad, in the open air as much as prudence will admit of in regard to weather, and well ventilated sleeping apartments should be secured. A perfect discipline should be observed as to taking food of nutritious quality in proper quantity — not quite satisfying the appetite — and at stated periods. Fruit to be taken only in the forenoon, and at most two hours after or before meals. Fluids to quench the thirst should be taken in exceedingly small quantities, and at no time for at least an hour before meals, and, if possible, only at meals. This is a general view of a suitable regime, to be modified and determined in detail by the attendant surgeon. And, as before stated, every effort should be made to compress the developing abscess by rollers, or lacings, over cotton batting, applying to the surface, and twice a day the following embrocation : —

   ℞.
    Lin. sapo. camph................ ℥ vj
    Tr. iodini.
    Tr. belladon.................. āā ℥ j
    Ol. cajeput..................... ʒ j
    Ft. emb.

The abscess should be carefully bathed, previous to each application of the embrocation, with warm water containing a drachm of common carbonate of potash to the quart, and diluted if, in some cases, too caustic. This solution affords much relief to distended abscesses when located about the body or limbs. A napkin or other soft material folded and kept moist with water should be applied to the affected part, when so sensitive as not to admit of compression. When in this condition, inflammation is not only confined to the abscess, but involves to a considerable extent the surrounding soft tissues, attended with increase of temperature and severe pain. For relief the evaporating lotion containing chlorate of ammonia, as before described, should be freely applied to the parts. Decided constitutional treatment is now, if not enforced earlier, imperative to the safety of the patient, and invites a careful investigation of the materia medica and its therapeutic tendencies to the resuscitation of organic vigor, to that of active assimilation, demanding an increase of nutrition to sustain the excessive drain that is now about to take place.

CHAPTER XIII.

TONICS, AND THEIR EFFECT UPON THE SYSTEM.

When Tonics are admissible or inadmissible. — Modus operandi of Tonics on the system. — Torpidity established from inability to move about or compulsory confinement. — Physical condition of the patient determines the therapeutic agents to be used. — Tonics materially increase the vital forces, if judiciously administered. — Abscesses may become incurable — SELECTION AND QUALITIES OF TONICS. — Exhilarating influence as a Tonic. — Pure Air as a Tonic. — MODIFYING INFLUENCES. — Well-ventilated apartments. — Temperature as a Tonic. — Medicinal agents as Tonics. — Treatment of chronic inflammation. — Diet. — Method of relieving the nerves and congestion of the bloodvessels — Means whereby inflammation is extended throughout the system. — Alterative treatment. — Application of Tonics to the various ailments treated in this work.

There are conditions of the system when tonics are inimical, as in cases of low degree of excitability, or deficiency of constitutional stamina, or from the abuse of stimulants. They tend to increase the depression by an exhausting influence — there being a deficiency of no latent power to excite into renewed action.

Tonics are inadmissible in indirect debility, as in an overwhelming congestion or inflammation of some important organ or tissue, there being such a concentration of blood and nervous force in one part that there is insufficient elsewhere to support the systemic actions generally, or, the organ affected is restricted so as to arrest its function and thus prostrate all dependent functions. Derivatives and diffusion of excitement are the remedies in such cases.

Tonics are only indicated as beneficial in cases of depression and debility, in which the excitability has not been exhausted by previous excessive exhaustion, in which the depression is neither the result of active congestion or irritation, nor sudden and transient as to call for stimulation more prompt than that which characterizes this class of remedies. Hence, they are most serviceable in slow impairment of health as in that of ordinary chronic ailments.

As to the *modus operandi* of tonics on the system, there is no well determined conclusion ; satisfactory results are well determined, and upon this we must rest satisfied for the present. They are not primary, but secondary ; the system must be in a suitable condition to be benefited by their use. They appear to act chiefly on the digestive organs; they promote the appetite, invigorate digestion, create a desire for more food to be eaten, and more thoroughly prepared for absorption and assimilation, and thus enrich the blood ; rendering it at once more stimulating to the functions and nutrition, exciting general tonic effects upon the system at large, and when alterative medicines are combined, greatly correct or modify morbid secretions. How they affect the ultimate organic constituents of the body is conjectural. They have been found by chemical investigation in the midst of the tissues. It is thought probable that each distinct function is performed through the instrumentality of a special power in the ultimate organic cells, nuclei, or molecules of the organ, and that tonics operate simply by stimulating this power into a somewhat increased activity. That there is a certain vital cohesion essential to the due performance of the function, and that tonics are moderate stimulants to this cohesion, is a reasonable conclusion. As, to this action, may be ascribed the great firmness of the tissues especially from the stimulant, the cause ceases and the system is left in a condition in which it can repair itself and resist the local irritant that is temporarily exhausting the system.

We have torpidity established from inability to move about — confined to a room — severe pain induced by motion and the surrounding monotonies. A child with morbus coxarius, or paralysis from *caries* of the spine, whose parents, greatly grieved at its condition, are from day to day entering the room with an unmistakable bearing of grief, soon becomes, as it were, partially paralyzed ; an apparent habit of insufficient action has been established, which the inherent force of the system cannot throw off, and it is disposed to continue in the condition in which it may have compulsorily continued for a considerable time. Here we have a torpidity of the mental and physical condition of the child, upon which special treatment for the ailment under the circumstances then existing can make no impression as to improvement, or even toward an arrest of progress to death. By giving the patient a change in the way of inviting scenery, tempting its thoughts to contemplate something more than its own condition, even this mental exercise will stimulate the weakest physical organization to a condition of susceptibility favorable to the influence of tonics. Here we perceive that home seclusion has a pernicious tendency; rendering treatment unavailable that would relieve the patient if favorably circumstanced, as he would be, in an institution designed for the purpose. When thus situated even the most gentle tonic would become a potent medicine, and essential to excite an impulse to the functions of assimilation, or that of a particular organ, most common in convalescence from acute disease.

General depression or debility may result from the torpidity of a particular function or organ, upon which impairment the general deficiency may react, so as to resist treatment for abscess or ulcer. A tonic suited to the condition of the patient would not only favor the treatment, but without other treatment than cleanliness cure the abscess or ulcer.

The patient's physical condition determines the therapeutic agents to be prescribed, with a special consideration of the expectation of results. The impairment of the physical condition of the patient from extraordinary supply to a morbid accumulation of pus and serum, demands a more than ordinary supply of sustenance, dependent upon an increase of assimilating power or functional effort to sustain more than a normal condition of the system. This must be the first effort, or the system will suffer from exhaustion. The second effort is to correct any morbid functional condition which may be favored by a normal tendency when the system is in a condition to assimilate nutrition. The supply of nutritious food may be all that could be desired, but the functional efforts insufficient to meet the demand. Then the indication is to increase the organic function of assimilating nutriment to a recuperative tendency.

Tonics, when properly prescribed and carefully administered, tend largely to an increase of the vital forces or recuperative power; however, there is no medicine in the materia medica that requires more practical judgment in the selection and use than tonics. If improperly used they induce a condition of indirect debility, to the great injury of the patient. By giving tone is to be understood as giving a certain vital cohesion of the living molecules in an organized tissue, which is essential to the due performance of its office; an augmentation of vital, cohesive tendency to increased energy of function, and thus only to the vital capacity, which has a limit. If called into excessive action it is proportionably exhausted; and in this state of exhaustion the ordinary healthful excitants have less than their normal effect, and depression, therefore, must follow.

Tonics have a relative capacity to the improvement of the patient's enfeebled condition and their suitable selection should be limited by the normal vital capacity. This view obviates the inference that tonics exhaust and depress the system. When tonics are properly selected they moderately and durably increase the vital action. Strength is dependent upon the normal state of the ultimate organic constituents of the tissues, which can be sustained by a due degree of all the vital processes which contribute to the nutrition or maintenance of parts, and tonics have the property of exciting these processes. It follows, when they are deficient, and debility has ensued as a result, tonics may prove not only excitants, but strengthening, providing the depressing causes have not exhausted the excitability of the organization.

Abscesses and secreting ulcers depressing the functional condition, tend to positive debility, and in degree to the power of resistance from stamina of the system, and may thus become incurable. They can only be considered in a state of permanency, however, when they overpower the vital energy of the system; and if timely aid is given by suitable tonics a resisting force may be induced, even to the recovery of the patient, by exciting the depressed functions and strengthening with nutrition the debilitated structures, and long before the excitability through which they operate has had time to suffer materially.

SELECTION AND QUALITIES OF TONICS.

EXHILARATING INFLUENCE AS A TONIC.

Favorable surroundings are of such importance as to ever give promise of success in the treatment of chronic diseases. The depressing emotions, as those of despondency, grief and fear, are resisting influences that will render the most skilfully prescribed medicines powerless in their operation as to the improvement of the patient's condition. Whatever in any manner contributes as an elevating influence to this condition of the feelings, must indirectly impart tone to the system, and although negative in its operation, must rank among tonic influences to the mental conditions, such as enlivening, elevating and pleasurable reflection, inducing an appreciation of the beautiful and the sublime in nature and art, tending to the enjoyment of all agreeable exercises of the intellectual faculties. For children, instrumental and vocal music in concert, the patient joining in the exercise, contributes most largely to the relief of depression of mind arising from long continued monotonous confinement. Games engaging reflection and calculation in order to win, are also useful, though this amusement must be limited, or the mind will become fatigued, if of a sedative character. Games engaging physical and mental energy, modified to the child's condition, are most salutary. Passive exercises exhaust physical energy without a compensatory increase, because of its sameness, having a depressing influence upon the mind, in a word, it is monotonous, and is objectionable from being so, however pleasing in appearance. Diversity and self-effort is essential to afford pleasure. Whatever is pleasurable to our mental and physical condition is a stimulant, and if limited to a healthful influence when in a state of debility, is positively tonic and restorative.

After a careful diagnosis of the ailment has been made, the patient's present or future condition for treatment is to be carefully considered and prescribed for, and in ratio to the fulfilment of the requirements to insure a state of social enjoyment to the patient, will be the promise of success, there being a sufficiency of vital energy to be made available from skilful treatment to the restoration of the patient.

The influence of the mind on the body is a matter of common observation. Reasonable excitement is known to increase the entire energies of the system, as is evidenced in the circulation of the blood, tending to an increase of functional activity involving nutrition and a desire for food, hence of primary importance as a therapeutic agent.

PURE AIR AS A TONIC.

Pure air is an invaluable tonic, and systems acquire renewed strength through the healthful agency of an uncontaminated atmosphere, well charged with electricity—the latter a most important consideration, as it demands a careful consideration of locality. When *static electricity* is not readily attained under an ordinary condition of favorable weather, depression of nervous energy prevails, unfavorable to health. Certain localities are thus conditioned, even within the limits of the radius of a mile. In a damp locality the excessive evaporation of moisture deprives the air of its electricity. The electrical machine, to obtain electricity from the atmosphere by friction and accumulated by insulation, readily determines the electrical condition of the atmosphere. It matters not how pure the air may be, even in dry, elevated localities, there are extraordinary conditions of the atmosphere at times, depriving the animal system of its normal quantum of energy — the influence extending to brutes; their depression being most manifest. They herd together, having no disposition to feed or to be aroused from their position. This influence is frequently noticed by persons who are wont to exclaim at such times: "What a depressing influence there is to day! I feel extremly languid, and yet the thermometer only indicates 80°." This condition of the atmosphere is most depressing only a few hours before the advent of a thunder storm, at which time not a spark of electricity can be obtained from the electrical machine, and in many instances for a period of twelve hours. After, and even during a thunder storm, however, electricity in abundance can be obtained, and the animal system is exhilarated; the brutes will feed, and be disposed to playful exercise — the temperature after the shower indicating 80° and upwards. This depressing influence is most worthy of notice, as it is continuous in certain localities, and its most pernicious influence is experienced in low, marshy districts of country and in which exist what are supposed to be malarious diseases, or those the result of malaria, the deficiency in electricity not being considered.

There are modifying influences that can be made available, such as those of carefully ventilating apartments by means of artificial heat, supplying patients with a continuous change of air, and avoiding excessive velocity that would exhaust the temperature of the body more rapidly than it could be generated.

TEMPERATURE AS A TONIC.

Temperature as a tonic influence claims special consideration. Cold is directly sedative, and serves an admirable purpose in allaying local inflammation, under careful observation. It does not, for a time, lessen power, while the excitability of the depressed part is increased by its suspension or rest; the consequence is, that when removed, the normal condition assumes an excitement beyond its former state. A cold breeze of air is exhilarating and tonic in effect as a negative influence, and often determines a stimulant effect to the circulatory system, by which injury is obviated. It is through the operation of these principles that reaction follows the first impression produced by cold. This reaction is not confined to the part first impressed alone, but extends throughout the system. Cold elevates the vital functions by a secondary influence, and thus it may be considered a tonic. Local and general debility is much relieved by a judicious exposure to cold air or water, and is one of the invigorating influences made available in traveling, when we have activity tending to an accumulation of strength. But here again there is danger. If too long continued it, at length, exhausts vital energy, and depression follows from inefficiency of excitability. If in excess, as regards the vital energies, it rapidly exhausts excitability — a vital organic condition. Reaction determines the avail ability of cold to the recuperative energy of the system, and must be most carefully observed in the use of cold baths. Cold must be carefully considered as an adjunct in the restoration of strength when there is a tendency to congestion of the brain or lungs, or disease of the heart. The first impression of cold to the surface repels the blood to the internal organs, and a pre-existing disposition to engorgement is fearfully compromised. The susceptibilities of individuals must be carefully observed. In case of a delicate child having a local excitement in a joint, great care must be taken in the application of volatile fluids or spirits to the part, because of its rapid evaporation, and exhaustion of the vital heat, as also when applied to the whole body, it has a most depressing and exhausting influence if reaction does not follow — which should be carefully noticed. The normal temperature being in direct ratio to the ability of the system to generate, and if delicate the loss is irreparable because of the depression on the vital forces.

TREATMENT OF CHRONIC INFLAMMATION.

In chronic ailments, as that of local inflammations resulting in the dissolution of the involved tissues, to the impairment of the general system — observable in cases of excessive drainage from open abscesses and ulcers — the pernicious influence appears to be diminished, elasticity or a depressing of vital cohesion from exhaustion. Normal organic function is impossible, because of deficiency in nutrition from excessive demand; the assimilating process being thus impaired, and the remedying of extraordinary excitement tending to increase the vital forces, which is done at the expense of exhausting excitability, if extraordinary recuperative power is not induced. Tonic medicines do not act directly upon the physical organization, but slowly through the vital functions as excitants. A proper cohesion of the living molecules in organized tissues is essential to the performance of normal functions, and a moderate augmentation of this vital cohesion may give increased energy to the function. It is simply through the vital powers that the several constituents of the body are enabled, under the influence of certain excitants to perform their functions duly, and extraordinarily, to the arrest of inimical influence, which is the tendency of the vital powers when sustained in energy. Tonics have no inherent quality of strength that can be imparted to the tissues; they are excitants having special tendencies that demand of the practitioner careful consideration in their selection to meet the carefully diagnosed indications of deficient energy of certain organic functions, and the extent of excitement the organs will bear without exhaustion or being overpowered by the excitant, or excess of excitant. There being a normal standard of excitability in each individual, the capacity must be carefully considered, as well as its relative condition when impaired by disease. Excesses from medication often do more injury than can be readily conceived of, especially when the friends of the patient are constantly insisting on having more done than the system will bear. In hospital practice the practitioner can offer no reasonable excuse for excessive medication, and hence the greater advantage to the patient from the practical experience of the skilful surgeon. So the judicious use of tonics in the treatment of chronic disease can only be determined by a long career of vigilant observation upon variously conditioned patients, and this knowledge only obtainable under the most favorable circumstances, not attainable in private practice, because of the indiscreet interference and inadvertent neglect usually attending practice in private families.

The sequence of chronic inflammation is most commonly that of abscess and ulceration. This invites attention, first, to that condition of inflammation that predisposes to these results; an inflammation under peculiar circumstances of health. As Mr. Erichsen remarks: " For the same treatment that would arrest inflammation in one form of the disease would certainly favor its progress in another." We do not have chronic inflammation in a healthy person; it is the result of some dyscrasia of the system, and that governs the treatment; as in cases where actual depletion is inadmissible, as it would increase the æsthenic condition of the patient. In these cases judicious treatment is not that intended to produce a great and sudden impression on the system, as we are required to do in the treatment of the acute condition. Attention to nutrition must be given in chronic inflammation, which can only be arrested to a tolerable condition by close attention to all the circumstances that tend to the improvement of the patient's general health, as well as by producing a favorable impression upon the part itself, by appropriate local means. Hence, in the treatment of chronic inflammation, hygienic measures are of primary consideration. In most cases nothing can be done without a proper régime, and much may be done by it that could not be effected by more direct medicinal means. This is our reason for first considering those available means of indirect tonic influences, as preparatory means for a more concise view of general principles in treatment, than was originally stated when treating upon special ailments, such as morbus coxarius, synovitis, and caries of the spine.

In the treatment of chronic inflammation, the diet is of much consequence as to the quantity and quality of nourishment, which must be carefully proportioned to the strength of the patient, and the taking of food carefully limited to stated periods, so as to afford ample time for digestion; restricting at all times the feeble patients from taking of fluids between meals, thus to prevent the dilution of the digestive fluids and an interruption to the normal process.

In the more active form of chronic inflammation, farinaceous food, with milk, may be allowed. The latter, however, is not suitable if the patient is subject to acid stomach; as lactic acid is most readily formed, and is a decided irritant to the mucous membrane of the stomach; under such circumstances, plain broth may be substituted, with that of carefully prepared bread.

In the less active forms, occurring in feeble constitutions, with depression of general power, animal food of nourishing quality may be given, such as roasted and broiled meats, accompanied with sweet, well-baked bread of a day's standing; milk twice a day, and fruit two or more hours afterward, but not in the latter part of the day. Mr. Erichsen has truly remarked: " Nothing requires greater nicety in practice than to proportion the diet, and determine the cases in which stimulants are necessary." It may be stated generally, that the more the disease assumes the asthenic and passive forms, the more are stimulants required; until, at last, in a truly adynamic type, our principal trust is in these agents, and large quantities of brandy and ammonia are required to maintain life. Brandy, in its raw state, is too much of an irritant to the stomach, and serves a much better purpose when combined with milk or other nutrient, as that of bread mucilage — stale bread having boiling water poured upon it, strained and sweetened. This makes a pleasant, nutritious beverage.

As favorable hygiene as possible having been obtained, our next effort should be to relieve the suffering condition of our patient. This invites attention to the cause of pain. Distention and pressure contribute most largely to this. Kolliker, Hasse, Burch, Paget and Martin Jones, agree that there is a dilatation of the blood-vessels. The phenomena of inflammation appear to be a stagnation or congestion of the capillaries from dilatation of the arteries; the immediate stagnation taking place in the capillaries, which are not in a direct line of passage from an artery into a vein, and the arrest taking place by the red corpuscles coalescing by mutual adhesion into masses, which, after being carried bodily up and down. more and more slowly, at last appear to block up the vessel, or partly, by overcrowding and distending it, and partly by becoming adherent to its walls, this adhesion usually commencing at the angle of union between two capillaries around the stagnant part of the vessel, or crowded by an aggregation of the red corpuscles, which appear to be more closely packed in consequence of the draining away of the liquor sanguin. The blood does not enter the part of the vessel in which stagnation has taken place, but passes off by a collateral branch. (Martin Jones.)

One of the most prominent symptoms is pain, supposed to be partly owing to increased sensibility of the nerves, but chiefly to the pressure made upon their terminal branches by dilatated blood-vessels, distended mainly by an excess of serum, and which manifests itself by the occurrence of pain. The condition is that of involving the superficial tissues, redness, swelling, from turgesence of the vessels, cognizable heat, a fulness or throbbing. The indications for relief appear to be very plain, relief to the congested condition of the parts primarily affected. The indication of deeper seated inflammation not presenting the superficial appearance requires the same means of relief; it is of the same pathological condition affecting the deeply located tissues, and, as yet, not involving the superficial integument, as when abscess is forming This, however, does not always follow, although its destructive influence may be very great as will be noticed as the results of inflammation.

Inflammation in its primary stage may be arrested, or terminate without interference, leaving no permanent impairment of the part affected, but an apprehension of unfavorable results, when it exists, prompts efforts to meet and terminate its tendency in the parts affected. From three of the terminations of inflammation serious consequences are to be apprehended. That is, when it terminates in the production of *pus,* the form being known as suppurative inflammation, when it terminates in an *ulcer,* and when the inflammatory action is of such destructive character as that of *gangrenous* inflammation. These conditions are to be avoided, if possible, in the incipient stage of inflammation. The disposition of inflammation is to extend and involve several tissues.

1. By spreading along the tissues affected in its *continuity,* as, for instance, along the skin, cellular or mucous membranes.

2. Where it spreads by *contiguity* of tissue, passing from one affected structure to an adjacent healthy one. Thus we see the opposite surface of an inflamed joint involved in disease at opposing points.

3. Inflammation extending to distant parts through the medium of the *blood,* this fluid being depraved, and increasing in liability to inflammation in other parts, as in some of the erratic forms of erysipelas, or conveying pus to a distance, as in phlebitis, and thus giving rise to numerous centres of inflammation.

4. Inflammation may be carried to distant parts by *metastasis,* or subside by *resolution.*

As we are limited to the treatment of inflammation with constitutional symptoms of the asthenic and irritative type, careful modification in agreement with the stamina of the patient is required. The local congestion, however, must be relieved in the early stage of the inflammation. If this is but slight, a gentle, saline purgative, and wet cloths at the temperature of the body, applied to the inflamed part, and the patient kept quiet for a few days, serves to afford relief. If not relieved in a reasonable time, a mild mercurial cathartic may be given, and a lotion of a drachm of ext. belladonna diffused in a quart of water should be applied to the painful part by means of thick folds of cloth; or, in place of belladonna, a similar preparation and quantity of stramonium. This treatment usually affords relief in cases of superficial inflammation. Deeper seated inflammation, and if located in a joint, requires a much more decided treatment; rest being enjoined, a brisk cathartic of calomel and scammony should be given, and a large sized plaster of cantharides applied from eight to ten hours over the joint, the blister being dressed with a large poultice of linseed meal, repeated every four hours for forty-eight hours. This decided treatment will usually relieve the patient of pain and tenderness in the affected joint. But the patient being constitutionally predisposed, from an asthenic condition, strumous or scorbutic, a constitutional course of treatment should not be neglected. Many, under these circumstances, are in a tolerable state of nutrition; to such, an alterative tonic, as that of hydrarg. bichlorid. and tr. cinch. comp., two grains of the former to six ounces of the latter, may be administered in half drachm doses, to be taken three times a day for a month or six weeks. This usually serves to protect the patient from future attacks of inflammation of the joint. Neuralgic pain in the limbs, after a few months, in many instances, alarm and annoy the patient; the arsenite of potassa solution in doses of four or five minims twice or three times a day soon affords relief.

In more feeble, attenuated patients, after the purgation and blistering, cod liver oil, two drachms a day, may be given, in divided doses, in conjunction with the alterative tonic.

Cod liver oil is only of service to the patient when thoroughly digested; hence the stools should be carefully examined to see that a portion or the entire amount of the oil has not passed, undigested, as is often the case when given in large doses. It is from the excess of oil that has passed undigested that injury may be apprehended, the excess imposed upon the stomach, in time, impairs the digestive functions. In these cases of feeble physical development and anæmic appearance, after a course of the alterative tonic, the preparations of iron may be employed, especially iron in combination with iodine, as that of the syr. ferri iod., or tr. ferri chlor., or the hypophosphate, carefully observed as to their influence upon the patient — a suitable quantity being decided upon. These serve a most admirable purpose, often greatly invigorating the feeble child to apparent good health. This end, however, is only attained where a proper regime is positively enforced to an increase of nutrition in which direction the iron greatly tends. In cases where the iron is inadmissible, and so determined from observation, the subnitrate of bismuth, in from four to five grain doses, three times a day, greatly serves to promote digestion, and thus increases the virtue of the cod liver oil.

In the advanced stage of inflammation, and in the induration of parts to the degree of abscess, the use of mercury must be most carefully prescribed. In a' gentle cathartic, it often times arrests excessive secretion from indurated tissues, and when in combination with a tonic, in a minute quantity, it is an invaluable thereapeutic agent in cachectic cases. It is not only serviceable in arresting the further progress of the disease, but especially in causing the absorption of some of the effusions that result from it, and in removing some of the other effects of the parts. It should be given in small doses for a considerable length of time, until the gums are slightly affected. In many cases of depressed power, it may be very advantageously conjoined with cinchona or sarsaparilla. The most useful preparations are calomel, in half grain doses, and the iodide of mercury in the same quantity, or, if a gradual and continued effect is desired, the bichloride of mercury, in doses of from a sixteenth to an eighth of a grain.

The iodide of potassium, as an alterative and absorbent, is of great value in some cases, especially in chronic inflammation of the fibrous and osseous tissues occurring in strumous constitutions, and mainly in those cases where mercury is inadmissible. In cases where mercury has been taken for some time without apparent benefit, its use has been followed with the best results. Some days, however, should be allowed to elapse after the mercury is discontinued before the iodide is given, otherwise profuse salivation, or even sloughing of the gums is apt to result.

Cod-liver oil is of very great value in the various strumous forms of inflammation in debilitated, emaciated, cachectic, and strumous subjects. In some cases it is advantageously conjoined with the iodide of potassium; or, in anæmic cases, the preparations of iron, the syr. ferri iod., being a most valuable one, or the sulphate, in one-fourth of a grain to one grain of ext. valerian in a pill, one to be taken every six hours where there is great nervous irritability and prostration, alternated with teaspoonful doses of the fresh, unbleached cod-liver oil. That the oil should be in this condition is an important consideration; as we have observed the fresh, unbleached cod-liver oil retained upon an irritable stomach of a child, when the manipulated or prepared oil in like quantities had been thrown off, from not being as palatable. But there are much more serious objections to the use of the bleached, or prepared oil, from a loss of its original constituents as cod-liver oil. In the dressing of leather, it is well known to the skilful mechanic that no other fish or other oils will serve him to perfect the dressing of a fine quality of leather. That its effects, when thus used, depends upon the same qualities that constitute it a medicine, we do not know nor are we prepared to advance that idea, but it certainly determines the fact that it possesses peculiar properties aside from its simple character as an oil. That these qualities may be lost from manipulation may be presumed from the fact that they are very readily neutralized — leaving simply a mass of ordinary rancid fish oil. This peculiar quality of cod-liver oil for the dressing of leather is thus spoken of by an elderly gentleman of our acquaintance, in the trade: "It has been used for the dressing of glove leather for all of a hundred years past, and no substitute of equal merit has been found in that time." It is said to have been used from time immemorial in the maritime districts of Holland, Germany, and the northern parts of Great Britain, as a popular remedy for rheumatism and rickets. It was first brought to the notice of the medical profession in the year 1782, by Dr. S. Percival, of England, and was afterward referred to by Dr. Bradsley, in his hospital reports, in 1807; but gained little attention until the publication of a paper by Schenck, in 1822, in Hufeland's Journal, containing a series of observations upon its efficacy in chronic rheumatism, particularly sciatica. After this time its employment was much extended in Germany and other parts of the Continent of Europe where its claims were set forth as of great value in the treatment of scrofula and tuberculosis. In 1841, it was most highly praised for its medicinal qualities by Dr. J. Hughes Bennett, Professor in the University of Edinburgh, which praises were confirmed by the ample experience of Dr. C. B. Williams of London, and, subsequently, by that of Dr. Walshe; and, both in that country and in our own, the use of the oil extended rapidly, until it has become a universal remedy for all classes of wasting disease, such as chronic debility, with impoverished blood, and defective nutrition or assimilation, not connected with inflammation or improper digestion. When in this condition, the subnitrate of bismuth serves an admirable purpose in promoting digestion, or, the nitrate of silver in one-fourth or one-eighth of a grain doses given half an hour after taking the oil. The nitrate of silver conjoined with the cod-liver oil is a most efficient remedy, where there is a tendency to chronic diarrhœa.

In this condition of the patient we have a low state of the vital forces and defective or depraved nutrition, where there is a tendency to the production — in various parts of the body, sometimes in one part and then in another, and in several parts at once — of a feeble, obstinate kind of scorbutic inflammation, strangely tending to the suppurative and ulcerative state and indisposed to a spontaneous cure, which is usually designated by writers as scrofulous inflammation. In the tuberculous form we have deposits remaining in a quiescent condition, but subject to irritation involving neighboring parts, gradually softening and then discharged in suppurations and ulcerations. Cod-liver oil has a most beneficial influence over the system when in this condition; generally controlling it in degree, often arresting for a time, if not determining a cure.

In the scorbutic diathesis, or scrofulous inflammation, without the tuberculous condition, free from the induration of parenchymatous structure, the oil will often effect complete cures.

When the system is characterized by the tuberculous deposit, it may be improved by the oil; but it is the opinion of some of our experienced teachers and practitioners in medicine, that the oil has no influence whatever over the tubercle when already formed, and that its regular course of degeneration, involving surrounding tissue, cannot be arrested. From this view it is obvious that the oil can only prove curative when employed either before the tuberculous deposit has taken place, or when it has occurred in situations, or in quantities, not necessarily destructive to life through disorganization of the tissues affected. Thus, when the tubercles are deposited in vital organs there is scarcely a chance of safety, because of their destructive influence when in a state of dissolution. The reverse may be said, however, when deposits are made in the external lymphatic glands, in the subcutaneous cellular tissues and within bones; there is reasonable hope of cure as the parts are less essential to life, and the destructive influence can generally be supported until the disintegration has passed off.

The effects of cod-liver oil are peculiarly obvious in the suppurative and ulcerative stage of the affection, whether in the lymphatic glands, subcutaneous tissues or the skin itself Scrofulous ulcers and abscesses of the neck, auxillæ, and groin, or of the skin and the areolar tissue in any part of the body, often rapidly improve and ultimately get well under its use. Large and exhausting abscesses of the lower extremities which have reduced the patient during months of suppuration to the lowest condition of emaciation and debility compatible with life, gradually improve; healthy granulations being presented, and the edges closing in by tender skin, pencilled with a weak solution of nitrate of silver, ten grains to the ounce of water, and for a dressing, fresh mutton tallow, in preference to the rancid ointments of the shops. The secretion being thin and watery, Turner's cerate serves a better purpose by giving tone to the rising granulations, or, a wash of acetate of lead and sulphate of zinc, two grains of each to the ounce of water, and when the ulcer is irritable, a drachm of the wine of opium added to the lotion. The mutton tallow containing a small portion of carb. acid, should be thinly spread upon a cloth as a covering. This simply sustains the recuperative tendency that the oil had produced.

During the treatment of this class of ailments, scrofulous ophthalmia presents in virulent form; demanding a gentle mercurial cathartic, and the above described lotion applied to the eyes to afford relief. The nitrate of silver may be applied to any existing ulcer of the cornea or other part of the conjunctival surface after the inflammation has subsided. Molasses dropped in the eye, or the cod-liver oil, greatly improves the ulcerated cornea.

Cutaneous eruptions, and even the obstinate *lupus* will yield to the oil and liq. potass. arsenitis in increasing doses, to as much as the stomach will bear. We have given, in obstinate cases, from ten to fifteen drops of the latter at a dose, three times a day, and this has been attended with the most happy results.

. In diseases of the bones and joints there is, probably, no other single remedy so efficacious as cod-liver oil, conjoined with alteratives and tonics. In advanced swellings of the hip, knee, and other joints, with or without abscess or caries. Though the cure is often protracted, yet its curative tendency is most satisfactorily demonstrated by the fact of the most extensive induration being healed, and the patient, although very decrepit, made to enjoy excellent health; and thus children are enabled to follow some occupation for self-support, even acquiring considerable muscular effort.

Enlargement of the mesenteric glands is a most common condition of children laboring under disease of the joints, and requires our special attention. Under such circumstances the oil is not admissible until the digestive functions are improved sufficiently to have it well digested. As we have before stated, the stomach must be in a condition to digest the oil, and in limited quantity, that the power of digestion may not be overburdened, which is a most weakening influence. The subnitrate of bismuth given in three or four grain doses, three times a day, an hour after meals — (the amount of fluids taken as drink being limited) — will most readily relieve this condition and pave the way for the exhibition of the oil, upon which the patient will improve, when alternated with the before mentioned tonic alterative of bichloride of mercury and tinct cinchona in small doses.

Some cachectic children have considerable abdominal distention and hardness, and, sometimes, peritoneal affections, enlarged liver, great emaciation, pallor and debility which have generally been ascribed to scrofulous disease of the mesenteric glands. Some of these cases yield quickly and most happily to cod-liver oil and cooperating treatment, while others are more or less obstinate, and not a few end fatally — as some of the cases are decidedly tuberculous. They enlarge and harden and are diffused throughout the abdomen, causing peritoneal inflammation. These are, in the general, incurable, except, perhaps, in a few instances, in which the tubercles — originally small in amount — are discharged from ulceration, or, possibly, undergo absorption from degeneration; as children under an enforced régime will recover from all the indications of a tuberculous diathesis. These affections of the liver and mesenteric glands are, in some cases, simply of an inflammatory character, and terminate favorably under skilful treatment.

Diarrhœa, with ulceration of the bowels, is not an unfrequent attendant on bone or joint disease. This sometimes depends upon tubercles in the bowels, leaving ulcers as they are discharged; in other instances, it has been found to be connected with tubercles. Scorbutic or scrofulous inflammation, without tubercles, when it attacks, exclusively, the mucous membrane, showing, when it does so, a tendency to affect the follicles, and to result in ulceration of these structures. The gastric mucous membrane may be invaded as well as the intestinal. In these cases, the nitrate of silver in one-fourth of a grain doses, given at an interval of an hour before or after taking food, three times a day, and a teaspoonful of cod-liver oil half an hour after. These are the remedies we most approve of; the cuticle being excited with the flesh brush, after cold sponging, once a day when about to retire for the night.

Chronic inflammation of the nostrils, or ozæna, chronic angina, and chronic laryngitis and bronchitis are all ailments that present in patients laboring under joint diseases, and yield tardily, but finally, to the use of cod-liver oil and alterative tonics.

RICKETS, being a disease, the result of impaired digestion, indicated by the light colored stools, capricious appetite — often voracious — is most readily relieved by the subnitrate of bismuth, tinct. cinchona, and hydrarg. bichlorid. This relief appears, however, to be only an arrest of the impairment of the digestive functions, and the cure is rendered more permanent by the use of cod-liver oil, which creates a most remarkable change toward improvement in the child's appearance; the enlarged joints assume normal form, and the sallow skin becomes clear and healthy in appearance.

It has been recommended to employ the oil externally, by friction, and most highly recommended for its constitutional impression upon children when administered in this form — the stomach being in a very irritable condition — and it may be worthy of a trial. Our practice has been not to attempt to give the oil before having allayed the irritability of the stomach and prepared the digestive organs for its proper digestion. In cases of irritable stomach it may prove beneficial from external application in the commencement of treatment, and thus limit the preparatory course tending to its assimilation when taken into the system, and worthy of favorable consideration in the treatment of patients, thus delicately conditioned

SENSIBLE PROPERTIES OF COD LIVER OIL.

In constitution cod liver oil is similar to ordinary fish oil. In its purest form it is of a color varying from a slightest tint of transparent yellow to a fine golden yellow; when less pure, of a light brown color, but still transparent; when most impure, dark brown and opaque in mass, though transparent in thin layers. Its odor and taste are quite peculiar, scarcely disagreeable in the finer kinds, but offensive in the most impure, and also somewhat acrid. The oil is injured by long exposure to the air. It contains a peculiar principle called *gaduin*, not known to have any medicinal virtue, various biliary principles, a little *iodine* and *bromine*, *olein* and *margarin*, and many other constituents of no special interest. Its most obvious characteristic properties are its odor and taste — quite different from that of ordinary fish oils — with which every practitioner should become familiar so as to be able to determine the pure from the adulterated oil; for adulteration of this article has become a very common practice. The odor strongly resembles that of shoe leather which owes its fineness of finish to the cod-liver oil used in its preparation. Another distinctive property derived from its biliary constituents, is that of assuming fine colors, or changes of colors, under the action of the mineral acids.

There has been, as yet, no substitute discovered that equals in medicinal properties the cod-liver oil. In Bennett's Practice of Medicine, page 184, we find the following impressive remarks upon this most valuable medicine : "Our present knowledge has led to a complete revolution in our practice. Thus, moderate exercise to stimulate respiration, cold sponging, nutritious diet, and a bracing system has been found more beneficial; at the same time avoiding anodynes and cough mixtures, which, by diminishing the appetite and inducing weakness, interfere with nutrition. Indeed, it has been proved that the best methods of lessening cough expectoration, and sweating, are the means which produce increase of general strength; so that if we can carry out the general indications, the local symptoms may be safely left to themselves.

"In doing this we have now the advantage of possessing a remedy which, in cases of tuberculosis, is of the highest nutritive importance, as it gives to the system that fatty element in which it is so defective, and in a form that is more easily assimilated, and more capable of adding to the molecular elements of the body than any other. I allude to *Cod-liver Oil.*

"And now, you cannot fail to perceive how the molecular doctrine of organization and of growth, not only explains the known facts in physiology and pathology, but constitutes the basis for a true therapeutics. Fatty particles, as we have seen, form the molecular fluid of chyle, while out of chyle blood, and through it, all the tissues are formed. Impairment of digestion in scrofula and tuberculosis renders chylification imperfect; the fatty constituents of food are not separated from it and assimilated; the blood consequently abounds in albuminous elements, and when exuded forms, as we have seen, tubercles. To induce health it is necessary to restore nutritive elements which are diminished, and that is done directly by adding a pure animal oil to the food. While an informatory exudation in previously healthy persons should be treated by supporting the vital powers generally, so as to permit its molecules going through the transformation necessary for their growth and elimination, in tuberculosis we add the constituent of food necessary for the formation of the molecules themselves. By so doing, we form good chyle and blood, we restore the balance of nutrition which has been disturbed; respiration again active in the excretion of carbonic acid gas; the tissues are more nourished from the blood, the elements so necessary for their sustenance. The entire economy is renovated, so that while the histogenetic processes are renewed, the hystolitic in the tubercle itself also are stimulated, and the whole disappears."

This is a very ingeniously devised theory, and acceptable to our expérience as to the modus operandi of cod liver oil — a medicine tending to the restoration of normal activity or progressive increase of vital energy in the animal economy, the very ultima thule of attainment in the treatment of chronic disease.

INDEX.

	PAGE.
Abdomen, relaxed	236
Abdominal hernia — its varieties	204
do its causes	205
do symptoms of	206
do divisions of the various	207
do diagnosis	209
do construction and application of trusses for relief of	222
Aberration of form, tendencies to	44
Abnormal position of bones, ligaments, and muscles in talipes varus	42
Absorption, promotion of	125
Accumulating static electricity, machine for	113
Action, alterative	124
sedative	124
Anchylosis, in its various conditions	298
treatment of	305
Ancient treatment of deformity	25
Ancient treatment of spinal deformities	28
Adventurous treatment, paralysis the result of	107
Air, pure, as a tonic	338
Alterative action of electricity	124
Amenorrhœa	137
Aneurisms	138
Anterio-posterior curvature, treatment of	192
Apparatus, Dr. Jerome Kidder's electrical	162
Appliances, mechanical	323
Applying electricity, modes of	140
Aponeurosis, contraction of the plantar	78
Arrangement of galvano-caustic battery and parts	158
Arteries, treatment of punctured	57
Bartlett's regulator	156
Batteries, portable galvanic	152
Battery, cabinet regulator and	155
the galvano-caustic	158
Bones, abnormal position of	42

	PAGE.
Bursæ	247
treatment of	249
Cabinet regulator and battery	155
Calcaneus, talipes	84
cause of talipes	85
treatment of congenital talipes	85
do non-congenital talipes	86
Capillary circulation	126
Caries of the cervical vertebræ	317
Caries of the vertebræ	310
do diagnosis	301
Caries of the last cervical and first dorsal vertebræ	106
spine, treatment of	319
cervical vertebræ	317
Cartilages, floating, in joints	309
Cases demanding tenotomy	102
Cases, prognosis of unfavorable	280
Causes of talipes valgus and condition of the bones	70
talipes equinus	80, 88
talipes calcaneus	85
necrosis	289
Caustic, galvano-battery, preparation of	158
do electrodes	160
do arrangement of battery and parts	158
Cautery, galvano	161
Chronic inflammation, treatment of	340
Circulation, capillary	126
Cod-liver oil, sensible properties of	351
Concussion, electric	118
Condition of the muscles in talipes varus	49
of the bones in talipes valgus	70
pathological, tending to lateral curvature	176
Conformation of the foot	40
Congenital luxation	23
Congenital talipes calcaneus, treatment of	85
Congestion, use of electricity in	127
Construction and application of trusses for the relief of abdominal hernia,	222
Contraction, organic	128
of hands, fingers, and toes	163
of the plantar aponeurosis	78
Contorted feet, surgical treatment of	50
do preparatory steps in treatment of	51
Contortions, impairment of tissues resulting in	37
of the foot, various forms of	39
resulting from paralysis	50
the sequence of infantile paralysis	88

INDEX. 357

	PAGE.
Contortions, special treatment of, the sequence of infantile paralysis	98
and ulceration of the toes	166
spastic, and their treatment	171
Counter-irritation	127
Coxarius, morbus	258
do special treatment of	273
Cure, means of, in first stage of talipes varus	63
tenotomy as a means of	184
Current, silent	117
Curvature, congenital lateral, of the spine	173
non-congenital, of the spine	174
do do diagnosis of	178
do do treatment of	180
pathological condition tending to lateral	176
treatment of anterio-posterior	192
of the spine, kyphosis, or posterior	313
Defective physical formation	7
Deformities, ancient treatment of spinal	28
Deformity, ancient treatment of	25
surgical means of redressing	30
Diagnosis, electrical	123
of lateral curvature of the spine	178
of caries of the vertebræ	311
of the various abdominal hernia	209
Digestion and menstruation, influence of electricity in	126
Diseases of sheaths of tendons	250
do do treatment of	257
of the joints, pathological consideration of	253
Divisions of the various abdominal hernia	207
Division of the sterno-cleido-mastoideus muscle	187
Dr. Jerome Kidder's electrical apparatus	162
Dr. Knight's modification of Scarpa's shoe, and mode of applying it in talipes varus	64
Ectropion vesicæ	235
Efforts, personal, in primary treatment	105
Electrical apparatus, Dr. Jerome Kidder's	162
Electrical induction	120
diagnosis	123
influence	124
rubefacient	120
sedative influence	120
Electricity as a therapeutic agent in treatment of paralysis	116
alterative action	124
as a counter-irritant	127
Electric concussion	118
revulsion	121

	PAGE.
Electric shock	118
sparks	118
Electricity, generators of galvanic	150
machine for accumulating static	113
modes of applying	140
promotion of nutrition by	125
do secretion by	125
do absorption by	125
results of static	129
reactive power of	124
sedative action	124
use of, in capillary circulation	126
digestion and menstruation	126
inflammation	126
congestion	127
vitalizing power of	123
Electrodes, galvano-caustic	160
Electro-magnetic machine, portable	151
Equinus, talipes	76
do causes of	80
do treatment of	81
do from various causes	83
Exercise, muscular	128
Exhilarating influence as a tonic	337
Extorsium, genu	196
do treatment of	198
Extreme cases of Talipes Valgus, treatment of	73
Fascia, severing the plantar	53
Feet, various forms of contortion of the	39
surgical treatment of contorted	50
preparatory steps in treatment of contorted	51
Femoral hernia, truss for the relief of reducible	222
Fingers, contraction of the	163
First dorsal vertebræ, paralysis the result of caries of	106
Flexum, posterior genu	109
Floating cartilages in joints	309
Foot, conformation of the	40
means of restoring the, to normal form	52
Form, tendencies to aberration of	44
Formation, defective physical	7
Galvanic batteries, portable	152
electricity, generators of	150
Galvano-caustic battery, the	158
do preparation of	158
do electrodes	160

INDEX.

	PAGE.
Galvano-cautery	161
Ganglion, diseases of sheaths of tendons	250
do do treatment of	251
Generators of galvanic electricity	150
General remarks on the treatment of talipes	60
Genu extorsium	196
do treatment of	198
Genu flexum, posterior	109
Genu valgum	200
do treatment of	201
Hands, contraction of the	163
Hernia	204
abdominal, its varieties	204
its causes	205
its symptoms	206
construction and application of trusses for the relief of,	222
complicated with a retained testicle	211
diagnosis of	209
divisions of the various	207
prognosis of the ailment	212
truss for the relief of reducible inguinal and femoral	227
Impairment of tissues resulting in contortions	37
Induction, electrical	120
Infantile paralysis	88
do contortions the sequence of	89
do prognosis in, the sequence of	92
do special treatment of contortion, the sequence of	98
Inflammation, use of electricity in	126
treatment of chronic	340
Influence, electrical	124
exhilarating, as a tonic	337
Influence of ligaments in talipes varus	49
of occupation tending to paralysis and the treatment	112
Inguinal hernia, truss for relief of reducible	227
Introductory remarks	1
Jerome Kidder's (Dr.) electrical apparatus	162
Joints, pathological consideration of diseases of the	253
floating cartilages in	309
Joint, synovitis of the knee	293
Kidder's (Dr. Jerome) electrical apparatus	162
Knight's (Dr.) modification of Scarpa's shoe, and mode of applying it in talipes varus	64
Knives, tenotomy	58
Kyphosis, or posterior curvature of the spine	313

INDEX.

	PAGE.
Last cervical vertebræ, paralysis the result of caries of	106
Lateral curvature of the spine, congenital	173
do do non-congenital	174
do do diagnosis of	178
do do treatment of	180
do do pathological condition tending to	176
Ligaments, abnormal position of, in talipes varus	42
influence of, in talipes varus	49
Luxation, congenital	23
Machine for accumulating static electricity	113
Magnetic, portable electro, machine	151
Maligna, onychia	169
Means of redressing deformity, surgical	30
of restoring the foot to normal form	52
of cure in first stage of talipes varus	63
of cure, tenotomy as a	184
Mechanical appliances	323
Menstruation, effect of electricity on	126
Modification of Scarpa's shoe (Dr. Knight's)	64
Morbus coxarius	258
Muscles, abnormal position of, in talipes varus	48
division of the sterno-cleido-mastoideus	187
Muscular exercise, use of electricity in	128
Necrosis, causes of	289
Non-congenital talipes calcaneus, treatment of	86
do lateral curvature of the spine	174
Nutrition, promotion of, by electricity	125
Occupations, influence of, tending to paralysis, and their treatment	112
Oil, cod liver, sensible properties of	351
Onychia Maligna	169
Organic contraction	128
Paralysis, contortion resulting from	50
infantile	88
prognosis in infantile	92
the sequence of unrelieved	93
special treatment of contortion, the sequence of infantile	98
the result of caries of the last cervical and first dorsal vertebræ	106
the result of adventurous treatment	107
electricity as a therapeutic agent in the treatment of	116
Pathological condition tending to lateral curvature	176
do of diseases of the joints	253
Personal efforts in primary treatment	105
Physical formation, defective	7
Plantar fascia, severing the	53

INDEX. 361

	PAGE.
Plantar aponeurosis, contraction of the	78
Portable electro-magnetic machine	151
galvanic batteries	152
Posterior genu flexum	109
curvature of the spine	313
Posterior tibial tendon, severing the	55
Power — reactive, in electricity	124
vitalizing, in electricity	123
Preparatory steps in treatment of contorted feet	51
Preparation of galvano caustic battery	158
Primary treatment, personal efforts in	105
Procidentia uteri	229
do treatment of	231
Projecting sternum	193
do treatment of	194
Prognosis of hernia	212
do unfavorable cases of morbus coxarius	280
Properties, sensible, of cod liver oil	351
Punctured arteries, treatment of	57
Pure air as a tonic	340
Rachitis	189
Reactive power of electricity	124
Redressing deformity, surgical means of	30
Reducible inguinal and femoral hernia	227
Regulator, Bartlett's	156
cabinet and battery	155
Relaxed abomen	236
Remarks, general, on treatment of talipes	60
introductory	1
Revulsion, electrical	127
Rubefacient, electrical	120
Scarpa's shoe, Dr. Knight's modification of, and mode of applying it in talipes varus	64
Second stage of talipes varus and treatment	66
Secretion, promotion of, by the use of electricity	125
Sedative action	124
Sedative influence of electricity	120
Sensible properties of cod liver oil	351
Severing the plantar fascia	53
the posterior tibial tendon	55
Sheaths of tendons (Ganglion), diseases of	250
Shock, electric	118
Sparks, electric	118
Spastic contortions and their treatment	171
Special treatment of contortion the sequence of infantile paralysis	98

	PAGE.
Special treatment of morbus coxarius	273
Spine, congenital lateral curvature of the	173
non-congenital lateral curvature of the	174
non-congenital lateral curvature of the diagnosis of	178
pathological condition, tending to lateral curvature	176
treatment of lateral curvature	180
treatment of caries of the	319
Static electricity, machine for accumulating	113
do results of	129
Sterno-cleido-mastoideus muscle, division of the	187
Sternum projecting	193
do treatment of	194
Surgical means of redressing deformity	30
treatment of contorted feet	50
Symptoms of abdominal hernia	206
Synovitis of the knee joint	293
System, tonics and their effects upon the	334
Talipes calcaneus	84
do causes of	85
do treatment of congenital	85
do treatment of non-congenital	86
Talipes equinus	76
do cause of	80
do treatment of	81
do from various causes	83
Talipes valgus — causes and condition of the bones	70
do treatment of	71
do extreme cases of	73
Talipes varus, abnormal position of bones, ligaments and muscles in	48
do the condition of the muscles in	48
do influence of ligaments in	49
do second stage of, and treatment	66
do third stage of	55, 65, 269
do general remarks on treatment of	60
do means of cure in the first stage of	63
do Dr. Knight's mode of applying his modification of Scarpa's shoe in	64
Taxis	214
Temperature as a tonic	339
Tendencies to aberration of form	44
Tendons, diseases of sheaths of	250
do treatment	251
severing the posterior tibial	55
Tenotomy knives	58
cases demanding	102
as a means of cure	184

INDEX. 363

	PAGE.
Testicle, retained, complicated with hernia.	211
prognosis of	212
Tibial tendon, severing the posterior	55
Tissues, impairment of, resulting in contortions	37
Toes, contraction of the	163
contortion and ulceration of the	166
Tonic, exhilarating influence as a	337
pure air as a	338
temperature as a	339
Tonics, selection and qualities of	334
and their effects upon the system	334
Torticollis, treatment of	186
Treatment of anchylosis in its various conditions	298
of anterio-posterior curvature	192
of bursæ	249
of caries of the spine	319
of chronic inflammation	340
of contorted feet, preparatory steps in	51
of congenital talipes calcaneus	85
of deformity, ancient	30
of diseases of sheaths of tendons	251
of genu extorsium	198
of genu valgum	201
of lateral curvature of the spine	180
and means of cure in the first stage of talipes varus	63
special, of morbus coxarius	273
of non-congenital talipes calcaneus	86
of diseases in occupations tending to paralysis	112
of paralysis, the result of adventurous	107
of paralysis, electricity as a therapeutic agent in	116
of procidentia uteri	229
of projecting sternum	194
of punctured arteries	57
of spinal deformities, ancient	28
surgical, of contorted feet	50
of talipes, general remarks on the	60
of talipes equinus	81
of talipes valgus	71
do extreme cases	73
in second stage of talipes varus	66
of third stage of talipes varus	69
of torticollis	186
of varicose ulcer	244
Truss for relief of abdominal hernia, construction and application	222
reducible inguinal and femoral hernia	227
Tumors, use of electricity in regard to	137

	PAGE.
Ulcer, varicose	243
treatment of	244
Unfavorable cases, prognosis of	280
Use of electricity, in capillary circulation	126
in congestion	127
in digestion and menstruation	126
in inflammation	126
in muscular exercise	128
in promotion of nutrition	125
do secretion	125
do absorption	125
Uteri, procidentia	229
treatment of	231
Valgum, genu, treatment of	201
Valgus, talipes, causes of, and condition of the bones	70
treatment of	71
extreme cases of	73
Varicose veins	240
ulcer	243
treatment of	244
Various forms of contortion of the feet	39
Varus, condition of muscles in talipes	48
abnormal position of bones, ligaments and muscles in talipes	49
first stage of talipes	63
second stage of talipes	66
third stage of talipes	55, 69
Vertebræ, caries of the	310
do diagnosis of	311
do cervical	317
last cervical and first dorsal, paralysis the result of caries of	106
Vesicæ, ectropion	235
Vitalizing power of electricity	128

www.ingramcontent.com/pod-product-compliance
Lightning Source LLC
Chambersburg PA
CBHW020314240426
43673CB00039B/801